Developing Climate-Resilient Crops

Footprints of Climate Variability on Plant Diversity

Series Editor: Shah Fahad

Climate Change and Plants: Biodiversity, Growth and Interactions
Shah Fahad, Osman Sönmez, Shah Saud, Depeng Wang, Chao Wu, Muhammad Adnan and Veysel Turan

Developing Climate-Resilient Crops: Improving Global Food Security and Safety
Shah Fahad, Osman Sönmez, Shah Saud, Depeng Wang, Chao Wu, Muhammad Adnan and Veysel Turan

Sustainable Soil and Land Management and Climate Change
Shah Fahad, Osman Sönmez, Veysel Turan, Muhammad Adnan, Shah Saud, Chao Wu and Depeng Wang

Developing Climate-Resilient Crops: Improving Global Food Security and Safety

Edited by
Shah Fahad, Osman Sönmez, Shah Saud, Depeng Wang,
Chao Wu, Muhammad Adnan, and Veysel Turan

CRC Press
Taylor & Francis Group
Boca Raton London New York

CRC Press is an imprint of the
Taylor & Francis Group, an **informa** business

First edition published 2022
by CRC Press
6000 Broken Sound Parkway NW, Suite 300, Boca Raton, FL 33487-2742

CRC Press is an imprint of Taylor & Francis Group, an Informa business

ISBN 13: 978-0-367-62347-0 (hbk)
ISBN 13: 978-0-367-62350-0 (pbk)

DOI: 10.1201/9781003109037

Library of Congress Cataloging-in-Publication Data
Names: Fahad, Shah (Assistant professor in agriculture), editor.
Title: Developing climate-resilient crops: improving global food security and safety / edited by Shah Fahad, Osman Sonmez, Shah Saud, Depeng Wang, Chao Wu, Muhammad Adnan, and Veysel Turan.
Description: First edition. | Boca Raton : CRC Press, 2021. | Series: Footprints of climate variability on plant diversity | Includes bibliographical references and index.
Identifiers: LCCN 2020058430 | ISBN 9780367623470 (hardback) | ISBN 9781003109037 (ebook)
Subjects: LCSH: Crops and climate. | Climate change. | Food security--Climatic factors.
Classification: LCC S600.7.C54 D46 2021 | DDC 630.2/515--dc23
LC record available at https://lccn.loc.gov/2020058430

Typeset in Times
by MPS Limited, Dehradun

Visit the Taylor & Francis Web site at
http://www.taylorandfrancis.com

and the CRC Press Web site at
http://www.crcpress.com

Contents

Acknowledgements

Words are bound and knowledge is limited to praise ALLAH, the Instant and Sustaining Source of all Mercy and Kindness, and the Sustainer of the Worlds. My greatest and ultimate gratitude is due to ALLAH (Subhanahu wa Taqadus). I thank ALLAH with all my humility, for everything that I can think of. His generous blessing and exaltation succeeded my thoughts and thrived my ambition to have the cherished fruit of my modest efforts in the form of this piece of literature from the blooming spring of blossoming knowledge. May ALLAH forgive my failings and weaknesses, strengthen and enliven my faith in HIM, and endow me with knowledge and wisdom. All praises and respects are for **Holy Prophet Muhammad** *Salle Allah Alleh Wassalam,* the greatest educator, the everlasting source of guidance and knowledge for humanity. He taught the principles of morality and eternal values and enabled us to recognize our Creator. I have a deep sense of obligation to my parents, my brothers, sisters, and son. Their unconditional love, care, and confidence in my abilities helped me achieve this milestone in my life. For this and much more, I am forever in their debt. It is to them that I dedicate this book. In this arduous time, I also appreciate the patience and serenity of my wife, who brought joy to my life in so many different ways. It is indeed on account of her affections and prayers that I was able to achieve something in my life.

Shah Fahad

Editors' Biographies

Dr Shah Fahad is an Assistant Professor in the Department of Agronomy, University of Haripur, Khyber Pakhtunkhwa, Pakistan. He obtained his PhD in Agronomy from Huazhong Agriculture University, China, in 2015. After doing his postdoctoral research in Agronomy at the Huazhong Agriculture University (2015–2017), he accepted the position of Assistant Professor at the University of Haripur. He has published over 190 peer-reviewed papers with more than 160 research and 30 review articles on important aspects of climate change, plant physiology and breeding, plant nutrition, plant stress responses and tolerance mechanisms, and exogenous chemical priming-induced abiotic stress tolerance. He has also contributed 20 book chapters to various book editions published by Springer, Wiley-Blackwell, and Elsevier. He has edited six book volumes, including this one, published by CRC Press, Springer, and Intech Open. He has won a Young Rice International Scientist award and distinguished scholar award in 2014 and 2015, respectively. He has worked and is presently continuing on a wide range of topics, including climate change, greenhouse emission gases, abiotic stresses tolerance, roles of phytohormones and their interactions in abiotic stress responses, heavy metals, and regulation of nutrient transport processes.

Prof. Dr. Osman Sönmez is a Professor in the Department of Soil Science, Faculty of Agriculture, Erciyes University, Kayseri, Turkey. He obtained his MS and PhD in Agronomy from Kansas State University, Manhattan-KS, USA, in 1996–2004. In 2014, he accepted the position of Associate Professor at the University of Erciyes. Since 2014, he has worked in the Department of Soil Science, Faculty of Agriculture at Erciyes University. He has published over 90 peer-reviewed papers, research, and review articles on soil pollution, plant physiology, and plant nutrition.

Dr. Veysel Turan is an Assistant Professor in the Department of Soil Science and Plant Nutrition, Bingöl University, Turkey. He obtained his PhD in Soil Science and Plant Nutrition from Atatürk University, Turkey, in 2016. After doing his postdoctoral research in the Department of Microbiology, University of Innsbruck, Austria (2017–2018), he is working in Bingöl University. He has worked and is presently continuing on a wide range of topics such as soil–plant interaction, heavy metal accumulation, bioremediation of soil by some plant, and soil amendment.

Dr. Muhammad Adnan is a Lecturer in the Department of Agriculture at the University of Swabi (UOS), Pakistan. He has completed his PhD (soil fertility and microbiology) from the Department of Soil and Environmental Sciences (SES), University of Agriculture, Peshawar, Pakistan, and the Department of Plant, Soil, and Microbial Sciences, Michigan State University, USA. He has received his MSc and BSc (Hons) in Soil and Environmental Sciences from the Department of SES, University of Agriculture, Peshawar, Pakistan.

Dr. Shah Saud received his PhD in turf grasses (Horticulture) from Northeast Agricultural University, Harbin, China. He is currently working as a Post Doctorate researcher in the Department of Horticulture, Northeast Agricultural University, Harbin, China. Dr. Shah Saud has published over 125 research publications in peer-reviewed journals. He has edited three books and written 25 book chapters on important aspects of plant physiology, plant stress responses, and environmental problems in relation to agricultural plants. According to Scopus®, Dr. Shah Saud's publications have received roughly 2500 citations with an h-index of 24.

Dr. Chao Wu engages in field-crop cultivation and physiology, and plant phenomics. He completed his PhD during 2013–2016 from Huazhong Agricultural University, Wuhan, China, and completed his post

PhD during 2017–2019 from Nanjing Agricultural University, Nanjing, China. Now, he is associate research fellow in the Guangxi Institute of Botany, Guangxi Zhuang Autonomous Region and the Chinese Academy of Sciences, Guilin, China. He chairs a Natural Science Foundation of Jiangsu Province and two Postdoctoral Science Foundation researches, and focuses mainly on physiological mechanisms of abiotic-stress tolerance (heat, drought) in crops and medicinal plants.

Dr. Depeng Wang completed his PhD in 2016 in the field of Agronomy and Crop Physiology from Huazhong Agriculture University, Wuhan, China. Presently, he is serving as a professor in the College of Life Science, Linyi University, Linyi, China. He is the principal investigator of Crop Genetic Improvement, Physiology & Ecology Center in Linyi University. His current research focuses on crop ecology, physiology, and agronomy including the key characteristics associated with high yielding crops, the effect of temperature on crop grain yield and solar radiation utilization, morphological plasticity to agronomic manipulation in leaf dispersion and orientation, and optimal integrated crop management practices for maximizing crop grain yield. Dr. Depeng Wang has published over 36 papers in reputed journals.

Contributors

Muhammad Aasim
Necmettin Erbakan University, Faculty of Science, Department of Biotechnology

Saghir Abbas
Department of Botany, Faculty of Life Sciences, Government College University, Faisalabad 38000, Pakistan

Yousef M Abu-Zaitoon
Department of Biological Sciences, Al Hussein Bin Talal University, P.O. Box 20, Ma'an, Jordan

Ishfaq Ahmad
Centre for Climate Research and Development (CCRD), COMSATS University Islamabad, Pakistan

Shakeel Ahmad
State Key Laboratory of Rice Biology, China National Rice Research Institute, Hangzhou, 310006, China

Muhammad Salim Akhter
Institute of Pure and Applied Biology, Bahauddin Zakariya University, Multan, Pakistan

Ezz Al-Dein-Al-Ramamneh
Department of Agricultural Sciences, AL-Shouback University College, Al-Balqa Applied University, AL-Shouback, Ma'an, Jordan

Muqadas Aleem
Nanjing Agricultural University, Nanjing 25000, China

Mohammad Alhawatema
Department of Applied Biological Science, Faculty of Science, Tafila Technical University, Tafila, Jordan.

Qasim Ali
Department of Botany, Faculty of Life Sciences, Government College University, Faisalabad 38000, Pakistan

Shafaqat Ali
Department of Environmental Sciences and Engineering, Government College University Faisalabad 38000, Pakistan

Zulfiqar Ali
Muhammad Nawaz Sharif, University of Agriculture, Multan 66000, Pakistan

Wafa'a A. Al-Taisan
Department of Biology, College of Science, Imam Abdulrahman Bin Faisal University, P.O. Box 1982, Dammam Saudi Arabia

Abdel Rahman M. Al-Tawaha
Department of Biological Sciences, Al Hussein Bin Talal University, P.O. Box 20, Ma'an, Jordan

Abdel Razzaq Al-Tawaha
Department of Crop Science, Faculty of Agriculture, Universiti Putra Malaysia, 43400 Serdang, Selangor, Malaysia

Amanullah
Agronomy Department, The University of Agriculture Peshawar, Pakistan

Amna
Department of Bioinformatics and Biotechnology, Government College University, Faisalabad-38000, Pakistan

Muhammad Anwar
College of Life Sciences and Oceanography, Shenzhen University, Shenzhen, China

Muhammad Arif
Department of Plant Protection, Faculty of Agricultural Sciences, Ghazi University, D. G. Khan, Pakistan

Muhammad Furqan Ashraf
College of Life Sciences, South China Agricultural University, Guangzhou, China

Mazen A. Ateyyat
Faculty of Agricultural Technology, Al Balqa Applied University

Muhammad Azeem
Department of Botany, Faculty of Life Sciences, Government College University, Faisalabad 38000, Pakistan

Ali Bakhsh
Department of Plant Breeding & Genetics, Faculty of Agricultural Sciences, Ghazi University, D. G. Khan, Pakistan

Usman Khalid Chaudhry
Department of Agricultural Genetic Engineering, Ayhan Şahenk Faculty of Agricultural Sciences and Technologies, Niğde Ömer Halisdemir University, Niğde, Turkey

Harpreet Kaur Cheema
Department of Plant Breeding and Genetics, Punjab Agricultural University, Ludhiana 141004, India

Marwa M. El-Deriny
Department of Nematodes Diseases and Central Lab of Biotechnology, Plant Pathology Institute, Agriculture Research Center (ARC), Egypt, 12619

Shah Fahad
Department of Agriculture, University of Haripur, Haripur, Khyber Pakhtunkhwa, Pakistan

Shahid Farooq
Department of Agronomy, Faculty of Agricultural Sciences, Ghazi University, D. G. Khan, Pakistan

Noreen Fatima
University of Agriculture, 38000 Faisalabad, Pakistan

Rida Fatima
University of Agriculture, 38000 Faisalabad, Pakistan

Ali Fuat Gökçe
Department of Agricultural Genetic Engineering, Ayhan Şahenk Faculty of Agricultural Sciences and Technologies, Niğde Ömer Halisdemir University, Niğde, Turkey

Farhana Gul
Department of Agriculture, University of Swabi, Pakistan

Mohamed Hashim
Botany department, Assiut University, Egypt

Mohammad Hassanzadeh
University of Mohaghegh Ardabili, Moghan Faculty of Agriculture and Natural Resources, Ardabil, Iran.

Nahid Hazrati
University of Mohaghegh Ardabili, Moghan Faculty of Agriculture and Natural Resources, Ardabil, Iran.

Junwei Huo
College of Horticulture & Landscape Architecture, Northeast Agricultural University, Harbin, Heilongjiang Province, China, 150030

Ansar Hussain
Department of Plant Breeding & Genetics, Faculty of Agricultural Sciences, Ghazi University, D. G. Khan, Pakistan

Mona Hussein
Biology department, Taibah university, Madinah, Saudi Arabia

Dina S. S. Ibrahim
Department of Nematodes diseases and Central Lab of Biotechnology, Plant Pathology Institute, Agriculture Research Center (ARC), Egypt, 12619

Muhammad Ijaz
College of Agriculture, BZU Bahadur Campus, Layyah, Pakistan

Imran
Agronomy Department, The University of Agriculture Peshawar, Pakistan

Muhammad Zaffar Iqbal
Agricultural Biotechnology Research Institute, Ayub Agricultural Research Institute Faisalabad 38000, Pakistan

Nusrat Jahan
Department of Botany, Faculty of Life Sciences Aligarh Muslim University, Aligarh-202002, UP, India

Shakra Jamil
Agricultural Biotechnology Research Institute, Ayub Agricultural Research Institute Faisalabad 38000, Pakistan

Muhammad Tariq Javed
Department of Botany, Faculty of Life Sciences, Government College University, Faisalabad 38000, Pakistan

Muhammad Daniyal Junaid
Department of Agricultural Genetic Engineering, Ayhan Şahenk Faculty of Agricultural Sciences and Technologies, Niğde Ömer Halisdemir University, Niğde, Turkey

Shah Khalid
Agronomy Department, The University of Agriculture Peshawar, Pakistan

Muhammad Ifnan Khan
Department of Plant Breeding & Genetics, Faculty of Agricultural Sciences, Ghazi University, D. G. Khan, Pakistan

Ekaterina Kozuharova
Medical University of Sofia, Faculty of Pharmacy, Department of Pharmacognosy

Shahzadi Mahapara
Department of Plant Breeding & Genetics, Faculty of Agricultural Sciences, Ghazi University, D. G. Khan, Pakistan

Seema Mahmood
Institute of Pure and Applied Biology, Bahauddin Zakariya University, Multan, Pakistan

Muhammad Mumtaz
Department of Management Sciences, University of Haripur, Pakistan

Huma Naz
Mohammad Ali Nazeer Fatima Degree College, Hardoi, U.P., India

Ali Noman
Department of Botany, Government College University, Faisalabad, Pakistan

Sibgha Noreen
Institute of Pure and Applied Biology, Bahauddin Zakariya University, Multan, Pakistan

Ali M. Qaisi, Department of Pharmaceutical Sciences, School of Pharmacy, University Of Jordan, Amman, Jordan

Dong Qin
College of Horticulture & Landscape Architecture, Northeast Agricultural University, Harbin, Heilongjiang Province, China, 150030

Sajid Ur Rahman
Agricultural Biotechnology Research Institute, Ayub Agricultural Research Institute Faisalabad 38000, Pakistan

Abdur Rauf
Department of Chemistry, University of Swabi, Anbar, Khyber Pakhtunkhwa, Pakistan

Ali Raza
Institute of Tropical Agriculture and Forestry, Hainan University, Haikou, China

Rokayya Sami
Department of Botany, Karnatak University, Dharwad, 580003, Karnataka, India

Jeyabalan Sangeetha
Department of Environmental Science, Central University of Kerala, Kasaragod, 671316, Kerala, India

Canan Sevinc
Necmettin Erbakan University, Faculty of Science, Department of Biotechnology

Mohd Abas Shah
Potato Research Station, Jalandhar, Punjab-144 003, India

Mohd Abas Shah
Department of Nutrition and Food Science, Taif University, Taif, Al-huwayah, 888, Kingdom of Saudi Arabia

Sana Shaheen
University of Agriculture, 38000 Faisalabad, Pakistan

Rahil Shahzad
Agricultural Biotechnology Research Institute, Ayub Agricultural Research Institute Faisalabad 38000, Pakistan

Hamna Shahzadee
University of Agriculture, 38000 Faisalabad, Pakistan

Iram Sharif
Cotton Research Station, Ayub agricultural Research Institute Faisalabad 38000, Pakistan.

Sonia Sheoran
ICAR-Indian Institute of Wheat and Barley Research, Karnal, India

Garima Singroh
ICAR-Indian Institute of Wheat and Barley Research, Karnal, India

Razia Sultana
Agricultural Biotechnology Research Institute, Ayub Agricultural Research Institute Faisalabad 38000, Pakistan

Devarajan Thangadurai
Department of Botany, Karnatak University, Dharwad, 580003, Karnataka, India

Devarajan Thangadurai
Department of Environmental Science, Central University of Kerala, Kasaragod, 671316, Kerala, India

Munir Turk
Department of Plant Production, Jordan University of Science and Technology, Irbid, Jordan

Ume Ummara
Institute of Pure and Applied Biology, Bahauddin Zakariya University, Multan, Pakistan

Muhammad Mohsin Waqas
Department of Agricultural Engineering, Khwaja Fareed University of Engineering and Information Technology, Rahim Yar Khan, Pakistan

Moh'd Ali Wedyan
Department of Biology and Biotechnology, The Hashemite University-Zarqa, Jordan

Fuchun Xie
College of Horticulture & Landscape Architecture, Northeast Agricultural University, Harbin, Heiongjiang Province, China, 150030

Madiha Zainab
College of Life Science, Fujian Agriculture and Forestry University, Fuzhou, China

1

Bimolecular Invention in Understanding Plant Adaptation to Climate Change

Nahid Hazrati and Mohammad Hassanzadeh
Moghan Faculty of Agriculture and Natural Resources, University of Mohaghegh Ardabili, Ardabil, Iran

1.1 Introduction

Increasing innovation in technologies that promote mitigation and adaptation is essential for encountering climate change. Such innovation may decline the costs of policy measures and improve more effectively responses to the complexities of climate changes. A great many literatures have examined the impact of innovation in climate policy (Popp 2010; Saud et al. 2013; Shah et al. 2013; Saud et al. 2014; Hafiz et al. 2016; Saud et al. 2016; Aziz et al. 2017a,b; Habib ur Rahman et al. 2017; Kamarn et al. 2017; Uz Zaman et al. 2017; Saud et al. 2017; Wajid et al. 2017; Yang et al. 2017; Zahida et al. 2017; Adnan et al. 2018a,b; Akram et al. 2018a,b; Depeng et al. 2018; Hafiz et al. 2018; Tariq et al. 2018; Adnan et al. 2019; Ahmad et al. 2019; Baseer et al. 2019; Hafiz et al. 2019; Muhammad et al. 2019; Sajjad et al. 2019; Wu et al. 2019; Adnan et al. 2020; Farhana et al. 2020; Farhat et al. 2020; Hafiz 2020; Hussain et al. 2020; Ilyas et al. 2020; Jan et al. 2019; Mubeen et al. 2020; Rehman et al. 2020; Saleem et al. 2020a,b,c; Shafi et al. 2020; Subhan et al. 2020; Wahid et al. 2020; Wu et al. 2020; Zafar-ul-Hye et al. 2020a,b).

Due to large number of fundamental and applied researches, significant progress has been made so far in understanding the mechanisms and processes of abiotic stress adaptation and defence in various plant species (Fraire-Velázquez et al. 2011; Fahad and Bano 2012; Fahad et al. 2013; Fahad et al. 2014a,b; Fahad et al. 2015a,b; Fahad et al. 2016a,b,c,d; Fahad et al. 2017; Fahad et al. 2018; Fahad et al. 2019a,b; Hesham and Fahad 2020). The sensing of abiotic or biotic stress circumstances induces signalling cascades that activate ion channels; kinase cascades; accumulation of hormones such as salicylic acid (SA), ethylene, jasmonic acid (JA), and abscisic acid; and production of reactive oxygen species (ROS). These signals ultimately induce expression of specific subsets of genes relating to defence that results in the assembly of the overall defence reaction (Jaspers and Kangasjärvi 2010; Pérez-Clemente et al. 2013).

In recent years, Omics technologies have been extensively used in the study of biotic and abiotic stress responses in plants. Omics enables a system biology approach towards understanding the complex interactions between genes, proteins, and metabolites within the resulting phenotype. This approach relies mainly on chemical analytical methods, bioinformatics and computational analysis, and many other disciplines of biology that lead to product conservation and improvement (Van Emon 2016).

Recent advances in genetics are expected to increase crop resilience to climate change by increasing the efficiency of plant breeding to improve agriculture under environmental constraints. It will be particularly useful for efficient exploitation of adaptation traits present in germplasm accessions (Huang and Han 2014) and to expand the gene pool for crop improvement (Brozynska et al. 2016).

In recent years, genomics knowledge based on next-generation sequencing (NGS), gene editing systems, gene silencing, and overexpression methods hasvesuggested a large amount of genetic information to help understanding the mechanisms of biotic and abiotic stress responses in plants (Luan et al. 2015; Ibraheem et al. 2018).

Proteomics is able to successfully distinguish and characterize many proteins in legumes and cereals responsible for stress responses and their regulations (Silva-Sanchez et al. 2015). Metabolomics focuses on the global profile of the metabolites with low molecular weight (<1000 Da) that are the end products of metabolisms in tissues and even whole organisms that are produced in response to abiotic stresses such as heavy metal toxicity, salinity, drought, and so on (Brosche et al. 2005; Ibraheem et al. 2018).

1.2 Genomics

1.2.1 Molecular Marker Resources

Although we have available tools, understanding of the genetic basis of adaptation is still an interesting task. We only have a basic understanding of most important traits, and the number of genes controlling each of these traits is large; therefore, the entire (and objective) approach would be understanding correlation between genes with phenotypes by scanning the whole genome with molecular markers (Howe and Brunner 2005). Molecular makers are used to detect the genetic variation caused by DNA polymorphisms in the DNA. The whole genome sequencing (WGS) provided the basis for the development of thousands of simple sequence repeat (SSR) markers and millions of single-nucleotide polymorphism (SNP) markers. The main method for the development of plant adaptation is molecular mapping and marker-assisted selection. In different species, amplified fragment length polymorphism (AFLPs), restriction fragment length polymorphism (RFLPs), RAPDs, microsatellites or simple sequence repeat (SSR), and SNPs are used for mapping of interesting traits (Iqbal 2019).

Since commercial SNP-genotyping platforms, such as Taqman, SNPlex, KASPar, Axiom Biobank, Infinium II, BioMark HD, GoldenGate, and iPlex, have been developed, the cost per data point for SNP-based genotyping has become cheaper than that for SSRs. However, the SNP information for target organisms is essential for commercial SNP-genotyping platforms, causing the increased costs and a longer experimental time (Kim et al. 2016; Chung et al. 2017).

Appearance of the next-generation sequencing (NGS) technologies has provided new opportunities for potent genotyping in various plant species. Recent improvements in high-throughput sequencing have enabled sequences to be used to detect and score single-nucleotide polymorphisms (SNPs) by shortening of the time-consuming process needed for marker development (Chung et al. 2017). Three main complexity reduction methods, namely reduced representation libraries (RRLs), restriction site associated DNA (RAD) sequencing, and genotyping-by-sequencing (GBS), are routinely used. Among these, GBS is a simple, robust, and affordable procedure for SNP distinguishing and mapping. Totally, this approach decreases complexity of genome with restriction enzymes (REs) in high diversity, and large genome species for efficient, high-throughput, and highly multiplexed sequencing. By using appropriate REs, repetitive regions of genomes can be inhibited and lower copy regions can be targeted, which decrease problems of alignments in genetically diverse species. This method was first explained by Elshire Robert et al. (2011). However, sequencing-based genotyping methods require computational expertise and a lot of time for analysing of data. This inhibits its use in marker-assisted breeding where timely selection is very crucial (Deshmukh et al. 2014).

Genome-wide association study (GWAS) is an observational study of a genome-wide set of genetic variants in different individuals to observe association of any variant with a trait. GWAS assesses the entire genome, opposite to methods that specifically test a small number of prespecified genetic regions. GWAS typically concentrates on correlations between single-nucleotide polymorphisms (SNPs) and traits. In general, where natural selection is acting and fixing the allele (eliminating other variants) of the SNPs, that constitutes the most favourable genetic adaptation (Barreiro et al. 2008).

Integrative investigations including GWAS, functional studies, selection scans, and fitness measurements in this regard have successfully distinguished loci for adaptation, demonstrated the molecular basis of genetic trade-offs, and showed that fitness can be predicted by polygenic effects of a number of loci related to local climate (Bamba et al. 2019).

1.2.2 QTL Mapping

Candidate genes can also be distinguished based on their positions on quantitative trait locus (QTL) maps or patterns of gene expression. Mapping quantitative trait loci (QTL) has become a routine tool for complex traits in functional genomic studies. Functional genomics is an important tool to find the correlation between phenotype and genome of an organism subjected to various environmental conditions (Soda et al. 2015; Ahmad et al. 2018). However, the construction of a high-density genetic linkage map is essential for genetic investigation of a target trait through quantitative trait locus (QTL) analysis. Moreover, high-resolution genetic linkage map is a crucial and powerful tool for positional cloning of genes, comparative genomic analysis, and scaffold sequence anchoring and genome assembly (Jones et al. 2009; Tang et al. 2019).

QTL analysis is complicated by the fact that one QTL region can have pleiotropic effects on a wide range of traits or consists of multiple QTLs. In addition, QTLs are frequently subjected to epistatic interactions, and their traits can be significantly influenced by the environment (Dixit et al. 2014; Tsaneva et al. 2019), so it is important to identify specific genes from new germplasm resources that are tolerant to multiple stresses. Exploit of the functions of genes that are responsible for drought will enable the plant biologists to use them in plant breeding programs to obtain cultivars resistant to drought stress (Ahmed et al. 2011). Realizing that mechanism of drought tolerance is quantitatively inherited and controlled by various genetic loci has led to the development of several drought-related QTLs (Sayed et al. 2012; Kalladan et al. 2013). Ahmad et al. (2018) have collected QTLs for different traits related to the adaptation of cereals to various stress conditions.

Abiotic stress induced through transcription elements, that is, dehydration-responsive element binding (DREB). *Dreb 1* genes are placed on 3A, 3B, and 3D chromosomes in wheat. Due to mapping, *Dreb B1* gene is placed between *Xmwg818* and *Xfbb117* on the 3BL chromosome. *Dreb B1* gene is responsible for salinity, drought, and heat tolerance in wheat. Another example is reduced height gene (Rht) in wheat that makes short stature. Dwarfing genes *Rht-B1b*, *Rht-D1b,* and *Rht8* are identified to have positively correlated with drought tolerance in wheat (Gasperini et al. 2012; Iqbal 2019).

Huang et al. (2018) carried out a study on QTL mapping of *fusarium* head blight and correlated agromorphological traits in an elite barley cultivar; the mapping population of recombinant inbred lines (RILs) was obtained from a cross between a moderately susceptible elite cultivar named *Rasmusson* and a highly susceptible Japanese landrace PI 383933 and evaluated for FHB severity, deoxynivalenol (DON) accumulation, and various agromorphological traits. QTL analysis using composite interval mapping (CIM) distinguished the main QTLs associated with FHB and DON on the centromeric region of chromosome 7 H, which was also correlated with spike length (SL), spike density (SD), and plant height (HT).

1.3 Transcriptomics

Transcriptomics is a powerful tool for understanding biological systems, and transcriptomic techniques such as sequencing-based approaches (RNA-seq)/next-generation sequencing (NGS) (Valdes et al. 2013), and microarrays methods (Johnson et al. 2014) provide capability of improved understanding of the functional elements of the genome. At the transcriptome level, technological innovations have made it feasible to overview the changes that occur at the transcriptomic level under different environmental circumstances. Microarrays and RNA sequencing techniques are employed to explain the differential expression of genes involved in biotic and abiotic stress adaptation in a variety of plant species (Budak and Akpinar 2015; Ibraheem et al. 2018).

Application of the next-generation sequencing (NGS) along with methods, such as restriction site-associated DNA sequencing (RADseq) or genotyping by sequencing GBS, specific length amplified fragment sequencing (SLAF-seq), double-digest RAD (ddRAD), and amplified-fragment single-nucleotide polymorphism and methylation (AFSM), is particularly promising. These technologies have successfully been applied to provide high-resolution genetic linkage maps for many crops (Bai et al. 2018; Hu et al. 2018; Huang et al. 2018; Tang et al. 2019). Above-mentioned methods allow deep coverage of the regions adjacent to restriction sites and provide high potential for high-throughput genotyping of entire genomes. Compared with microarray analyses, these methods present advantages of the flexibility and the lack of requirements to refer to a set of predetermined polymorphisms.

Microarray is considered to be a high-throughput technology where thousands of probes representing different genes are hybridized with RNA samples. Gene expression is calculated using hybridization signal level. The Affymetrix Gene Chip representing 61 K probe sets is routinely used for transcriptome profiling under different abiotic stresses (Le et al. 2012). The normalized expression data generated by the Affymetrix Gene Chip can be used to compare plants' experiments performed all around the world (Deshmukh et al. 2014). Gene expression profiling using microarrays not only provides a prevalent sensitive tool for investigating the phenotypic plasticity, but also provides an entry into revelation of the actual genes responsible for natural variation in adaptive traits. However the extent and adaptive significance of phenotypic plasticity remains somewhat understood, it is clear that phenotypic plasticity may be a powerful mean of adaptation in which large-scale modifications occur rapidly in gene expression in response to environmental changes (Howe and Brunner 2005).

Zhu et al. (2013) applied a comparative microarray analysis approach to study the transcriptome changes of cotton plant under five abiotic stresses. Their investigation unravelled the functional genes and stress-related pathways, and also suggested a cross-talk of responsive genes or pathways to various abiotic stresses in seedlings of this plant.

The relative accuracy of microarrays and RNA-seq has been evaluated using proteomics, and it has been revealed that RNA-seq provides a better assessment of absolute expression levels. RNA-seq is used for transcription start site mapping, small RNAs' characterization, gene fusion detection, and detection of alternative splicing events so that application of RNA-seq can give an enhanced understanding of molecular regulations of adaptation process (Ozsolak and Milos 2010; Deshmukh et al. 2014).

Mahdavi Mashaki et al. (2018) have performed RNA sequencing in roots and shoots of two chickpea genotypes under water-deficit conditions in early flowering stage. Gene Ontology (GO) analysis revealed several subcategories related to the stress, including response to drought in the tolerant genotype as compared with the sensitive one under drought stress. In addition, they distinguished several transcription factors (TFs) in major metabolic pathways such as ABA, proline, and flavonoid biosynthesis. Drought tolerance dissection in the genotypes revealed that the genes and pathways involved in shoots of tolerant genotype were the most important factor to make a difference between the genotypes for drought tolerance. In a research, RNA-seq analysis of *Chenopodium quinoa* under four water treatments (field capacity to drought) revealed an overlap between drought stress tolerance and other abiotic stress mechanisms (Raney et al. 2014).

By application of the genomic techniques such as high-throughput analysis of expressed sequence tags (EST), the knowledge about plant response to abiotic stresses is enhanced. Expressed sequence tags, which are short, single-pass sequence reads from cDNA, have proven to be an efficient and rapid strategy to distinguish novel genes involved in tolerance to environmental changes. ESTs, along with advances in parallel designs for gene expression analysis, with large-scale EST projects provide new insights into understanding the molecular basis of important traits in crop plants. EST sequence projects have been completed for some plant species. These projects have provided useful tools for intragenomic and intergenomic comparisons, molecular marker identification, microarray development, gene discovery, and polyploid species' genomic resource development (Manickavelu et al. 2012).

Different investigations illustrate that plant's defence response genes are transcriptionally activated by pathogens and different types of abiotic stress. Induction of specific defence genes in response to some pathogens has been shown to be related to specific environmental conditions and indicates a complex signalling network that helps plant withstand various environmental stresses (Jaspers and Kangasjärvi 2010; Pérez-Clemente et al. 2013). On the other hand, kinase cascades of the mitogen-activated protein kinase (MAPK) class play an important role in plant signalling of a variety of biotic and abiotic stresses, and it is an essential step in the establishment of resistance to pathogens. It has been reported that in

Arabidopsis, MEKK1 and ANP1 act in the environmental stress response, and MPK3, MPK4, and MPK6 are involved in different signalling pathways responding to various external stimuli including abiotic stresses, pathogens, and oxidative stress (Pitzschke et al. 2014; Takáč et al. 2016). In addition, reactive oxygen species (ROS) and lipid-derived pathways have been involved in plant adaptation to environmental stresses (Garg et al. 2014; Golldack et al. 2014).

Another important factor is calcium ions that play a crucial role in several signal transduction cascades in plants, especially abiotic stress signalling. Calcineurin B-like proteins (CBLs) function as calcium sensor and modulate the activity of CBL-interacting protein kinases (CIPKs). The CBL–CIPK network helps maintaining proper ion balances during adverse conditions (Garg et al. 2014).

Golldack et al. (2014) and Nakashima et al. (2014) studies illustrated the main role of transcription factors (TFs) in abiotic stress response and tolerance mechanisms and provided important insights into signalling mechanism and transcriptional regulatory network. Nakashima et al. (2014) suggested that TFs function in cross-talk among various abiotic stress responses and are utilized to improve abiotic stress tolerance in different crops (Garg et al. 2014).

Transcriptomic technologies have the capability of providing deep coverage and unbiased representation of transcript abundance, which is very important in non-model plants lacking genome sequence information. However, the frequent incongruity between the abundance of cognate gene transcripts and protein levels suggested the need of complementary analysis of the proteome for further validation of candidate genes and pathways (Soda et al. 2015).

1.4 Proteomics

Although most stress-responsive genes are regulated at the transcriptional level, in many cases, it has been shown that changes at the transcriptional level do not always necessarily correspond to the changes in protein content, which suggests that the transcriptome needs to be assessed in conjunction with the proteome to link the phenotype of stress adaptation (Li et al. 2018). Anderson and Seilhamer (1997) showed that the correlation coefficient in quantity between mRNA and protein abundance is relatively low. Protein level may be modulated by changing either the rate of synthesis or the stability of the messenger RNA (mRNA or transcript), or the synthesis or stability of the protein itself (Komili and Silver 2008; Soda et al. 2015), so the survival of cells exposed to stress conditions requires rapid and efficient reprogramming of messenger RNA (mRNA) translation, the process aimed at the conservation of energy and repair of stress-induced damage. Such changes in stress-responsive protein synthesis target both global translation and the translation of specific mRNAs (Advani and Ivanov 2019). Over the past decade, there has been a rapid increase in the understanding of the role of translation as a key regulatory node in gene expression and plant adaptation. Thus, the development of methods to infer the translatome, which indicates the entirety of mRNAs associated with ribosomes for protein synthesis, has facilitated the discovery of new principles and mechanisms of translation and expanded our view of protein synthesis (King and Gerber 2016).

Although modulation of translation can happen at any step of protein synthesis, most regulatory events occur in the initiation step. As a general rule, once translation enters the elongation phase, ribosomes are responsible for complete synthesis of a nascent polypeptide to avoid formation of a harmful incomplete protein. Translation rates are not uniform among different mRNAs and are commonly determined by the proteins bonded to them to form messenger ribonucleoprotein complexes (mRNPs). Specific protein factors within mRNPs determine the translational status (e.g. stimulation or inhibition of translation) and localization (e.g. sequestration into RNA granules) of mRNAs (Hershey et al. 2018). Applying of such factors into mRNPs is determined by the presence of specific sequences or structural motifs in each mRNA, which interact with specific RNA-binding proteins (RBPs). Other factor that determines translational rates is the availability of transfer RNAs (tRNAs) and codon composition of an mRNA (Kuscu et al. 2018; Advani and Ivanov 2019).

Since proteomics is a powerful tool to investigate how plants adjust their physiology and metabolism in response to an altered environment, proteomic experiments have been performed several times in different stress conditions in tissues, organelles, and plant species (Rodziewicz et al. 2014;

Janmohammadi et al. 2015; Jorrín-Novo et al. 2015; Hoang et al. 2017; Lu et al. 2017; Carrera et al. 2018; Kosová et al. 2018; Qin et al. 2019).

Mass spectrometry (MS) plays an important role in enabling the analysis of proteomes and typically is the method for identifying proteins present in biological systems, and several methodologies based on MS have been developed for the analysis of proteomes. Due to the complexity of the biological systems, proteome should be separated before analysis. Both gel chromatography- and liquid chromatography-based separations have been found to be useful in this regard. Typically, after these extensive separations, proteins are characterized by MS analysis (Hixson et al. 2017). MS breaks molecules into their ions and records their mass to charge ratio, which is then calculated by the time it takes for the ions to travel through the flight tube. Detecting and identifying of the proteins present in very small quantities in biological samples have been easier using new developments in soft ionization MS technology. MALDI (matrix-assisted laser desorption ionization) and ESI (electrospray ionization) are two new methods for preparing the protein for ionization. In electrospray ionization, the protein sample in solution is sprayed into a tube and the molecules are induced by an electric field to pick up extra protons from the solvent, but in MALDI method, the protein sample is adsorbed onto a solid matrix, and protonation is achieved with a laser beam (Soderberg 2019).

Additional experimental procedures such as tandem MS (MS/MS) are employed when confident identifications cannot be made by mass measurements alone, particularly in lower resolution instruments. Also, several procedures with or without stable isotopic labelling have been proposed to facilitate protein quantitation (e.g. characterize changes in protein abundances between given biological substances) (Hixson et al. 2017). SDS-PAGE technique separates proteins by size. These proteins are trapped inside a gel but can be transferred into nitrocellulose for Western blot analysis. In the Western blot technique, a protein of interest can be visualized by adding labelled antibodies. The Western blot can be used to distinguish the relative abundance of the given protein. High -performance liquid chromatography (HPLC) is another useful tool for separating proteins. This method retains the proteins in a liquid and the column materials separate the proteins from the mixture (Clark and Pazdernik 2016). Developed techniques and equipment for separation and detection such as high-pressure nano-HPLC and new techniques for multidimensional HPLC separation enable proteomics to experience dynamic growth and enter new pathways (Mitulović 2014).

Since microarrays are used for mass screening of nucleic acids, protein arrays can be used for global protein analysis. They consist of arrays with many different proteins applied as spots on a solid support. One type of protein array contains antibodies for screening. The other type represents the proteome and consists of many different cellular proteins. Such arrays may be screened for binding of a variety of labelled molecules to the proteins. These involve other proteins, enzyme substrates, and RNA (Clark and Pazdernik 2016).

Transcriptional regulation of plant genes is controlled by networks of transcription factors together with transcription factor-binding sites (TFBS) (Chaves and Oliveira 2004). Transcription factors are proteins containing a DNA domain that binds to cis-acting elements that are present in the upstream region of all gene promoters (Ciarmiello et al. 2011). Furthermore, TFs activate or repress the activity of RNA polymerase, leading to gene regulation (Kimotho et al. 2019). Based on the specific motifs appearing in their protein structure, the TFs are classified into different families. Regarding TFs, it has been reported that around 5–7% of coding sequences present in a plant genome are used for making this protein group (Udvardi et al. 2007).

We will briefly discuss several widely studied and researched TF families involved in abiotic stress regulation mechanisms, together with their regulons.

1.5 The Ethylene-Responsive Element-Binding Factors (ERFs)

ERFs are versatile TFs that are involved in development as well as responses to different types of stresses. Usually, ERFs affect cross-talk between different responses acting at the interface between different signalling pathways. The RNA levels of specific *ERF* genes are regulated by heat, drought, cold, pathogen infection, wounding, or treatment with ethylene, SA, or JA (Onate-Sánchez and Singh 2002; Cheng et al. 2013). Based on the number of AP2/ERF domains and other signatures present in the protein, this superfamily is divided into five categories, known as (i) RAV (related to ABI3/VP1) TFs with an AP2/ERF domain and a B3 DNA-binding domain, (ii) AP2 TFs carrying two AP2/ERF domains, (iii) DREB (dehydration-responsive element-binding proteins)/CBF (CRT (C-repeat)-binding factor) with an AP2/ ERF domain but being different in amino acid sequence from that of the AP2/ERF domain of ERF TFs, (iv) ERF (ethylene-responsive element-binding factor) with an AP2/ERF domain, and (v) Soloist (Licausi et al. 2013). The last subgroup has been recently added to classify any members that possess AP2 domain but display a more diverse sequence and structure in comparison with the ERF TFs (Nakano et al. 2014; Hoang et al. 2017).

It has been shown that ERF proteins function as either activators or repressors of transcription, which is of great relevance in all processes related to plant development and its responses to adverse growing conditions and in regulating plant adaptive mechanisms. According to some investigations, AP2 TFs are suggested to bind to an AT-rich DNA domain or bind to sequence GACC (A/G)N (A/T)TCCC(A/G) ANG(C/T) (Dinh et al. 2012). Regarding DREB TFs, previous studies have shown that dehydration-responsive element-binding protein 1 (DREB1)/C repeat-binding factor (CBF) and DREB2 regulons function in an ABA-independent manner, and they are able to bind to the DRE (dehydration-responsive element)/CRT regions that are commonly found in the promoters of genes which respond to salinity, drought, and adverse temperature conditions (Lakhwani et al. 2016). *DREB* genes are usually divided into six subgroups (A-1 to A-6). The general trend observed in *Arabidopsis* is that *DREB1/CBF* (A-1) genes are induced by low temperature, while *DREB2* genes (A-2) are involved in responses to osmotic stress (salinity and dehydration) (Nakashima et al. 2009). Conversely, ABI4, as the only member of the A-3 subgroup, is involved in lipid mobilization in embryos, ABA and sugar signalling, chloroplast functioning, retrograde signalling, and germinating seeds (Shkolnik-Inbar and Bar-Zvi 2011). Investigations suggest that A-4 genes are involved in the cross-talk between abiotic and biotic stresses response through connecting DRE- and ERE- (ethylene-responsive element) mediated signalling pathways. Moreover, members in *DREB-A4* family such as HARDY (*HRD*) also positively regulate drought and salt tolerance (Xie et al. 2019). Genes of the A-5 subgroup from *Arabidopsis* such as *RAP2.1* exhibit induction by drought and cold stress (Dong and Liu 2010). In group A-6, *RAP2.4* gene is most strongly expressed in dry seeds and roots but can be strongly induced throughout the plant by abiotic stress treatments, such as drought and salt, or by the abscisic acid (Phuong and Hoi 2015) and *RAP2.4B* (heat responsive) (Figueroa-Yañez et al. 2016).

1.6 NAC Transcription Factors

NAC TFs (ATAF, CUC, and NAM) are plant-specific multifunctional transcription factors and involve in wide range of processes relating to biotic and abiotic stress responses, flowering, anther dehiscence, lateral root development, etc. They act downstream of auxin and ethylene pathways and also ABA signalling pathway (He et al. 2005) and are mostly involved in salinity tolerance (Gujjar 2016). The TFs belonging to the NAC family share a highly conserved N-terminus made up of nearly 150–160 amino acids, a DNA-binding domain that carries five subdomains (A–E) and a varying C-terminal (Hu et al. 2008; Kimotho et al. 2019). It has been observed that overexpression of a stress-responsive NAC transcription factor gene *ONAC022* and overexpression of *NAC67* transcription factor from finger millet (*Eleusine coracana* L.) confer tolerance to drought and salinity stresses in rice (Hong et al. 2016; Rahman et al. 2016). Furthermore, Fang et al. (2015) stated that stress-responsive NAC transcription factor *SNAC3* confers heat and drought tolerance through modulation of five reactive oxygen species (ROS) in rice. It has been shown that *PbeNAC1* transcription factor derived from *Pyrus betulifolia* improved cold and drought tolerance through interacting with *PbeDREB*s and activated the expression of stress-responsive genes (Jin et al. 2017) Also, it was found that *CarNAC4*, which is considered NAC-type chickpea transcription factor, enhanced drought and salt stress tolerances in *Arabidopsis*. Overexpression of an *NAC* gene *MlNAC5* derived from *Miscanthus lutarioriparius* confers enhanced drought and cold tolerance in *Arabidopsis* through transcriptionally activating stress-responsive genes and increase in hypersensitivity to abscisic acid (Yang et al. 2015). Overexpression of *BoNAC019*, which is an NAC transcription factor obtained from *Brassica oleracea*, negatively regulates the dehydration response and anthocyanin biosynthesis in *Arabidopsis*; the expression of *BoNAC019* was induced by dehydration, salt, abscisic acid, and H_2O_2 treatments, and overexpression of *BoNAC019* reduced drought tolerance (Wang et al. 2018).

1.7 bZIP Transcription Factors

Another important family of transcription factors is the so-called Basic-Domain Leucine-Zipper (bZIP), which is considered to be regulator of important plant processes. It was found that bZIP is involved in many important biological activities, such as histological tissue differentiation, metabolic activity, cell elongation, seed storage protein gene regulation, embryogenesis, and seed maturation. At the same time, the bZIP is involved in the response to biotic and abiotic stresses including photoreaction, salt and drought tolerance, hormone and sugar signalling, and pathogen defence (Liu et al. 2019). Thc bZIP proteins are characterized by a 40- to 80-amino-acid-long conserved domain (bZIP domain) that is composed of two motifs: a leucine zipper required for TF dimerization and a basic region responsible for specific binding of the TF to its target DNA (Gujjar 2016). This is the explanation why some of the bZIP TFs are best regulators for responses of plant to stress on account of their capacity of binding specifically to ABA-responsive element (ABRE) sequence, which is the cis-acting sequence found within promoter regions of many stress-responsive genes controlled by ABA, especially under osmotic conditions. The expression of *bZIP* gene family in signal transduction, stress responses, and development processes is mainly regulated by abscisic acid (ABA). ABA – as a plant hormone – plays significant role in regulating gene expression and related physiological processes in abiotic stress response (Kaifa et al. 2012).

1.8 WRKY Transcription Factors

WRKY proteins represent the largest superfamily of TFs, which are only found in plants. It has been shown that numerous WRKY TFs from a wide range of plant species participate in plant growth and development, metabolism, and responses to environmental ingredients (Phukan et al. 2016; Chen et al. 2017; Jiang et al. 2017). WRKY proteins participate in metabolism through the synthesis of phytohormones such as jasmonic acid (JA), salicylic acid (SA), or phytoalexins and other defence-related chemicals that emphasize major role of the WRKY family in plant immunity (Liang et al. 2017; Chen et al. 2019). WRKY family members demonstrate enhanced expression and/or DNA-binding activity following the induction by a range of defence signals and wounding (Rushton et al. 2010). WRKY proteins also interact with other proteins, such as kinases, receptors, and other TFs, and are in close relation with transcriptional regulatory networks (Lal et al. 2018). Functional analysis of *ZmWRKY33* under different abiotic stresses revealed that this gene is activated by cold, dehydration, ABA, and salt treatments (Li et al. 2013). In general, WRKY TFs regulate target genes through binding W-box (TTGACY, with the core sequence TGAC) cis-elements in gene promoters. However, some WRKY TFs recognize cis-elements without the W-box core sequence (Chen et al. 2019).

1.9 MYB Transcription Factors

Another important group of regulators in plant defence pathways against abiotic stresses are MYB TFs, which can be distinguished from members of other TF families by their specific binding to the MYB-binding sites (MBS) (Riechmann and Ratcliffe 2000). MYB family proteins, which are characterized by up to four repeat sequences (R) and depending on the number of repeats, can be classified into four different classes, R1, R2R3, R3, and R4. Among these four classes, the R2R3 class of MYB proteins is plant specific and forms the largest group of MYB TFs functioning in response to biotic and abiotic stresses along with primary and secondary metabolisms (Ambawat et al. 2013; Ng et al. 2018).

AtMYB96 acts through the ABA signalling cascade to regulate drought stress and disease resistance (Seo and Park 2010). AtMYB33 and AtMYB101 are involved in ABA-mediated responses to environmental signals (Pérez-Clemente et al. 2013). It was found that transcript levels of *PtsrMYB* were up-regulated by abiotic stresses such as salt, dehydration, cold, and ABA treatments. It has been observed that overexpression of *PtsrMYB* in tobacco confers enhanced dehydration tolerance, as indicated by less water loss, lower levels of malondialdehyde, and reactive oxygen species (Sun et al. 2014). Shukla et al. (2015) stated that in response to stresses such as desiccation, salinity, high temperatures, abscisic acid, and salicylic acid treatments, *SbMYB44* transcript from a succulent halophyte showed up-regulation. SbMYB44 recombinant protein showed binding to dehydration-responsive cis-elements (RD22 and MBS-1), which suggests its possible role in stress signalling. Lv et al. (2017) revealed that overexpression of *OsMYB30* in rice caused increased cold sensitivity, while the *osmyb30* knockout mutant resulted in increased cold tolerance. Also, microarray and quantitative real-time polymerase chain reaction analyses showed that a few β-amylase (*BMY*) genes were down-regulated by *OsMYB30*. It has been found that overexpression of *OsMYB6*-a *MYB* family gene increased tolerance to drought and salinity stresses in transgenic rice which had been evaluated by higher proline content, higher CAT and SOD activities, lower REL and MDA content under drought and salt stress conditions in addition to the remarkable increase in expression of abiotic stress-responsive genes in *OsMYB6* these plants (Tang et al. 2019). Overexpression of the *OsMYB2* resulted in higher accumulation rates of soluble sugars and proline, and less amounts of H_2O_2 and malondialdehyde than wild-type plants, and there was higher up-regulation of stress-related genes, including *OsRab16A*, *OsLEA3*, and *OsDREB2A,* and enhanced activities of antioxidant enzymes including peroxidase, superoxide dismutase, and catalase which show interactions of TFs in stress signalling pathways (Yang et al. 2012).

1.10 MYC Transcription Factors

MYC proteins are involved in the response of plants to unfavourable environmental conditions. MYC and MYB recognition sites are present in the promoter region of *RD22* gene (Abe et al. 2003). This transcription factor family plays important role in induction of the apoptosis, which includes response to pathogen attack (Pérez-Clemente et al. 2013). Especially, several *MYC* genes were reported to correlate with JA (jasmonate) signal. For example, *Arabidopsis MYC2*, *MYC3*, *MYC4*, and *MYC5* are master regulators in JA signal pathway (Chen et al. 2019). Meanwhile, *MYC* genes function in various molecular and physiological processes, especially in growth and development. In *Arabidopsis thaliana*, *MYC2* functions synergistically with *MYC3* and *MYC4* in regulating leaf senescence (Tiancong et al. 2015a), root elongation (Gasperini et al. 2015), stamen development (Tiancong et al. 2015b), seed formation, and accumulation of seed storage protein (Gao et al. 2016; Chen et al. 2019). Furthermore, expression of the *AtMYC2* is induced by drought and salt stresses, and overexpression of this gene enhances osmotic stress tolerance (Shinozaki and Yamaguchi-Shinozaki 2007). ICE1, a MYC-type transcription factor, plays an important role in the induction of *CBF3/DREB1A* to control cold-responsive genes and cold tolerance. Moreover, two MYC-type transcription factors, *MYC67* and *MYC70*, were identified as ICE1-interacting proteins. The MYC mutants were more tolerant to freezing temperatures than wild type (Ohta et al. 2018).

1.11 Metabolomics

Since metabolites are in close relation to phenotype and are the actual representatives of any visible change, metabolomic profiling of crops will provide useful insights into genetic mechanisms of plants adaptation in response to climate change. The information resulted from the systematic metabolomic investigations of the response of plants to climate change can be used to identify candidate genes responsible for conferring specific traits of adaptation (Kumar et al. 2014). The analysis of the metabolome using complementary mass spectrometry techniques (GC-MS and LC-Orbitrap-MS) is believed to allow the reliable quantification of metabolites; is comparatively rapid and cheaper; and has wide range of applications than other OMICS approaches. Capillary electrophoresis mass spectrometry (CE-MS) and nuclear magnetic resonance (NMR) are also used in plants sciences.

The chemical interaction between plants and their environment is mainly mediated by the biosynthesis of secondary metabolites, which perform their biological roles as a plastic adaptive response to their environment. Such chemical interaction often involves variations in synthesis of the plant metabolites (Pavarini et al. 2012; Miranda et al. 2015; Leite Sampaio et al. 2016). When a plant encountered environmental stress and perceived it, multiple signalling cascades are activated. These include intricate cross-talk between the different plant hormones and other signalling pathways involving phosphatases and kinases, ROS, calcium, and lipids. Metabolic studies also have revealed the variable response of plant metabolism in different developmental stages and degrees of dehydration. In this sense, most amino acids are accumulated in severely desiccated leaves but decreased in mild conditions.

Previous studies in *Pinus radiata* revealed that salicylic acid and abscisic acid are crucial factors in the initial response to heat stress (Escand6n et al. 2016) and may be attributed to the plant tendency to quickly regulate stomatal closure as observed in other species (Acharya and Assmann 2009). Other phytohormones such as cytokinins (CKs) and indoleacetic acid (IAA) seem to be more important for recovery of the plant. Nevertheless, how these hormones interact and trigger changes in the plant metabolome to cope with heat stress remains unknown (Escand6n et al. 2018). Analysis of the distinguished metabolites and metabolic pathways across the time of heat exposure by Escand6n et al. (2018) reveals the dynamics of the metabolome correlated with high temperature response in *P. radiata* and determined the existence of a turning point at which *P. radiata* plants are transformed from an initial stress response programme to an acclimation phase.

Verma et al. (2016) stated that under abiotic stress conditions, stress hormones such as abscisic acids and ethylene are induced, whereas production and distribution of growth-promoting hormones such as gibberellic acid (GA), brassinosteroid (BR), and cytokinin are modified to provide optimum responses. Regulation of plant hormone signalling during abiotic stresses is partly mediated by hormone and stress-sensitive transcription factors (Bechtold and Field 2018).

1.12 Conclusion

Molecular mechanisms in plant adaptation provide a comprehensive perspective of a wide variety of various mechanisms underlying adaptation to challenges for plant survival. Molecular mechanisms by exploring the latest technological advances used in plant adaptation researches provide readers with an overview of potent technologies and their applications.

Plant modifies its Omics profiles to cope with the changes in environment for survival. The main objective of the Omics approach is to realize the molecular interactions, their relationships with the signalling cascade, and to process the information that links specific signals with specific responses at molecular level.

Plants are inevitably exposed to the changing environmental conditions; hence, they should cope with them in order to minimize the loss of fitness. The acclimation and adaptation responses have extensively been investigated, but little is known about the complex underlying mechanisms. Increased knowledge about pathways and regulated proteins at the cellular levels would facilitate efforts to optimize rational breeding.

REFERENCES

Abe H, Urao T, Ito T, Seki M, Shinozaki K, Yamaguchi-Shinozaki K (2003) Arabidopsis AtMYC2 (bHLH) and AtMYB2 (MYB) function as transcriptional activators inabscisic acid signaling. Plant Cell 15:63–78.

Acharya BR, Assmann SM (2009) Hormone interactions in stomatal function. Plant Mol Biol 69:451–462. https://doi.org/10.1007/s11103-008-9427-0.

Adnan M, Fahad S, Khan IA, Saeed M, Ihsan MZ, Saud S, Riaz M, Wang D, Wu C (2019) Integration of poultry manure and phosphate solubilizing bacteria improved availability of Ca bound P in calcareous soils. 3 Biotech 9:368.

Adnan M, Fahad S, Muhammad Z, Shahen S, Ishaq AM, Subhan D, Zafar-ul-Hye M, Martin LB, Raja MMN, Beena S, Saud S, Imran A, Zhen Y, Martin B, Jiri H, Rahul D (2020) Coupling phosphate-solubilizing bacteria with phosphorus supplements improve maize phosphorus acquisition and growth under lime induced salinity stress. Plants 9:900. https://doi.org/10.3390/plants9070900.

Adnan M, Shah Z, Sharif M, Rahman H (2018b) Liming induces carbon dioxide (CO_2) emission in PSB inoculated alkaline soil supplemented with different phosphorus sources. Environ Sci Poll Res Int 25:9501–9509.

Adnan M, Zahir S, Fahad S, Arif M, Mukhtar A, Imtiaz AK, Ishaq AM, Abdul B, Hidayat U, Muhammad A, Inayat-Ur R, Saud S, Muhammad ZI, Yousaf J, Amanullah, Hafiz MH, Wajid N (2018a) Phosphate-solubilizing bacteria nullify the antagonistic effect of soil calcification on bioavailability of phosphorus in alkaline soils. Sci Rep 8:4339. https://doi.org/10.1038/s41598-018-22653-7.

Advani VM, Ivanov P (2019) Translational control under stress: reshaping the translatome. BioEssays 41:1900009. https://doi.org/10.1002/bies.201900009.

Ahmad HM, Ansari M-u-R, Azeem F, Tahir N, Iqbal MS (2018) QTL mapping for crop improvement against abiotic stresses in cereals. J Anim Plant Sci 28:1558–1573. https://www.researchgate.net/publication/329012134.

Ahmad S, Kamran M, Ding R, Meng X, Wang H, Ahmad I, Fahad S, Han Q (2019) Exogenous melatonin confers drought stress by promoting plant growth, photosynthetic capacity and antioxidant defense system of maize seedlings. PeerJ 7:e7793. http://doi.org/10.7717/peerj.7793.

Ahmed MS, Khaliq I, Farooq J, Awan SI, Ahmed N, Awan FS (2011) Assessment of the combining ability and authentication of F1 hybrids using SSR markers in wheat (*Triticum aestivum* L.). Front Agric China 5:135–140.

Akram R, Turan V, Hammad HM, Ahmad S, Hussain S, Hasnain A, Maqbool MM, Rehmani MIA, Rasool A, Masood N, Mahmood F, Mubeen M, Sultana SR, Fahad S, Amanet K, Saleem M, Abbas Y, Akhtar HM, Waseem F, Murtaza R, Amin A, Zahoor SA, ul Din MS, Nasim W (2018a) Fate of organic and inorganic pollutants in paddy soils. In: Hashmi MZ, Varma A (eds) Environmental pollution of paddy soils, soil biology. Springer International Publishing AG, Switzerland, pp 197–214.

Akram R, Turan V, Wahid A, Ijaz M, Shahid MA, Kaleem S, Hafeez A, Maqbool MM, Chaudhary HJ, Munis, MFH, Mubeen M, Sadiq N, Murtaza R, Kazmi DH, Ali S, Khan N, Sultana SR, Fahad S, Amin A, Nasim W (2018b) Paddy land pollutants and their role in climate change. In: Hashmi MZ, Varma A (eds) Environmental pollution of paddy soils, soil biology. Springer International Publishing AG, Switzerland, pp 113–124.

Ambawat S, Sharma P, Yadav NR, Yadav RC (2013) MYB transcription factor genes as regulators for plant responses: an overview. Physiol Mol Biol Plants 19:307–321.

Anderson L, Seilhamer JA (1997) Comparison of selected mRNA and protein abundances in human liver. Electrophoresis 18:533–537.

Aziz K, Daniel KYT, Fazal M,Muhammad ZA, Farooq S, Fan W, Fahad S, Ruiyang Z (2017a) Nitrogen nutrition in cotton and control strategies for greenhouse gas emissions: a review. Environ Sci Pollut Res 24:23471–23487. https://doi.org/10.1007/s11356-017-0131-y.

Aziz K, Daniel KYT, Muhammad ZA, Honghai L, Shahbaz AT, Mir A, Fahad S (2017b) Nitrogen fertility and abiotic stresses management in cotton crop: a review. Environ Sci Pollut Res 24:14551–14566. https://doi.org/10.1007/s11356-017-8920-x.

Bai B, Wang L, Zhang YJ, Lee M, Rahmadsyah R, Alfiko Y, Qing Ye B, Purwantomo S, Suwanto A, Chua NH, Yue GH (2018) Developing genome-wide SNPs and constructing an ultrahigh-density linkage map in oil palm. Sci Rep 8:691. https://doi.org/10.1038/s41598-017-18613-2.

Bamba M, Kawaguchi YW, Tsuchimatsu T (2019) Plant adaptation and speciation studied by population genomic approaches. Develop Growth Differ 61:12–24. https://doi.org/10.1111/dgd.12578.

Barreiro LB, Laval G, Quach H, Patin E, Quintana-Murci L (2008) Natural selection has driven population differentiation in modern humans. Nat Genet 40: 340–345. https://doi.org/10.1038/ng.78. PMID 18246066.

Baseer M, Adnan M, Fazal M, Fahad S, Muhammad S, Fazli W, Muhammad A, Jr. Amanullah, Depeng W, Saud S, Muhammad N, Muhammad Z, Fazli S, Beena S, Mian AR, Ishaq AM (2019) Substituting urea by organic wastes for improving maize yield in alkaline soil. J Plant Nutrition https://doi.org/10.1080/01904167.2019.1659344.

Bechtold U, Field B (2018) Molecular mechanisms controlling plant growth during abiotic stress. J Exp Bot 69:2753–2758. https://doi.org/10.1093/jxb/ery157.

Brosche M, Vinocur B, Alatalo ER, Lamminmaki A, Teichmann T, Eric AO, Dimitar D, Dany A, Marie-Béatrice Bogeat T, Arie A, Andrea P, Erwin D , Stephen R, Lars P, Petri A, Jaakko K (2005) Gene expression and metabolite profiling of *Populus euphratica* growing in the Negev desert. Genome Biol 6:R101. https://doi.org/10.1186/gb-2005-6-12-r101.

Brozynska M, Furtado A, Henry RJ (2016) Genomics of crop wild relatives: expanding the gene pool for crop improvement. Plant Biotechnol J 14:1070–1085. https://doi.org/10.1111/pbi.12454.

Budak H, Akpinar BA (2015) Plant miRNAs: biogenesis, organization and origins. Funct Integr Genomics 15:523–531.

Carrera DA, Oddsson S, Grossmann J, Trachsel C, Streb S (2018) Comparative proteomic analysis of plant acclimation to six different long-term environmental changes. Plant Cell Physiol 59:510–526. https://doi.org/10.1093/pcp/pcx206.

Chaves MM, Oliveira MM (2004) Mechanisms underlying plant resilience to water deficits: prospects for water-saving agriculture. J Exp Bot 55:2365–2384.

Chen F, Hu Y, Vannozzi A, Wu K, Cai H, Qin Y, Mullis A, Lin Z, Zhang L (2017) The WRKY transcription factor family in model plants and crops. Crit Rev Plant Sci 36:311–335.

Chen X, Li C, Wang H, Guo Z (2019b) WRKY transcription factors: evolution, binding, and action. Phytopathology Res 1:13.

Chen S, Zhao H, Luo T, Liu Y, Nie X, Li H (2019a) Characteristics and expression pattern of MYC genes in *Triticum aestivum*, *Oryza sativa*, and *Brachypodium distachyon*. Plants 8:274. https://doi.org/10.3390/plants8080274.

Cheng, MC, Liao PM, Kuo WW, Lin TP (2013) The Arabidopsis ETHYLENE RESPONSE FACTOR1 regulates abiotic stress-responsive gene expression by binding to different cis-acting elements in response to different stress signals. Plant Physiol 162:1566–1582. https://doi.org/10.1104/pp.113.221911.

Chung YS, Choi SC, Jun TH, Kim C (2017) Genotyping-by-sequencing: a promising tool for plant genetics research and breeding. Hortic Environ Biotechnol 58:425–431. https://doi.org/10.1007/s13580-017-0297-8.

Ciarmiello LF, Woodrow P, Fuggi A, Pontecorvo G, Carillo P (2011) Plant genes for abiotic stress, abiotic stress in plants - mechanisms and adaptations. InTech. DOI: 10.5772/22465. http://www.intechopen.com/books/abioticstress-in-plants-mechanisms-and-adaptations/plant-genes-for-abiotic-stress.

Clark DP, Pazdernik NJ (2016) Proteomics. In: Biotechnology, 2nd edn. Elsevier, pp 295–333. https://doi.org/10.1016/B978-0-12-385015-7.00009-0.

Depeng W, Fahad S, Saud S, Muhammad K, Aziz K, Mohammad NK, Hafiz MH, Wajid N (2018) Morphological acclimation to agronomic manipulation in leaf dispersion and orientation to promote "Ideotype" breeding: evidence from 3D visual modeling of "super" rice (*Oryza sativa* L.). Plant Physiol Biochem 135:499–510. https://doi.org/10.1016/j.plaphy.2018.11.010.

Deshmukh R, Sonah H, Patil G, Chen W, Prince S, Mutava R, Vuong T, Valliyodan B, Nguyen HT (2014) Integrating omic approaches for abiotic stress tolerance in soybean. Front Plant Sci 5:244. https://doi.org/10.3389/fpls.2014.00244.

Dinh, TT, Girke, T, Liu, X, Yant, L, Schmid, M, Chen, X (2012) The floral homeotic protein APETALA2 recognizes and acts through an AT-rich sequence element. Development 139:1978–1986.

Dixit S, Huang BE, Cruz MTS, Maturan PT, Ontoy JCE, Kumar A (2014) QTLs for tolerance of drought and breeding for tolerance of abiotic and biotic stress: an integrated approach. PLOS One 9:e109574

Dong CJ, Liu JY (2010) The Arabidopsis EAR-motif-containing protein RAP2.1 functions as an active transcriptional repressor to keep stress responses under tight control. BMC Plant Biol 10:47

Elshire Robert J, Glaubitz JC, Sun Q, Poland JA, Kawamoto K, Buckler ES, Mitchell SE (2011) A robust, simple genotyping-by-sequencing (GBS) approach for high diversity species. PLOS One 6:e19379. https://doi.org/10.1371/journal.pone.0019379.

Escandón M, Cañal MJ, Pascual J, Pinto G, Correia B, Amaral J, Meijón M (2016) Integrated physiological and hormonal profile of heat-induced thermotolerance in *Pinus radiata*. Tree Physiol 36:63–77. https://doi.org/10.1093/treephys/tpv127.

Escandón M, Meijón M, Valledor L, Pascual J, Pinto G, Cañal MJ (2018) Metabolome integrated analysis of high-temperature response in *Pinus radiata*. Front Plant Sci 9:485. https://doi.org/10.3389/fpls.2018.00485.

Fahad S, Adnan M, Hassan S, Saud S, Hussain S, Wu C, Wang D, Hakeem KR, Alharby HF, Turan V, Khan MA, Huang J (2019b) Rice responses and tolerance to high temperature. In: Hasanuzzaman M,

Fujita M, Nahar K, Biswas JK (eds) Advances in rice research for abiotic stress tolerance. Woodhead Publ Ltd, Cambridge, England, pp 201–224.

Fahad S, Bajwa AA, Nazir U, Anjum SA, Farooq A, Zohaib A, Sadia S, NasimW, Adkins S, Saud S, Ihsan MZ, Alharby H, Wu C, Wang D, Huang J (2017) Crop production under drought and heat stress: plant responses and management options. Front Plant Sci 8:1147. https://doi.org/10.3389/fpls.2017.01147.

Fahad S, Bano A (2012) Effect of salicylic acid on physiological and biochemical characterization of maize grown in saline area. Pak J Bot 44:1433–1438.

Fahad S, Chen Y, Saud S, Wang K, Xiong D, Chen C, Wu C, Shah F, Nie L, Huang J (2013) Ultraviolet radiation effect on photosynthetic pigments, biochemical attributes, antioxidant enzyme activity and hormonal contents of wheat. J Food Agri Environ 11:1635–1641.

Fahad S, Hussain S, Bano A, Saud S, Hassan S, Shan D, Khan FA, Khan F, Chen Y, Wu C, Tabassum MA, Chun MX, Afzal M, Jan A, Jan MT, Huang J (2014a) Potential role of phytohormones and plant growth-promoting rhizobacteria in abiotic stresses: consequences for changing environment. Environ Sci Pollut Res 22:4907–4921. https://doi.org/10.1007/s11356-014-3754-2.

Fahad S, Hussain S, Matloob A, Khan FA, Khaliq A, Saud S, Hassan S, Shan D, Khan F, Ullah N, Faiq M, Khan MR, Tareen AK, Khan A, Ullah A, Ullah N, Huang J (2014b) Phytohormones and plant responses to salinity stress: A review. Plant Growth Regul 75:391–404. https://doi.org/10.1007/s10725-014-0013-y.

Fahad S, Hussain S, Saud S, Hassan S, Chauhan BS, Khan F et al (2016a) Responses of rapid viscoanalyzer profile and other rice grain qualities to exogenously applied plant growth regulators under high day and high night temperatures. PLoS One 11:e0159590. https://doi.org/10.1371/journal.pone.0159590.

Fahad S, Hussain S, Saud S, Hassan S, Ihsan Z, Shah AN,Wu C, Yousaf M, Nasim W, Alharby H, Alghabari F, Huang J (2016c) Exogenously applied plant growth regulators enhance the morphophysiological growth and yield of rice under high temperature. Front Plant Sci 7:1250. https://doi.org/10.3389/fpls.2 016.01250.

Fahad S, Hussain S, Saud S, Hassan S, Tanveer M, Ihsan MZ, Shah AN, Ullah A, Nasrullah KF, Ullah S, AlharbyH NW, Wu C, Huang J (2016d) A combined application of biochar and phosphorus alleviates heat-induced adversities on physiological, agronomical and quality attributes of rice. Plant Physiol Biochem 103:191–198.

Fahad S, Hussain S, Saud S, Khan F, Hassan Amanullah S Jr, Nasim W, Arif M, Wang F, Huang J (2016b) Exogenously applied plant growth regulators affect heat-stressed rice pollens. J Agron Crop Sci 202:139–150.

Fahad S, Hussain S, Saud S, Tanveer M, Bajwa AA, Hassan S, Shah AN, Ullah A,Wu C, Khan FA, Shah F, Ullah S, Chen Y, Huang J (2015a) A biochar application protects rice pollen from high-temperature stress. Plant Physiol Biochem 96:281–287.

Fahad S, Muhammad ZI, Abdul K, Ihsanullah D, Saud S, Saleh A, Wajid N, Muhammad A, Imtiaz AK, Chao W, Depeng W, Jianliang H (2018) Consequences of high temperature under changing climate optima for rice pollen characteristics-concepts and perspectives. Arch Agron Soil Sci 64:1479–1488. https://doi.org/10.1080/03650340.2018.1443213.

Fahad S, Nie L, Chen Y, Wu C, Xiong D, Saud S, Hongyan L, Cui K, Huang J (2015b) Crop plant hormones and environmental stress. Sustain Agric Rev 15:371–400.

Fahad S, Rehman A, Shahzad B, Tanveer M, Saud S, Kamran M, Ihtisham M, Khan SU, Turan V, Rahman MHU (2019a) Rice responses and tolerance to metal/metalloid toxicity. In: Hasanuzzaman M, Fujita M, Nahar K, Biswas JK (eds) Advances in rice research for abiotic stress tolerance. Woodhead Publ Ltd, Cambridge, England, pp 299–312.

Fang Y, Liao K, Du H, Xu Y, Song H, Li X, Xiong LA (2015) Stress-responsive NAC transcription factor SNAC3 confers heat and drought tolerance through modulation of reactive oxygen species in rice. J Exp Bot 66:6803–6817.

Farhana G, Ishfaq A, Muhammad A, Dawood J, Fahad S, Xiuling L, Depeng W, Muhammad F, Muhammad F, Syed AS (2020) Use of crop growth model to simulate the impact of climate change on yield of various wheat cultivars under different agro-environmental conditions in Khyber Pakhtunkhwa, Pakistan. Arabian J Geosci 13:112. https://doi.org/10.1007/s12517-020-5118-1.

Farhat A, Hafiz MH, Wajid I, Aitazaz AF, Hafiz FB, Zahida Z, Fahad S, Wajid F, Artemi C (2020) A review of soil carbon dynamics resulting from agricultural practices. J Environ Manage 268:110319

Figueroa-Yañez L, Pereira-Santana A, Arroyo-Herrera A, Rodriguez-Corona U, Sanchez-Teyer F, Espadas-Alcocer J, Espadas-Gil F, Barredo-Pool F, Castaño E, Rodriguez-Zapata LC (2016) RAP2.4a is transported through the phloem to regulate cold and heat tolerance in papaya tree (*Carica papaya* cv. Maradol): implications for protection against abiotic stress. PLOS One 11:e0165030. https://doi.org/10.1371/journal.pone.0165030.

Fraire-Velázquez S, Rodríguez-Guerra R, Sánchez- Calderón L (2011) Abiotic and biotic stress response crosstalk in plants. In: Abiotic stress response in plants—physiological biochemical and genetic perspectives. InTech, Rijeka, Croatia, pp 3–26. https://doi.org/10.5772/23217.

Gao C, Qi S, Liu K, Li D, Jin C, Li Z,Huang G, Hai J, Zhang M, Chen M (2016) MYC2, MYC3, and MYC4 function redundantly in seed storage protein accumulation in Arabidopsis. Plant Physiol Biochem 108:63–70.

Garg R, Varshney RK, Jain M (2014) Molecular genetics and genomics of abiotic stress responses. Front Plant Sci 5:398. https://doi.org/10.3389/fpls.2014.00398.

Gasperini D, Chételat A, Acosta IF, Goossens J, Pauwels L, Goossens A, Dreos R, Alfonso E, Farmer EE (2015) Multilayered organization of jasmonate signalling in the regulation of root growth. PLos Genet 11:e1005300

Gasperini D, Greenland A, Hedden P, Dreos R, Harwood W, Griffiths S (2012) Genetic and physiological analysis of rht8 in bread wheat: an alternative source of semidwarfism with a reduced sensitivity to brassinosteroids. J Exp Bot 63:4419–4436.

Golldack D, Li C, Mohan H, Probst N (2014) Tolerance to drought and salt stress in plants: unraveling the signaling networks. Front Plant Sci 5:151. https://doi.org/10.3389/fpls.2014.00151.

Gujjar RS (2016) Transcription factors in abiotic stress tolerance. In: Yadav P, Kumar S, Jain V (eds) Recent advances in plant stress physiology. Daya Publishing House, New Delhi, pp 49–67.

Gupta B, Sengupta A, Saha J, Gupta K (2013) Plant abiotic stress: 'Omics' approach. J Plant Biochem Physiol 1:3. http:/dx.doi.org/10.4172/2329-9029.1000e108.

Habib ur Rahman M, Ashfaq A, Aftab W, Manzoor H, Fahd R, Wajid I, Md Aminul I, Vakhtang S, Muhammad A, Asmat U, Abdul W, Syeda RS, Shah S, Shahbaz K, Fahad S, Manzoor H, Saddam H, Wajid N (2017) Application of CSM-CROPGRO-Cotton model for cultivars and optimum planting dates: evaluation in changing semi-arid climate. Field Crops Res 238:139–152. http://dx.doi.org/10.1016/j.fcr.2017.07.007.

Hafiz MH, Abdul K, Farhat A, Wajid F, Fahad S, Muhammad A, Ghulam MS, Wajid N, Muhammad M, Hafiz FB (2020b) Comparative effects of organic and inorganic fertilizers on soil organic carbon and wheat productivity under arid region. Commun Soil Sci Plant Anal 51:1406–1422. https://doi.org/10.1080/00103624.2020.1763385.

Hafiz MH, Farhat A, Ashfaq A, Hafiz FB, Wajid F, Carol Jo W, Fahad S, Gerrit H (2020a) Predicting kernel growth of maize under controlled water and nitrogen applications. Int J Plant Prod 14:609–620. https://doi.org/10.1007/s42106-020-00110-8.

Hafiz MH, Farhat A, Shafqat S, Fahad S, Artemi C, Wajid F, Chaves CB, Wajid N, Muhammad M, Hafiz FB (2018) Offsetting land degradation through nitrogen and water management during maize cultivation under arid conditions. Land Degrad Dev 1–10. https://doi.org/10.1002/ldr.2933.

Hafiz MH, Muhammad A, Farhat A, Hafiz FB, Saeed AQ, Muhammad M, Fahad S, Muhammad A (2019) Environmental factors affecting the frequency of road traffic accidents: a case study of sub-urban area of Pakistan. Environ Sci Pollut Res 26:11674–11685. https://doi.org/10.1007/s11356-019-04752-8.

Hafiz MH, Wajid F, Farhat A, Fahad S, Shafqat S, Wajid N, Hafiz FB (2016) Maize plant nitrogen uptake dynamics at limited irrigation water and nitrogen. Environ Sci Pollut Res 24:2549–2557. https://doi.org/10.1007/s11356-016-8031-0.

He JX, Mu RL, Cao WH, Zhang ZG, Zhang JS, Chen SY (2005) AtNAC2, a transcription factor downstream of ethylene and auxin signaling pathways, is involved in salt stress response and lateral root development. Plant J 44:903–916.

Hershey JWB, Sonenberg N, Mathews MB (2018) Principles of translational control: an overview. Cold Spring Harb Perspect Biol 12. https://doi.org/10.1101/cshperspect.a032607.

Hesham FA, Fahad S (2020) Melatonin application enhances biochar efficiency for drought tolerance in maize varieties: modifications in physio-biochemical machinery. Agron J 112:2826–2847.

Hixson KK, Lopez-Ferrer D, Robinson EW, Paša-Tolić L (2017) Proteomics. In: Encyclopedia of spectroscopy and spectrometry, 3rd edn. Elsevier, pp 766–773. https://doi.org/10.1016/B978-0-12-803224-4.00061-3.

Hoang XLT, Nhi DNH, Thu NBA, Thao NP, Tran LP (2017) Transcription factors and their roles in signal transduction in plants under abiotic stresses. Current Genomics 18:483–497. https://doi.org/10.2174/1389202918666170227150057.

Hong Y, Zhang H, Huang L, Li D, Song F (2016) Overexpression of a stress-responsive NAC transcription factor gene ONAC022 improves drought and salt tolerance in rice. Front Plant Sci 7:4

Howe GT, Brunner AM (2005) An evolving approach to understanding plant adaptation. New Phytologist 167:1–5. https://doi.org/10.1111/j.1469-8137.2005.01469.x.

Hu H, You J, Fang Y, Zhu X, Qi Z, Xiong L (2008) Characterization of transcription factor gene SNAC2 conferring cold and salt tolerance in rice. Plant Mol Biol 67:169–181.

Hu XH, Zhang SZ, Miao HR, Cui FG, Shen Y, Yang WQ, Xu TT, Chen N, Chi XY, Zhang ZM, Chen J (2018) High-density genetic map construction and identification of QTLs controlling oleic and linoleic acid in peanut using SLAF-seq and SSRs. Sci Rep 8:5479

Huang Y, Haas M, Heinen S, Steffenson BJ, Smith KP, Muehlbauer GJ (2018) QTL mapping of Fusarium head blight and correlated agromorphological traits in an elite barley cultivar rasmusson. Front Plant Sci 9:1260. https://doi.org/10.3389/fpls.2018.01260.

Huang X, Han B (2014) Natural variations and genome-wide association studies in crop plants. Annu Rev Plant Biol 65:531–551. https://doi.org/10.1146/annurev-arplant-050213-035715.

Hussain MA, Fahad S, Rahat S, Muhammad FJ, Muhammad M, Qasid A, Ali A, Husain A, Nooral A, Babatope SA, Changbao S, Liya G, Ibrar A, Zhanmei J, Juncai H (2020) Multifunctional role of brassinosteroid and its analogues in plants. Plant Growth Regul 92:141–156. https://doi.org/10.1007/s10725-020-00647-8.

Ibraheem O, Adigun RO, Olatunj IT (2018) Omics technologies in unraveling plant stress responses; using sorghum as a model crop, how far have we gone? Vegetos 31:2. https://doi.org/10.4172/2229-4473.1000405.

Ilyas M, Mohammad N, Nadeem K, Ali H, Aamir HK, Kashif H, Fahad S, Aziz K, Abid U (2020) Drought tolerance strategies in plants: a mechanistic approach. J Plant Growth Regul https://doi.org/10.1007/s00344-020-10174-5.

Iqbal J (2019) Morphological, physiological and molecular markers for the adaptation of wheat in drought condition. Asian J Biotechnol Genet Eng 2:1–13. https://doi.org/10.9734/AJBGE/2019/46253.

Jan M, Anwar-ul-Haq M, Noor Shah A, Yousaf M, Iqbal J, Li X, Wang D, Fahad S (2019) Modulation in growth, gas exchange, and antioxidant activities of salt-stressed rice (Oryza sativa L.) genotypes by zinc fertilization. Arabian J Geosci 12:775. https://doi.org/10.1007/s12517-019-4939-2.

Janmohammadi M, Zolla L, Rinalducci S (2015) Low temperature tolerance in plants: changes at the protein level. Phytochemistry 117:76–89.

Jaspers P, Kangasjärvi J (2010) Reactive oxygen species in abiotic stress signaling. Physiologia Plantarum 138:405–413.

Jiang J, Ma S, Ye N, Jiang M, Cao J, Zhang J (2017) WRKY transcription factors in plant responses to stresses. J Integr Plant Biol 59:86–101.

Jin C, Li KQ, Xu XY, Zhang HP, Chen HX, Chen YH, Hao J, Wang Y , Huang XS, Zhang SL (2017) A novel NAC transcription factor, PbeNAC1, of *Pyrus betulifolia* confers cold and drought tolerance via interacting with PbeDREBs and activating the expression of stress-responsive genes. Front Plant Sci 8:1049. http://doi.org/10.3389/fpls.2017.01049.

Johnson SM, Fei-Ling L, Aliza F, Hillel F, Antoni RS, Knight MR (2014) Transcriptomic analysis of sorghum bicolor responding to combined heat and drought stress. BMC Genomics 15:456

Jones N, Ougham H, Thomas H, Pasakinskiene I (2009) Markers and mapping revisited: finding your gene. New Phytol 183:935–966. https://doi.org/10.1111/j.1469-8137.2009.02933.x.

Jorrín-Novo JV, Pascual J, Sánchez-Lucas R, Romero-Rodríguez MC, Rodríguez-Ortega MJ, Lenz C, Valledor L (2015) Fourteen years of plant proteomics reflected in proteomics: moving from model species and 2DE-based approaches to orphan species and gel-free platforms. Proteomics 15:1089–1112.

Kaifa W, Juan C, Yanmei W, Yanhui C, Shaoxiang C, Yina L, Si P, Xiaojun Z, Daoxin X (2012) Genome-wide analysis of bZIP-encoding genes in maize. DNA Res 19:463–476.

Kalladan R, Worch S, Rolletschek H, Harshavardhan VT, Kuntze L, Seiler C, Sreenivasulu N, Röder MS (2013) Identification of quantitative trait loci contributing to yield and seed quality parameters under terminal drought in barley advanced backcross lines. Mol Breed 32:71–90.

Kamarn M, Wenwen C, Irshad A, Xiangping M, Xudong Z, Wennan S, Junzhi C, Shakeel A, Fahad S, Qingfang H, Tiening L (2017) Effect of paclobutrazol, a potential growth regulator on stalk mechanical strength, lignin accumulation and its relation with lodging resistance of maize. Plant Growth Regul 84:317–332. https://doi.org/10.1007/s10725-017-0342-8.

Kim C, Guo H, Kong W, Chandnani R, Shuang L-S, Paterson AH (2016) Application of genotyping by sequencing technology to a variety of crop breeding programs. Plant Sci 242:14–22.

Kimotho RN, Baillo EH, Zhang Z (2019) Transcription factors involved in abiotic stress responses in maize (*Zea mays* L.) and their roles in enhanced productivity in the post genomics era. PeerJ 7:e7211. https://doi.org/10.7287/peerj.preprints.27549v1.

King HA, Gerber AP (2016) Translatome profiling: methods for genome-scale analysis of mRNA translation. Brief Funct Genomics 15:22–31. https://doi.org/10.1093/bfgp/elu045.

Komili S, Silver PA (2008) Coupling and coordination in gene expression processes: a systems biology view. Nat Rev Genet 9:38–48.

Kosová K, Vítámvás P, Urban MO, Prášil IT, Renaut J (2018) Plant abiotic stress proteomics: the major factors determining alterations in cellular proteome. Front Plant Sci 9:122. https://doi.org/10.3389/fpls.2018.00122.

Kumar A, Kage U, Mosa K, Dhokane D (2014) Metabolomics: a novel tool to bridge phenome to genome under changing climate to ensure food security. Med Aromat Plants 3:4. https://doi.org/10.4172/2167-0412.1000e154.

Kuscu C, Kumar P, Kiran M, Su Z, Malik A, Dutta A (2018) tRNA fragments (tRFs) guide Ago to regulate gene expression post-transcriptionally in a Dicer-independent manner. RNA 24:1093–1105. https://doi.org/10.1261/rna.066126.118.

Lakhwani D, Pandey A, Dhar YV, Bag SK, Trivedi PK, Asif MH (2016) Genome-wide analysis of the AP2/ERF family in Musa species reveals divergence and neofunctionalisation during evolution. Sci Rep 6:18878

Lal NK, Nagalakshmi U, Hurlburt NK, Flores R, Bak A, Sone P, Ma X, Song G, Walley J, Shan L, He P, Casteel C., Fisher AJ, Dinesh-Kumar SP (2018) The receptor-like cytoplasmic kinase BIK1 localizes to the nucleus and regulates defense hormone expression during plant innate immunity. Cell Host Microbe 23:485–497.

Le DT, Nishiyama R, Watanabe Y, Tanaka M, Seki M, Ham LH, Yamaguchi-Shinozaki K, Shinozaki K, Tran LSP (2012) Differential gene expression in soybean leaf tissues at late developmental stages under drought stress revealed by genome-wide transcriptome analysis. PLOS One 7:e49522. https://doi.org/10.1371/journal.pone.0049522.

Leite Sampaio B, Edrada-Ebel RA, Da Costa FB (2016) Effect of the environment on the secondary metabolic profile of *Tithonia diversifolia*: a model for environmental metabolomics of plants. Sci Rep 6:29265. https://doi.org/10.1038/srep29265.

Li H, Gao Y, Xu H, Dai Y, Deng DQ, Chen JM (2013) ZmWRKY33, a WRKY maize transcription factor conferring enhanced salt stress tolerances in Arabidopsis. Plant Growth Regul 70:207–216.

Li YF, Zheng Y, Vemireddy LR, Panda SK, Jose S, Ranjan A, Panda P, Govindan G, Cui J, Wei K, Yaish MW, Naidoo GC, Sunkar R (2018) Comparative transcriptome and translatome analysis in contrasting rice genotypes reveals differential mRNA translation in salt-tolerant Pokkali under salt stress. BMC Genomics 19:935. https://doi.org/10.1186/s12864-018-5279-4.

Liang X, Chen X, Li C, Fan J, Guo Z (2017) Metabolic and transcriptional alternations for defense by interfering OsWRKY62 and OsWRKY76 transcriptions in rice. Sci Rep 7:2474

Licausi F, Ohme-Takagi M, Perata P (2013) APETALA2/Ethylene responsive factor (AP2/ERF) transcription factors: mediators of stress responses and developmental programs. New Phytol 199:639–649.

Liu M, Wen Y, Sun W, Ma Z, Huang L, Wu Q, Tang Z, Bu T, Li C, Chen H (2019) Genome-wide identification, phylogeny, evolutionary expansion and expression analyses of bZIP transcription factor family in tartaty buckwheat. BMC Genomics 20:483. https://doi.org/10.1186/s12864-019-5882-z.

Lu Y, LiR, Wang R, Wang X, Zheng W, Sun Q, Tong S, Dai S, Xu S (2017) Comparative proteomic analysis of flag leaves reveals new insight into wheat heat adaptation. Front Plant Sci 8:1086. https://doi.org/10.3389/fpls.2017.01086.

Luan Y, Cui J, Zhai J, Li J, Han L, Meng J (2015) High-throughput sequencing reveals differential expression of miRNAs in tomato inoculated with phytophthora infestans. Planta 241:1405–1416.

Lv Y, Yang M, Hu D, Yang Z, Ma S, Li X et al (2017) The OsMYB30 transcription factor suppresses cold tolerance by interacting with a JAZ protein and suppressing β-amylase expression. Plant Physiol 173:1475–1491. https://doi.org/10.1104/pp.16.01725.

Mahdavi Mashaki K, Garg V, Nasrollahnezhad Ghomi AA, Kudapa H, Chitikineni A, Zaynali Nezhad K, Yamchi A, Soltanloo H, Kumar Varshney R, Thudi M (2018) RNA-Seq analysis revealed genes associated with drought stress response in kabuli chickpea (*Cicer arietinum* L.). PLOS One 13:e0199774. https://doi.org/10.1371/journal.pone.0199774.

Manickavelu A, Kawaura K, Oishi K, Shin-I T, Kohara Y, Yahiaoui N, Keller B, Abe R, Suzuki A, Nagayama T, Yano K, Ogihara Y (2012) Comprehensive functional analyses of expressed sequence tags in common wheat (*Triticum aestivum*). DNA Res 19:165–177. https://doi.org/10.1093/dnares/dss001.

Miranda MA, Varela RM, Torres A, Molinillo JMG, Gualtieri SCJ, Macías FA (2015) Phytotoxins from *Tithonia diversifolia*. J Nat Prod 78:1083–1092.

Mitulović G (2014) New HPLC techniques for proteomics analysis: a short overview of latest developments. J Liq Chrom Relat Tech 38:390–403. https://doi.org/10.1080/10826076.2014.941266.

Mubeen M, Ahmad A, Hammad HM, Awais M, Hafiz UF, Mazhar S, Sami ul Din M, Asad A, Amjed A, Fahad S, Wajid N (2020) Evaluating the climate change impact on water use efficiency of cotton-wheat in semi-arid conditions using DSSAT model. J Water Clim Change 11:1661–1675. https://doi.org/10.2166/wcc.2019.179/622035/jwc2019179.pdf.

Muhammad Z, Abdul MK, Abdul MS, Kenneth BM, Muhammad S, Shahen S, Ibadullah J, Fahad S (2019) Performance of Aeluropus lagopoides (mangrove grass) ecotypes, a potential turfgrass, under high saline conditions. Environ Sci Pollut Res 26:13410–13421. https://doi.org/10.1007/s11356-019-04838-3.

Nakano T, Fujisawa M, Shima Y, Ito Y (2014) The AP2/ERF transcription factor SlERF52 functions in flower pedicel abscission in tomato. J Exp Bot 65:3111–3119.

Nakashima K, Ito Y, Yamaguchi-Shinozaki K (2009) Transcriptional regulatory networks in response to abiotic stresses in Arabidopsis and grasses. Plant Physiol 149 88–95.

Nakashima K, Yamaguchi-Shinozaki K, Shinozaki K (2014) The transcriptional regulatory network in the drought response and its crosstalk in abiotic stress responses including drought, cold, and heat. Front Plant Sci 5:170. https://doi.org/10.3389/fpls.2014.00170.

Ng DWK, Abeysinghe JK, Kamali M (2018) Regulating the regulators: the control of transcription factors in plant defense signaling. Int J Mol Sci 19:3737. https://doi.org/10.3390/ijms19123737.

Ohta M, Sato A, Renhu N, Yamamoto T, Oka N, Zhu JK, Tada Y, Suzaki T, Miura K (2018) MYC-type transcription factors, MYC67 and MYC70, interact with ICE1 and negatively regulate cold tolerance in Arabidopsis. Sci Rep 8:11622. https://doi.org/10.1038/s41598-018-29722.

Onate-Sánchez L, Singh KB (2002) Identification of Arabidopsis ethylene-responsive element binding factors with distinct induction kinetics after pathogen infection. Plant Physiol 128 1313–1322.

Ozsolak F, Milos PM (2010) RNA sequencing: advances, challenges and opportunities. Nat Rev Genet 12:87–98. https://doi.org/10.1038/nrg2934.

Pavarini DP, Pavarini SP, Niehues M, Lopes NP (2012) Exogenous influences on plant secondary metabolite levels. Anim Feed Sci Technol 176:5–16.

Phukan UJ, Jeena GS, Shukla RK (2016) WRKY transcription factors: molecular regulation and stress responses in plants. Front Plant Sci 7:760

Phuong ND, Hoi PX (2015) Isolation and characterization of a OsRap2.4A transcription factor and its expression in Arabidopsis for enhancing high salt and drought tolerance. Curr Sci India 108:51–62.

Pitzschke A, Datta S, Persak H (2014) Salt stress in Arabidopsis: lipid transfer protein AZI1 and its control by mitogen-activated protein kinase MPK3. Mol Plant 7:722–738.

Popp D (2010) Innovation and climate policy. NBER Working. http://www.nber.org/papers/w15673. Accessed 10 August 2019.

Pérez-Clemente RM, Vives V, Zandalinas SI, López-Climent MF, Muñoz V, Gómez-Cadenas A (2013) Biotechnological Approaches to Study Plant Responses to Stress. Hindawi Publishing Corporation. http://dx.doi.org/10.1155/2013/654120.

Qin L, Walk TC, Han P, Chen L, Zhang S, Li Y, Hu X, Xie L, Yang Y, Liu J, Lu X, Yu C, Tian, Shaff JE, Kochian LV, Liao X, Liao H (2019) Adaption of roots to nitrogen deficiency revealed by 3D quantification and proteomic analysis. Plant Physiol 179:329–347. https://doi.org/https://doi.org/10.1104/pp.18.00716.

Rahman H, Ramanathan V, Nallathambi J, Duraialagaraja S, Muthurajan R (2016) Over-expression of a NAC 67 transcription factor from finger millet (*Eleusine coracana* L.) confers tolerance against salinity and drought stress in rice. BMC Biotechnol 16:7–20.

Raney JA, Reynolds DJ, Elzinga DB, Page J, Udall JA, Jellen EN, Bonfacio A, Fairbanks DJ, Maughan PJ (2014) Transcriptome analysis of drought induced stress in *Chenopodium quinoa*. Am J Plant Sci 5:338–357. http://dx.doi.org/10.4236/ajps.2014.53047.

Rehman M, Fahad S, Saleem MH, Hafeez M, Habib ur Rahman M, Liu F, Deng G (2020) Red light optimized physiological traits and enhanced the growth of ramie (*Boehmeria nivea* L.). Photosynthetica 58: 922–931.

Riechmann JL, Ratcliffe OJ (2000) A genomic perspective on plant transcription factors. Curr Opin Plant Biol 3:423–434.

Rodziewicz P, Swarcewicz B, Chmielewska K, Wojakowska A, Stobiecki M (2014) Influence of abiotic stresses on plant proteome and metabolome changes. Acta Physiol Plant 36:1–19.

Rushton PJ, Somssich IE, Ringler P, Shen QXJ (2010) WRKY transcription factors. Trends Plant Sci 15:247–258.

Sajjad H, Muhammad M, Ashfaq A, Waseem A, Hafiz MH, Mazhar A, Nasir M, Asad A, Hafiz UF, Syeda RS, Fahad S, Depeng W, Wajid N (2019) Using GIS tools to detect the land use/land cover changes during forty years in Lodhran district of Pakistan. Environ Sci Pollut Res 27:39676–39692. https://doi.org/10.1007/s11356-019-06072-3.

Saleem MH, Fahad S, Adnan M, Mohsin A, Muhammad SR, Muhammad K, Qurban A, Inas AH, Parashuram B, Mubassir A, Reem MH (2020a) Foliar application of gibberellic acid endorsed phytoextraction of copper and alleviates oxidative stress in jute (*Corchorus capsularis* L.) plant grown in highly copper-contaminated soil of China. Environ Sci Pollution Res 27:37121–37133. https://doi.org/10.1007/s11356-020-09764-3.

Saleem MH, Fahad S, Shahid UK, Mairaj D, Abid U, Ayman ELS, Akbar H, Analía L, Lijun L (2020c) Copper-induced oxidative stress, initiation of antioxidants and phytoremediation potential of flax (Linum usitatissimum L.) seedlings grown under the mixing of two different soils of China. Environ Sci Poll Res 27:5211–5221.https://doi.org/10.1007/s11356-019-07264-7.

Saleem MH, Rehman M, Fahad S, Tung SA, Iqbal N, Hassan A, Ayub A, Wahid MA, Shaukat S, Liu L, Deng G (2020b) Leaf gas exchange, oxidative stress, and physiological attributes of rapeseed (*Brassica napus* L.) grown under different light-emitting diodes. Photosynthetica 58:836–845.

Saud S, Chen Y, Fahad S, Hussain S, Na L, Xin L, Alhussien SA (2016) Silicate application increases the photosynthesis and its associated metabolic activities in Kentucky bluegrass under drought stress and post-drought recovery. Environ Sci Pollut Res 23:17647–17655. https://doi.org/10.1007/s11356-016-6957-x.

Saud S, Chen Y, Long B, Fahad S, Sadiq A (2013) The different impact on the growth of cool season turf grass under the various conditions on salinity and drought stress. Int J Agric Sci Res 3:77–84.

Saud S, Fahad S, Cui G, Chen Y, Anwar S (2020) Determining nitrogen isotopes discrimination under drought stress on enzymatic activities, nitrogen isotope abundance and water contents of Kentucky bluegrass. Sci Rep 10:6415. https://doi.org/10.1038/s41598-020-63548-w.

Saud S, Fahad S, Yajun C, Ihsan MZ, Hammad HM, Nasim W, Amanullah Jr, Arif M, Alharby H (2017) Effects of nitrogen supply on water stress and recovery mechanisms in kentucky bluegrass plants. Front. Plant Sci 8:983. https://doi.org/10.3389/fpls.2017.00983.

Saud S, Li X, Chen Y, Zhang L, Fahad S, Hussain S, Sadiq A, Chen Y (2014) Silicon application increases drought tolerance of Kentucky bluegrass by improving plant water relations and morph physiological functions. SciWorld J 2014:1–10. https://doi.org/10.1155/2014/368694.

Sayed MA, Schumann H, Pillen K, Naz AA, Léon J (2012) AB-QTL analysis reveals new alleles associated to proline accumulation and leaf wilting under drought stress conditions in barley (*Hordeum vulgare* L.). BMC Genet 13:61. https://doi.org/10.1186/1471-2156-13-61.

Seo PJ, Park CM (2010) MYB96-mediated abscisic acid signals induce pathogen resistance response by promoting salicylic acid biosynthesis in Arabidopsis. New Phytologist 186:471–483.

Shafi MI, Adnan M, Fahad S, Wahid F, Khan A, Yue Z, Danish S, Zafar-ul-Hye M, Brtnicky M, Datta R (2020) Application of single superphosphate with humic acid improves the growth, yield and phosphorus uptake of wheat (*Triticum aestivum* L.) in calcareous soil. Agron 10:1224. https://doi.org/10.3390/agronomy10091224.

Shah F, Lixiao N, Kehui C, Tariq S, Wei W, Chang C, Liyang Z, Farhan A, Fahad S, Huang J (2013) Rice grain yield and component responses to near 2°C of warming. Field Crop Res 157:98–110.

Shinozaki K, Yamaguchi-Shinozaki K (2007) Gene networks involved in drought stress response and tolerance. J Exp Bot 58:221

Shkolnik-Inbar D, Bar-Zvi D (2011) Expression of Abscisic acid insensitive 4 (ABI4) in developing Arabidopsis seedlings. Plant Signal Behav 6:694–696.

Shukla PS, Agarwal P, Gupta K, Agarwal PK (2015) Molecular characterization of an MYB transcription factor from a succulent halophyte involved in stress tolerance. AoB PLANTS 7:plv054. https://doi.org/10.1093/aobpla/plv054.

Silva-Sanchez C, Li H, Chen S (2015) Recent advances and challenges in plant phosphoproteomics. Proteomics 15:1127–1141.

Soda N, Wallace S, Karan R (2015) Omics study for abiotic stress responses in plants. Adv Plants Agric Res 2:28–34

Soderberg T (2019) Mass spectrometry - Chemistry LibreTexts. https://chem.libretexts.org › Chapter_04:_Structure_Determination_I › 4.1. Accessed 15 August 2019.

Subhan D, Zafar-ul-Hye M, Fahad S, Saud S, Martin B, Tereza H, Rahul D (2020) Drought stress alleviation by ACC deaminase producing *Achromobacter xylosoxidans* and *Enterobacter cloacae*, with and without timber waste biochar in maize. Sustain 12 https://doi.org/10.3390/su12156286.

Sun P, Zhu X, Huang X, Liu JH (2014) Overexpression of a stressresponsive MYB transcription factor of *Poncirus trifoliata* confers enhanced dehydration tolerance and increases polyamine biosynthesis. Plant Physiol Biochem 78:71–79.

Takáč T, Vadovič P, Pechan T, Luptovčiak I, Šamajová O, Šamaj J (2016) Comparative proteomic study of Arabidopsis mutants mpk4 and mpk6. Sci Rep 6:28306. https://doi.org/10.1038/srep28306.

Tang Y, Bao X, Zhi Y, Wu Q, Guo Y, Yin X, Zeng L, Li J, Zhang J, He W, Liu W, Wang Q, Jia C, Li Z, Liu K (2019a) Overexpression of a MYB family gene, OsMYB6, increases drought and salinity stress tolerance in transgenic rice. Front Plant Sci. https://doi.org/10.3389/fpls.2019.00168.

Tang YQ, Xia ZQ, Ding ZT, Ya-Cao Ding YC, Liu Z, Ma X, Liu JP (2019b) Construction of a high-density linkage map and QTL mapping for important agronomic traits in *Stylosanthes guianensis* (Aubl.) Sw. Sci Rep 9:3834. https://doi.org/10.1038/s41598-019-40489-7.

Tariq M, Ahmad S, Fahad S, Abbas G, Hussain S, Fatima Z, Nasim W, Mubeen M, Habib ur Rehman M, Khan MA, Adnan M. (2018). The impact of climate warming and crop management on phenology of sunflower-based cropping systems in Punjab, Pakistan. Agri and Forest Met 256–257:270–282.

Tiancong Q, Huang H, Susheng S, Daoxin X (2015b) Regulation of jasmonate-mediated stamen development and seed production by a bHLH-MYB complex in Arabidopsis. Plant Cell 27:1620–1633.

Tiancong Q, Jiaojiao W, Huang H, Bei L, Hua G, Yule L, Susheng S, Daoxin X (2015a) Regulation of jasmonate-induced leaf senescence by antagonism between bHLH subgroup IIIe and IIId factors in Arabidopsis. Plant Cell 27:1634–1649.

Tsaneva M, De Schutter K, Verstraeten B, Van Damme EJM (2019) Lectin sequence distribution in QTLs from rice (*Oryza sativa*) suggest a role in morphological traits and stress responses. Int J Moleculer Science 20:437. https://doi.org/10.3390/ijms20020437.

Udvardi MK, Kakar K, Wandrey M, Montanari O, Murray J, Andriankaja A, Zhang JY, Benedito V, Hofer JMI, Chueng F, Town CD (2007) Legume transcription factors: global regulators of plant development and response to the environment. Plant Physiol 144: 538–549. https://doi.org/10.1104/pp.107.098061.

Uz Zaman Q, Zubair A, Muhammad Y, Muhammad ZI, Abdul K, Fahad S, Safder B, Ramzani PMA, Muhammad N (2017) Zinc biofortification in rice: leveraging agriculture to moderate hidden hunger in developing countries. Arch Agron Soil Sci 64:147–161. https://doi.org/10.1080/03650340.2017.1338343.

Valdes A, Ibanez C, Simo C, Garcia-Canas V (2013) Recent transcriptomics advances and emerging applications in food science. Trends Anal Chem 52:142–154.

Van Emon JM (2016) The Omics revolution in agricultural research. J Agric Food Chem 64:36–44. https://doi.org/10.1021/acs.jafc.5b04515.

Verma V, Ravindran P, Kumar PP (2016) Plant hormone-mediated regulation of stress responses. BMC Plant Biol 16:86. https://doi.org/10.1186/s12870-016-0771-y.

Wahid F, Fahad S, Subhan D, Adnan M, Zhen Y, Saud S, Manzer HS, Martin B, Tereza H, Rahul D (2020) Sustainable management with mycorrhizae and phosphate solubilizing bacteria for enhanced phosphorus uptake in calcareous soils. Agri 10:334. https://doi.org/10.3390/agriculture10080334.

Wajid N, Ashfaq A, Asad A, Muhammad T, Muhammad A, Muhammad S, Khawar J, Ghulam MS, Syeda RS, Hafiz MH, Muhammad IAR, Muhammad ZH, Habib ur Rahman M, Veysel T, Fahad S, Suad S, Aziz K, Shahzad A (2017) Radiation efficiency and nitrogen fertilizer impacts on sunflower crop in contrasting environments of Punjab. Pakistan Environ Sci Pollut Res 25:1822–1836. https://doi.org/10.1007/s11356-017-0592-z.

Wang J, Lian W, Cao Y, Wang X, Wang G, Qi C, Liu L, Qin S, Yuan X, Li X, Ren S, Guo YD (2018) Overexpression of BoNAC019, a NAC transcription factor from *Brassica oleracea*, negatively regulates the dehydration response and anthocyanin biosynthesis in Arabidopsis. Sci Rep 8:13349. https://doi.org/10.1038/s41598-018-31690-1.

Wu C, Kehui C, She T, Ganghua L, Shaohua W, Fahad S, Lixiao N,Jianliang H, Shaobing P, Yanfeng D (2020) Intensified pollination and fertilization ameliorate heat injury in rice (*Oryza sativa* L.) during the flowering stage. Field Crops Res 252:107795.

Wu C, Tang S, Li G, Wang S, Fahad S, Ding Y (2019) Roles of phytohormone changes in the grain yield of rice plants exposed to heat: a review. PeerJ 7:e7792. https://doi.org/10.7717/peerj.7792.

Xie Z, Nolan TM, Jiang H, Yin Y (2019) AP2/ERF transcription factor regulatory networks in hormone and abiotic stress responses in Arabidopsis. Front Plant Sci 10:228. https://doi.org/10.3389/fpls.2019.00228.

Yang A, Dai X, Zhang WH (2012) A R2R3-type MYB gene, OsMYB2, is involved in salt, cold, and dehydration tolerance in rice. J Exp Bot 63:2541–2556. https://doi.org/10.1093/jxb/err431.

Yang X, Wang X, Ji L, Yi Z, Fu C, Ran J, Hu R, Zhou G (2015) Overexpression of a *Miscanthus lutarioriparius* NAC gene MlNAC5 confers enhanced drought and cold tolerance in Arabidopsis. Plant Cell Rep 34:943–958.

Yang Z, Zhang Z, Zhang T, Fahad S, Cui K, Nie L, Peng S, Huang J (2017) The effect of season-long temperature increases on rice cultivars grown in the central and southern regions of China. Front Plant Sci 8:1908. https://doi.org/10.3389/fpls.2017.01908.

Zafar-ul-Hye M, Muhammad N, Subhan D, Fahad S, Rahul D, Mazhar A, Ashfaq AR, Martin B, Jiˇrí H, Zahid HT, Muhammad N (2020a) Alleviation of cadmium adverse effects by improving nutrients uptake in bitter gourd through cadmium tolerant rhizobacteria. Environ 7:54. https://doi.org/10.3390/environments7080054.

Zafar-ul-Hye M, Tahzeeb-ul-Hassan M, Abid M, Fahad S, Brtnicky M, Dokulilova T, Datta R, Danish S (2020b) Potential role of compost mixed biochar with rhizobacteria in mitigating lead toxicity in spinach. Scientific Rep 10:12159. https://doi.org/10.1038/s41598-020-69183-9.

Zahida Z, Hafiz FB, Zulfiqar AS, Ghulam MS, Fahad S, Muhammad RA, Hafiz MH,Wajid N, Muhammad S (2017) Effect of water management and silicon on germination, growth, phosphorus and arsenic uptake in rice. Ecotoxicol Environ Saf 144:11–18.

Zhu YN, Shi DQ, Ruan MB, Zhang LL, Meng ZH, Liu J, Yang WC (2013) Transcriptome analysis reveals crosstalk of responsive genes to multiple abiotic stresses in cotton (*Gossypium hirsutum* L.). PLOS One 8(11):e80218.

2

Physiological Ecology

Mohammad Hasanzadeh and Nahid Hazrati
Moghan Faculty of Agriculture and Natural Resources, University of Mohaghegh Ardabili, Ardabil, Iran

2.1 Introduction

Physiological ecology, or in other words, ecophysiology, deals with physiological events that occur in living organisms in response to environmental factors. Generally speaking, all features that an organism reflects are direct or indirect consequences of environment loads. Plants, opposite to animals, do not possess ability to escape from undesirable circumstances so they need more potent mechanisms to cope with such conditions. These may be better understood by studying ecological events that plants suffer from. To explain these, we should first be familiar with ecology that illustrates correlations between living creatures and their environment. Production and decomposition, mass–energy flow, bio-geochemical circles, and food chains are among main topics in ecological studies.

Considering the vast diversity in growth habitats for plants, complicated ecological impacts would be expected. For example, C3 and C4 plants do respond to cool and warm climates by different growing habits, or specific reaction against salinity has been shown by halophytes and glycophytes. What makes such differences relates to the physiological pathways plants acquire to survive against severe ecological conditions. In other words, over the millions of years, plants have been selected by surrounding environments. Different physiological responses may be caused as a result of some ecological aspects. Among them is photosynthesis, which is affected by light; oxygen; CO_2 and humidity; environmental stresses like salinity, drought, cold, and heat; heavy metal stresses; water–soil–plant relations; and nutrition.

2.2 Photosynthesis

Photosynthesis is the consequence of all biochemical pathways in plants that terminate to assimilate production. It has been estimated that photosynthesis is responsible for converting about 200 billion tons of CO_2 into complexes of organic compounds along with releasing of nearly 140 billion tons of oxygen into atmosphere annually. This phenomenon acts globally as the energy flow in food chains (Johnson 2016).

Leaf is the main site in plants for photosynthesis and contains chloroplasts. These organelles having found to be transferred from cyanobacteria kind of bacteria with photosynthetic potential to eukaryotes as endosymbiosis during the early evolution and encoded in their genome. These lens-shaped organelles contain outer and inner boundary membranes, and similarity of lipid composition of outer membrane to those of eukaryotes while inner membrane to those of prokaryotes could be a confirmation for abovementioned fact.

During photosynthetic pathway, solar energy, as direct or reflected light, results in breaking down of water molecules to produce protons along with electrons (light reactions) in organelles called grana which are composed of thylakoids. Produced electrons are transferred to $NADP^+$ to form NADPH. Also, using protons, phosphorylation of ADP into ATP is performed. Whole process occurs in two photosystems: the first is called photosystem I (PSI) with a maximum radiation adsorption in 700 nm wavelength. The other is called photosystem II (PSII), which is active in 680 nm wavelength. Abovementioned compounds are shifted to another pathway named Calvin–Benson cycle – a complicated network of different biochemical reactions – needed for converting CO_2 into carbohydrates and organic acids (dark reaction). This cycle occurs in a plasmatic matrix named stroma of chloroplast. Stroma contains different enzymes involved in photosynthesis. Also, starch globules accumulate there.

For CO_2 assimilation, this gas is combined with ribulose 1,5-bisphosphate (RUBP) to be carboxylated by RUBP carboxylase (Rubisco) that is abundant in leaves. The product is a six-carbon compound that soon decomposes to yield two 3-phosphoglyceric acid (3-PGA). Since this substance contains three carbon atoms, the plants involving this pathway are called C3. Eventually, a substance named 3-phosphoglycer aldehyde is produced which is considered as a precursor for different carbohydrates, starch, carboxylic acids, and amino acids. For producing one mole of carbohydrate, two moles of NADPH and three moles of ATP are used.

In some plants, the first assimilation product is a di-carboxylic acid with four carbon atoms rather than a three-carbon atom substance. In this pathway, CO_2 enters a combination with phosphoenolpyruvate (PEP). Plants with such pathway are called C4 plants. These plants mainly belong to warm climates, produce higher yields than C3 plants, and can efficiently use low CO_2 concentrations in the atmosphere. Some important cereals like sugarcane, corn, sorghum, and millet belong to C4 plants, whereas most important cool season plants such as wheat, barley, sugar beet, bean, lentil, and chickpea are classified as C3 plants.

2.2.1 Factors Influence Photosynthesis

2.2.1.1 Light

Chloroplasts absorb light needed for photosynthesis. The light used in the photosynthesis covers a range of 380 nm (violet) through 760 nm (red). Out of this range, light has no efficiency for photosynthesis. In wavelengths less than 380 nm, absorbed energy may damage cells, whereas in rates more than 760 nm, such a weak energy absorbed could practically not be effective in photosynthesis. In addition to light wavelength, light intensity is the other main factor in photosynthesis. In a mid-day of summer, the highest light intensity of about 100,000 lux had been calculated to reach ground surface, but this is very high amount for almost all plant species; hence, 1/4 of total radiation (25,000 lux) seems to be ideal for maximum photosynthesis. In shade plants, this rate is very low and reaches about 4000–5000 lux. Imagine the canopy of a tree where the leaves in outermost part are exposed to direct light with the highest radiation intensity, while the leaves in inner parts are well adapted to indirect light and harvest yellow to green wavelength of the sunlight spectrum. Very little amount of captured light energy is consumed in photosynthesis, and the majority is turned into temperature, fluorescent, and phosphorescent.

2.2.1.2 Oxygen

Oxygen in two forms influences this phenomenon: Complete lack of oxygen may suppress the photosynthesis activity, while excess amount of this gas could reduce its ratio so much that some plants lose their photosynthesis capacity after only two hours, while some others exposed to more than 15 hours of oxygen-free environment are not able to perform photosynthesis. This has not been a constant finding, and some opposite findings show that oxygen may not be required for photosynthesis. As purple bacteria have ability to perform photosynthesis like plants, it has been found that abovementioned bacteria survive only under severe anaerobic environment. Also, it has been found that some plants can withstand up to 24 hours of anaerobic atmosphere without permanent damages.

2.2.1.3 Carbon Dioxide

In most cases, there is a limit for carbon dioxide adsorption to be sufficient for maximum photosynthesis. Prevalently, CO_2 constitutes about 0.03% of the atmosphere, but this rate is progressively increased due to industrial combustion of hydrocarbons. This has positively affected dry matter production in plants, that is, increase in CO_2 rates up to 0.1% results in higher photosynthesis, but excess of CO_2 may decrease the process of photosynthesis. Naumburg and Ellsworth (2000) reported no increase in photosynthesis rate in elevated CO_2. A probable cause is acidification of the cell contents by disturbing buffering capacity. Also, it has been found that at elevated CO_2, mesophyll conductance is decreased (Flexas et al. 2007). On the other hand, in some cases (radish leaves), the maximum photosynthesis is obtained in higher rates of about 5%.

2.2.1.4 Carbohydrates

There is a consensus on photosynthesis decline in case of carbohydrate accumulation, especially after prolonged illumination in many plants. This could be attributed to blocking of the enzymes involved in photosynthesis, or by diversion of light energy to so-called photooxidative route in which reverse reactions are accelerated to consume produced carbohydrates. In contrast, there exist opinions that claim plants left in dark and exposed to starvation for several hours are less efficient in photosynthesis than normal plants. It has been suggested that existence of carbohydrates results in higher rates of respiration and consequently higher amounts of photosynthesis.

2.2.1.5 Available Water

Since water is a constituent of all cells, water availability is more effective on cell wall shrinking and turgor rather than direct effect on photosynthesis mechanism, and in case low rates of water are absorbed or when cell loses water, its osmotic potential becomes more negative and, hence, water could not properly flow inside the colloidal cell sap. Also, enzyme activity involved in photosynthesis is reduced. Water deficit indirectly restricts water and gas exchange between cell and environment by closing of the stomata. These organs tend to close at high temperatures to inhibit water loss and consequently result in decline in photosynthesis.

2.3 Stresses

Plants survive under very different environmental circumstances that most of which influence plant growth negatively, named stresses. In other words, in most cases, environmental factors are not mainly optimal for a healthy growth and potent yield. Plants either tolerate or escape from stress conditions by different ways. Prevalent stresses that plants encounter include salinity, drought, heat, chilling, heavy metals, and pathogens.

2.3.1 Salinity Stress

This stress mainly occurs in arid and semi-arid regions in which annual evaporation exceeds total precipitation. In such soils, electrical conductivity (EC) of soil solution reaches higher than 4 dS m^{-1}, which is mainly caused by NaCl, but other ions also contribute to this phenomenon (Fahad and Bano 2012). In recent years, overapplication of chemical fertilizers with different shapes of nitrogen has led to penetration of excess amounts of these substances into the underground waters. During evaporation, these forms of nitrogen are elevated to the soil surface by capillary micro tubes and accumulate there after drying as a white layer. In addition, some water flows by passing through saline soils, dissolves different salty ions, and deposits them in other places, and hence, salinity may spread. Also, some subsurface waters have salinity rates to some extent, and by using these waters as field irrigation, salinity reaches the non-saline soils.

Considering the global warming and decline in precipitation, unfavourable perspective in terms of soil salinity is expected. High solubility of normal salts results in their absorption in toxic rates by plants so that salinity influences plant growth negatively, and resulted injuries depend on different factors like climatic conditions, plant species, and soil conditions (Tang et al. 2015). In terms of tolerance against salinity, plants mainly are divided into glycophytes and halophytes. Unfortunately, majority of the crop plants include the first category, and it has been found that growth of most crop plants is inhibited and/or prevented in the presence of NaCl concentrations up to 100–200 mM (Munns and Termaat 1986), but halophytes can resist up to 300–500 mM concentration of the mentioned salt (Parida and Das 2005; Flowers and Colmer 2015).

Salinity exhibits many features by which plants are affected. Among them are direct and indirect impacts. When the concentration of a salt increases in the root zone and we ignore their uptake by root system, no matter these salts are chlorides of alkaline metals like sodium, potassium, magnesium, etc. or excess amounts of macro (urea, ammonium nitrate, etc.) and micro (iron, copper, zinc sulphates, etc.) nutrients, water absorption is restricted due to the higher osmotic potential and plant growth is exposed to risk, but when we speak about direct impact, this matter becomes more complicated. Excess uptake of deleterious ions like Na^+ and Cl^- may interfere with uptake of nutrient ions like potassium and calcium (Marschner 1995). Also, overaccumulation of salt ions may impose toxic effects on plant cells. Hernández et al. (1993) found that as a result of salinity stress, reactive oxygen species (ROS) lead to oxidative stress. Also during salt stress, leaves may be deformed and stay smaller than normal, which in turn negatively affects photosynthesis. Necrotic spots may be observed in some species as well. Moreover, in case of high rates of salinity accumulated in forage plants, their consumption as livestock feed may be restricted.

Some plants possess abilities to cope with salinity damage. For instance, a halophyte plant may grow in saline soils without or with less injury. Such plant benefits from some adaptive mechanisms help it survive even at salinity levels lethal for other plants. For example, these plants can accumulate salts in their cell sap in osmotic potential rates higher than that of soil solution. Also, protoplast of such plants resists salinity. Another way is to accumulate salts in vacuoles and expel them outside without contacting the cells (Badr and Shafei 2002). Totally, there exist different protective ways by which plants tolerate salinity such as morphological changes, water relations, hormonal activities, changes in photosynthesis, distribution of salts in different parts of plant, and biochemical pathways.

2.3.2 Drought Stress

Undoubtedly, drought stress is considered to be the most important environmental factor limiting the crop yield all round the world. Unfortunately, this stress is going to spread to regions that did not experience it before. Threats of this event could not be expected due to its dependence on various factors such as precipitation occurrence and distribution, evaporation, global warming, and soil capacity of water storage (Wery et al. 1994; Fahad et al. 2013; Saud et al. 2013; Fahad et al. 2014a,b; Saud et al. 2014; Saud et al. 2016; Saud et al. 2017; Fahad et al. 2019a,b; Hesham and Fahad 2020; Saud et al. 2020). Each year, this stress leads to thousands of tons of yield loss in different crops, and as a result, in some regions cropping system of high water demand crops like rice has been changed with those of lower requirements.

Like other stresses, this one is responded by plants in different ways. The first impact of drought is appeared in leaf gas and water exchange rates. When the humidity in the air decreases, guard cells placed on the leaf surface close the stomata to restrict water loss from the leaf. This in turn leads to decline in CO_2 absorption. Furthermore, CO_2 diffusion into chloroplast is decreased (Pinheiro and Chaves 2011), and hence, photosynthesis is reduced. In such conditions, cell membrane system tends to be adjusted (Krasensky and Jonak 2012). Another defence mechanism is accumulation of some protective solutes such as proline. This substance has been found to stabilize proteins that are sensitive to dehydration. Also, proline may facilitate water uptake by adjusting impact on cytoplasmic osmotic potential. In addition, removing of ROS, as a substance produced under different stresses like water deficit, is at least partly attributed to proline production (Valliyodan and Nguyen 2006); however, plants benefit from antioxidant enzymatic defence system like superoxide dismutase (SOD) and peroxidase (POD) to cope with drought stress (Farooq et al. 2009; Sharma et al. 2012).

2.3.3 Cold Stress

Cold stress includes temperatures under which growth and development of each growing stage of a plant are reduced and/or suppressed. As we know, plants are totally divided into cool season, moderate, and temperate plants, and each category needs definite range of temperature for a maximal growth and reproduction. For example, small seeded cereals like wheat, barley, oat, and rye are amongst cool season plants and may tolerate low temperatures especially in earlier growth stages, as wheat can germinate at 3–5 °C, which seems to be too low for large seeded cereals like maize, which is considered to be a temperate crop. Other examples of crops that are sensitive to cold temperatures are soybean, cotton, sugarcane, cucumber, tomato, and watermelon. Although wheat can withstand low temperatures near zero or even lower in early growing season, this does not mean that the plant could resist low temperatures throughout growing period. For instance, this plant is impaired severely by temperatures optimal for germination. In most plants, flowering/reproduction stage is considered so sensitive to low temperatures that if a tree in flowering stage encounters low to freezing temperatures, most probably the great injury will be occurred in terms of yield loss. Even in some cases, it has been observed that whole walnut tree has dried when faced with snowfall during the flowering stage. Symptoms of injury usually appears about 48–72 hours after exposure to cold. The most important thing is the temperature drop rate; when a mid-sensitive to cold plant is gradually exposed to low/reduced temperatures, there may be acclimation to cold stress that is called chilling tolerance and/or cold acclimation, but if the rate of exposure is rapidly occurred, reverse results might be achieved. Based on the definition, it has been accepted that plant tolerance against temperature range of 0–15 °C without injury or damage is called chilling tolerance (Somerville 1995), while cold acclimation is known as the tolerance to the physical and physiochemical events caused by freezing stress (Guy 1990; Thomashow 1999; Khan et al. 2016). Generally, this feature is seen in cool and moderate season plants and very scarcely occurs in temperate crops. Cold acclimation and chilling tolerance are complicated events and depend on biochemical, molecular, and metabolic processes (Kotak et al. 2007; Zhu et al. 2007). During rapid fall of the temperature to the point of freezing, existing water within the cells turns to the ice crystals, resulting in the severe damage to them and leakage of cell sap, while in case of gradual exposure to such temperatures, protective substances and osmolytes such as sugars, amino acids, organic acids, lipids, and polyamines (Nayyar et al. 2005; Farooq et al. 2009) accumulate in cells and lower the freezing temperature of water inside. This phenomenon is well understood in wheat leaves in the rosette stage under the layer of snow that acts as an insulator for freezing. Today, thanks to the genes transfer techniques, many efforts have been made to transfer cold tolerance genes to the crops.

2.3.4 Heat Stress

Global warming is increased progressively and based on the report of Intergovernmental Panel on Climate Change (IPCC 2007), an increase of 0.2 °C in air temperature is predicted per decade; hence, by the year 2100, about 1.8 to 4 °C higher temperatures than the current level are expected. In such a rising temperature, exposure of cool season plants to high temperatures would result in yield loss. Negative impacts of this stress depend on different factors like intensity and duration of high temperature, plant type, growth stage, and nutrition. Like cold stress, high temperatures may result in ROS, which cause oxidative stress (Hasanuzzaman et al. 2013a; Fahad et al. 2016a,b,c,d; Fahad et al. 2017; Fahad et al. 2018). Heat stress may cause alterations in gene expression relating to the production of substances including osmoprotectants, regulatory proteins (heat shock proteins), and oxidative enzymes. Some substances that may impact on heat stress tolerance include phytohormones (brassinosteroids, abscisic acid, jasmonic acid, and salicylic acid), signalling molecules (nitric oxide), trace elements (selenium and silicon), and nutrients (phosphorus, potassium, and calcium) (Barnabás et al. 2008; Hasanuzzaman et al. 2013a).

At high temperatures, cells death or imposed injuries may occur in few minutes. Denaturation of proteins may occur at high temperatures, and enzymes may show dysfunction. In addition, membranes and RNA may be negatively affected. Below we can find information on high temperature impacts on some crops in terms of heat rate, growth stage, and major effect, respectively:

Pepper: 38 °C – Reproductive, maturity and harvesting time – Reduced fruit weight/Rice: Above 33 °C – Heading stage – Reduced spikelet fertility/Wheat: 37 °C – Seed filling and maturity – Reduced leaf size, reduced seed number, seed yield loss/Sorghum: 40 °C – Maturity – Decreased chlorophyll content, increased ROS content, yield loss/Soybean: 38 °C – Flowering – Decreased stomatal conductance, reduced activities of antioxidant enzymes (Hasanuzzaman et al. 2013b).

2.3.5 Heavy Metal Stress

Soil, as the bed for plant root, continuously encountering chemical pollutions, either biological or mineral substances, may negatively affect plant growth. There are elements in the periodic table whose lethal effects could intensely limit crop performance. Some of these elements include nickel (Ni), cadmium (Cd), chromium (Cr), vanadium (V), lead (Pb), mercury (Hg), tin (Sn), copper (Cu), and cobalt (Co). These are mainly found in soils by application of industrial sewage sludge and wastes, phosphate fertilizers, and irrigation with waters originating from naturally enriched with abovementioned elements (Bell et al. 2001; Schwartz et al. 2001; Passariello et al. 2002). Despite some of these elements such as zinc and copper are considered to be nutrient elements, but in excess rates, these become toxic to plants.

Cadmium amounts in soil have been reported to be near 100 ppm, but it may rise due to industrial activities (Salt and Rauser 1995). Toxic levels of this element could reduce photosynthesis and water uptake and disturb proper uptake of nutrient elements. Apparent symptoms of this element on plants include chlorosis and browning of root tips (Wójcik and Tukiendorf 2004; Mohanpuria et al. 2007). Zinc concentration up to 150–300 ppm has been found to have toxic effect on plants (de Vries et al. 2002; Warne et al. 2008), and this effect may appear as chlorosis in young leaves which can extend to older ones (Ebbs and Kochian 1997; Shah et al. 2020). Zinc toxicity also may inhibit proper uptake of manganese (Mn) and Cu. Exposure of plants to excess rates of Copper may cause oxidative stress and ROS production (Stadtman and Oliver 1991). Mercury, due to its unique form and extreme toxicity to animals, humans, and plants, should be precisely monitored in croplands as this element can accumulate in lipid structures without removal of body, and cropping in contaminated soils should be avoided. This element can bind to the proteins, induce closure of stomata, and interfere with water flow in crop (Zhang and Tyerman 1999). Chromium has been found to be toxic for many plants. For example, in some cereals like barley, seedling growth is inhibited up to 40% in concentrations of about 100 μM (Skeffington et al. 1976). Lead is a widespread soil pollutant and can adversely impact on water balance, membrane permeability, and enzyme activity (Sharma and Dubey 2005). Excess amounts of Cobalt in soil may decrease biomass, Fe content, and catalase enzyme activity in some plants (Chatterjee and Chatterjee 2000). Excess rates of nickel can cause nutrient imbalance and disorder of cell membrane functions. For instance, plasma membrane activity has been found to be negatively affected by this element in rice shoots (Ros et al. 1992).

Despite plant biochemical protective reactions against heavy metal stress, the best way of avoiding such damages is planting in non-polluted soils, irrigating with low impurity values, and isolating croplands from industrial zones.

2.4 Water–Soil–Plant Relations

Water–soil–plant relations are of great importance in crop production and cover different disciplines which their understanding would help to better manage field.

Water, as a unique substance, has great ecologic impact due to its specific density, polarity, cohesion and adhesion properties, freezing and boiling points, and also solvent features. All the reactions and processes in plant directly or indirectly relate to water. For instance, respiration rate in immature seeds is very high but over the maturity – while the seeds tend to lose their water content – its rate severely is decreased and could result in seed survival for a long time in nature. Water possesses some phenomena which include (i) structural role: more than 80–90% of forage crops and up to 50% of woody plants structure consist of water, (ii) solvent role: gases, minerals, and organic substances are dissolved and transferred between environment and plant through water. Also, rather high permeability of cell membrane would supply continuous flow of different materials inside the plants, (iii) chemical role: many important processes such as photosynthesis, respiration, sugar hydrolysis, enzymatic protective pathways, etc. take place in wet beds, (iv) turgor role: cell growth and development, stomata closure, leaf twisting, flower opening, etc. are directly related to water availability.

Soil texture and structure extremely vary from region to region, and different soils owing the variable abilities to water retention keep plant diversity all around the world. Some soil features include consistency, plasticity, strength, compactibility, penetrability, and trafficability. Rate of water per mass or volume unit of soil is called 'moisture content,' and physiochemical situation of water in soil is defined as 'water potential.'

2.4.1 Soil Specifications

Below is a list of soil specifications by which water moisture content could be calculated:

2.4.1.1 Particle Density

This specification varies between 2.6 and 2.7 g cm^{-3} and indicates density of solid particles in soil. Sometimes, this is demonstrated as 'specific gravity.'

2.4.1.2 Dry Bulk Density

Its rate is lower than the abovementioned specification and ranges from 1.3 to 1.4 g cm^{-3}.

2.4.1.3 Wet Bulk Density

It is also called 'total bulk density' and completely depends on soil humidity.

2.4.1.4 Specific Volume of Dry Soil

This is one of the important soil indices describing the compaction rate of the soil.

2.4.1.5 Porosity

This specification varies between 30 and 60% in croplands. With the finest soil texture, the higher rates of porosity are observed.

2.4.1.6 Mass and Volume Humidity of the Soil

This specification is expressed as mass humidity ratio or volume humidity ratio.

2.4.1.7 Saturation Degree

This indicates a ratio of humidity volume to pores volume and varies from 0% in dried soil to 100% in saturated soil.

Water content of the soil may be calculated directly or indirectly. Direct method includes sampling of the soil, determining wet (fresh) weight, and drying in oven to measure dry weight, but indirect methods include neutron meter, gamma radiation, and plaster block. The final goal in each method is to plan a suitable programme for irrigation.

2.4.2 Water Potential

There are some water potentials in soil and plant by which water uptake and flow in soil–plant system can be described to some extent. These potentials are as follows:

2.4.2.1 Gravitational Potential

It originates from earth gravity, and in this, potential water penetrates due to its weight downwards.

2.4.2.2 Matric Potential

In unsaturated soil, water molecules adsorb to the soil particles, which is known as matric potential. Rate of this potential in saturated soil which water can freely move is zero. In this situation, this potential is replaced with 'pressure potential.'

2.4.2.3 Osmotic Potential

It is a potential that relates to the dissolved substances in water. Dissolved salts and/or organic matters in water induce a potential to absorb more humidity in order to balance its concentration. With the presence of more dissolved matter, the higher tendency appears to uptake water. By using this phenomenon through accumulation of different solutes in cell sap, some plants having high water use efficiency save their water content under high evaporation conditions and retain water absorption even in presence of salts.

Water absorption by plants comprises many mechanisms by which water may elevate to the heights of up to 100 meters in some trees. Water absorption from soil is mainly divided into active and non-active pathways, but in both of which environmental factors such as air temperature, precipitation rate, soil composition, rate of photosynthesis and respiration, plant nutrition, and plant age are effective. When soil is properly wet and transpiration rate of crop is low, water inside is under positive pressure which is called root pressure. It seems that this pressure depends on cells' activity and, thus, is defined as active absorption, but in plants with rapid transpiration, water movement is performed as mass flow and is called non-active absorption. When water is evaporated from the leaves, drop of water potential results in moving of the water from xylems into the leaves. Such trend is repeated to the roots, and thus, continuous water flow is provided.

Although water loss through transpiration may result in wilting especially at mid-day, this is necessary for plant growth and ensures optimum yields.

Different plants possess various abilities to absorb and use water. Majority of plants like wheat, sunflower, beans, etc. are considered, have moderate water use efficiency and relative to C4 crops, exhibit either impotent water uptake like fibrous roots or fast water loss during the warm days of summer while plants like maize or millet, possessing CO_2 concentration systems, inhibit water loss from stomata and spend more water for photosynthesis than temperature exchange and balance. The most potent plants in this regard are those that belong to *crassulaceae* family, which retain water in plant body even in high temperatures without rainfall for several months; however, their yield rate may be lower than C4 plants. Some other plants exhibit rather high water use efficiency while being C3, like alfalfa, which can absorb water from high depths of the soil thanks to their potent root systems. Some other plants, most of which found in deserts, have specific anatomic features like deep stomata, low surface of leaf, fluffy leaves, etc. to save water.

2.5 Nutrition

Nutrient elements are materials needed for plant growth, development, and yield and divided in two categories including macro (nitrogen (N), phosphorous (P), potassium (K), calcium (Ca), magnesium (Mg), and sulphur (S)) and micro (iron (Fe), zinc (Zn), copper (Cu), boron (B), molybdenum (Mo), cobalt (Co), selenium (Se), manganese (Mn), and sometimes sodium (Na)). Among macro elements, N possesses widespread usage, followed by P and K. N helps better growth, yield, and protein increase, but it decreases plant tolerance against disease and pests. On the other hand, P and K are essential for healthy growth and tolerance against pathogens and even physiological disorders like lodging caused by overirrigation, heavy rainfall, wind, or high N application.

A large amount of applied fertilizers relate to mineral resources or industrially synthesized (like nitrogen), and very little part includes organic substances like animal manure, crop residues, and composts. Regardless of the source of nutrient, type of element, growth stage, environmental conditions, type of culture, etc. play important role in mineral applying. For example, when a hydroculture is concerned, stringent and precise measurements in terms of nutrient calculation in irrigated solution are needed for gaining maximum yields. In such conditions, little deviation from the proper concentration of nutrients may result in high yield loss as plant has no choice to cope with the nutrient deficiencies other than received solution, but in nature, there are different ways to meet the plant requirements for nitrogen such as atmospheric nitrogen fixation, symbiosis with various microorganisms to provide elements like phosphorous, etc.

2.5.1 Shapes and Usages of Commercial Macro Nutrients

2.5.1.1 Nitrogen

This element comprises nearly 1–5% of plant dry matter and is utilized to form proteins. Also, it is part of chlorophyll, and leaf area production mainly depends on it (Ahmad, et al. 2009). Although N comprises nearly 78% of atmosphere as N_2 gas, plants are not capable to use it. The proper forms of this element be absorbed by plants are NO^{3-} and NH^{4+}. N has various types of fertilizers including urea (46% N), ammonium nitrate (34% N), monoammonium phosphate (11% N), diammonium phosphate (18% N), ammonium sulphate (21% N), potassium nitrate (13% N), ammoniac (82% N), and biologically fixed nitrogen.

Due to high solubility of most nitrogen compounds and also high rates of this element needed by plants, application of this nutrient is increasingly widespread in most croplands, resulting in many environmental problems; however, N is the most applied fertilizer all around the world. In the case of excess rates of nitrogen application, especially under high rates of precipitation or overirrigation, nitrogen leakage into ground waters as NO_3 can occur. Another form of N loss is through volatilization of NH_3. Under anaerobic conditions, part of the N may be lost in its oxide forms by activity of microorganism as denitrification. Also, it has been found that during application of N at excess rates, nitrous oxide emission is increased (Erickson et al. 2001), which is harmful for human and impacts global warming. Another disadvantage of overdose of N appears as nitrate accumulation in some crops like potato, spinach, and lettuce, which can negatively impact health. This element can easily be absorbed by soil application, topdressing, and foliar spraying, and if humidity is insufficient, excess rates of N may result in damage to plant. In addition, in such situations, some forms like urea tend to be partly lost in gaseous form. This element efficiently is absorbed by leaves; hence, this can be applied as auxiliary fertilization. Almost in all crops, it is preferred to divide total application of N into at least three or more stages rather than using only in the seeding stage.

2.5.1.2 Phosphorous

This element is less abundant than N and K in environment and in plant parts, and constitutes nearly 0.2% of dry weight. It is important in optimum reproductive growth, as it transfer energy as ATP, ADP, and AMP. Also, glycophosphates are essential components of DNA and RNA. Furthermore, compounds such as phospholipids are the key substances in cell membranes. For calculations, this element usually is presented as anhydride phosphoric acid (P_2O_5). Fertilizer types such as monoammonium phosphate (61% P), diammonium phosphate (46%), simple superphosphate (22% P), triple superphosphate (46% P), and phosphoric acid (54 and 85%) are the well-known forms of phosphorous. Plants can absorb this element from the soil solution as $H_2PO_4^-$ and HPO_4^{2-}.

Availability of this element is restricted in acidic soil (pH below 6) due to formation of low soluble compounds of phosphorous and aluminium and iron (precipitated as Al and Fe phosphates). Also, its availability is limited in alkaline soils because of creating ligands with calcium to yield low soluble phosphorous (Marschner 1995). As its main source in soil as rock granules, symbiosis with microorganisms (such as VAM: vesicular arbuscular mycorrhiza) results in dissolution of phosphorous with abovementioned microorganisms, which can then be absorbed by plant roots; hence, the runoff rate of this element is very rare. Also, in some plants, especially those that grow in low P contents, root exudates including organic acids (i.e. citrate) and enzymes result in dissolution and availability of this element from Ca phosphate minerals in soil (Hoffland et al. 1989). Organic P accounts for about 50% of P found in surface of soil granules. In the presence of soil microorganisms, mineralization of organic P provides labile phosphorous for plant use. In soil application, only 10–20% of P can be utilized by plant in that year, and the rest is tended to be fixed (immobilized) rapidly in soil (Holford 1997). In many croplands, P fertilizers are applied in soil in autumn while preparing seed bed, whereas seeding is performed in spring. In such conditions, majority of the element is converted into insoluble forms over the autumn and winter and may not be efficiently adsorbed. Even in time of seeding when granules are placed in long distances from the seeds, chance of the maximum benefit is reduced. For this reason, topdress application is not advised. Instead, placing fertilizer granules just adjacent to seed is preferred. However, foliar spraying has no such limits.

2.5.1.3 Potassium

After N, K is found in high concentrations in plant parts than any other nutrients. This element is responsible for water relations and gas exchange (charge balance) between cells and environment. Some clays like illite and vermiculite tend to contain high levels of K, and this element may be released into the soil by weathering process. In areas with heavy precipitation, leakage may result in K deficiency. K can displace readily by other cations in soil solution. Like P, K availability decreases as soil pH drops. This in part relates to the competition between K and Al, with solubility of the latter increasing in acidic soils. K is found in five shapes: organic K, primary minerals like feldspar and mica, non-exchangeable K, exchangeable K on surface of minerals, and K^+ in soil solution. Types of this element as fertilizer are as follows: potassium chloride (60–62% K), potassium nitrate (46% K), potassium sulphate (44% K), and potassium magnesium sulphate (24% K). These are the most common sources of potassium. Potassium chloride is a cheap and highly soluble K source, but due to damages of Cl^- anion to most plants as salinity, its application is restricted. Instead, potassium sulphate is a better choice.

This element has been found to alleviate adverse effect of salinity in plants (Garg and Gupta 1998) and to enhance plant tolerance against environmental stresses (Römheld and Kirkby 2010). One of the important effects of potassium on plant is osmoregulation (Cakmak 2010).

2.5.1.4 Calcium and Magnesium

These elements belong to alkaline soil metals and often are found in high concentrations in soils of arid and semi-arid regions. Needed amount of Ca is nearly two times that of Mg. Availability of these elements is reduced as pH drops below 6. Although they are base elements and may dissolve in soil acidity, they held onto the cation exchange capacity (CEC), and in such circumstances, Al_3^+ competes with them to occupy the CEC positions. Thus, dissolved Ca and Mg ions are exposed to leach from soil solution. Application of lime may lead to increase available Ca and Mg, but higher than pH 8 rates may cause decrease of the mentioned elements due to precipitation with carbonates. These elements are absorbed from the soil solution through mass flow. Sulphates of Ca and Mg are considered the available sources of these elements.

Ca is essential component of cell wall and also involves in transportation of assimilates and nutrient inside the plant. Mg is a component of chlorophyll and ribosomes. Also, it is required for transferring phosphate from ATP. Opposite to Ca, Mg deficiency is seen in old leaves first. Because of Mg role in chlorophyll formation, its deficiency symptoms are the same to that of N as interveinal chlorosis.

2.5.1.5 Sulphur

This element is essential component of some amino acids and also found as complex substances in *brassicaceae* family.

The main form of the absorption of this element is SO_4^{2-}, and it is available in many macro and micro nutrient fertilizers such as ammonium sulphate, potassium sulphate, and iron sulphate; hence, sole application of this element is not prevalent. However, granulated sulphur (S) may be used as well. Biological substances also contain rates of S. Other source of this element is H_2SO_4, which is not considered fertilizer but, due to its application in improving soil pH, can increase S in soil.

REFERENCES

Ahmad S, Ahmad R, Ashraf MY, Ashraf M, Waraich EA (2009) Sunflower (*Helianthus annuus* L.) response to drought stress at germination and seedling growth stages. Pak J Botany 41:647–654

Badr MA, Shafei AM (2002) Salt tolerance in two wheat varieties and its relation to potassium nutrition. Al-Azhar J Agric Res 35:115–128.

Barnabás B, Jäger K, Fehér A (2008) The effect of drought and heat stress on reproductive processes in cereals. Plant Cell Environ 31:11–38.

Bell FG, Bullock SET, Halbich TFJ, Lindsay P (2001) Environmental impacts associated with an abandoned mine in the Witbank Coalfield, South Africa. Int J Coal Geol 45:195–216.

Cakmak I (2010) Potassium for better crop production and quality. Plant Soil 335:1–2.

Chatterjee J, Chatterjee C (2000) Phytotoxicity of cobalt, chromium and copper in cauliflower. Environ Pollut 109:69–74.

de Vries W, Lofts S, Tipping E, Meili M, Groenenberg JE, Schütze G (2002) Impact of soil properties on critical concentrations of cadmium, lead, copper, zinc, and mercury in soil and soil solution in view of ecotoxicological effects. Rev Environ Contam Toxicol 191:47–89.

Ebbs SD, Kochian LV (1997) Toxicity of zinc and copper to *Brassica* species: implications for phytoremediation. J Environ Qual 26:776–781.

Erickson H, Keller M, Davidson EA (2001) Nitrogen oxide fluxes and nitrogen cycling during post-agricultural succession and forest fertilization in the humid tropics. Ecosystems 4:67–84.

Fahad S, Adnan M, Hassan S, Saud S, Hussain S, Wu C, Wang D, Hakeem KR, Alharby HF, Turan V, Khan MA, Huang J (2019b) Rice responses and tolerance to high temperature, In: Hasanuzzaman M, Fujita M, Nahar K, Biswas JK (eds) Advances in rice research for abiotic stress tolerance. Woodhead Publ Ltd, Cambridge, England, pp 201–224.

Fahad S, Bajwa AA, Nazir U, Anjum SA, Farooq A, Zohaib A, Sadia S, NasimW, Adkins S, Saud S, Ihsan MZ, Alharby H,Wu C,Wang D, Huang J (2017) Crop production under drought and heat stress: plant responses and management options. Front Plant Sci 8:1147. https://doi.org/10.3389/fpls.2017.01147.

Fahad S, Bano A (2012) Effect of salicylic acid on physiological and biochemical characterization of maize grown in saline area. Pak J Bot 44:1433–1438.

Fahad S, Chen Y, Saud S,Wang K, Xiong D, Chen C,Wu C, Shah F, Nie L, Huang J (2013) Ultraviolet radiation effect on photosynthetic pigments, biochemical attributes, antioxidant enzyme activity and hormonal contents of wheat. J Food Agri Environ 11:1635–1641.

Fahad S, Hussain S, Bano A, Saud S, Hassan S, Shan D, Khan FA, Khan F, Chen Y, Wu C, Tabassum MA, Chun MX, Afzal M, Jan A, Jan MT, Huang J (2014a) Potential role of phytohormones and plant growth-promoting rhizobacteria in abiotic stresses: consequences for changing environment. Environ Sci Pollut Res 22:4907–4921. https://doi.org/10.1007/s11356-014-3754-2.

Fahad S, Hussain S, Matloob A, Khan FA, Khaliq A, Saud S, Hassan S, Shan D, Khan F, Ullah N, Faiq M, Khan MR, Tareen AK, Khan A, Ullah A, Ullah N, Huang J (2014b) Phytohormones and plant responses to salinity stress: A review. Plant Growth Regul 75:391–404. https://doi.org/10.1007/s10725-014-0013-y.

Fahad S, Hussain S, Saud S, Hassan S, Chauhan BS, Khan F et al (2016a) Responses of rapid viscoanalyzer profile and other rice grain qualities to exogenously applied plant growth regulators under high day and high night temperatures. PLoS One 11:e0159590. https://doi.org/10.1371/journal.pone.0159590.

Fahad S, Hussain S, Saud S, Hassan S, Ihsan Z, Shah AN, Wu C, Yousaf M, Nasim W, Alharby H, Alghabari F, Huang J (2016c) Exogenously applied plant growth regulators enhance the morphophysiological growth and yield of rice under high temperature. Front Plant Sci 7:1250. https://doi.org/10.3389/fpls.2016.01250.

Fahad S, Hussain S, Saud S, Hassan S, Tanveer M, Ihsan MZ, Shah AN, Ullah A, Nasrullah KF, Ullah S, AlharbyH NW, Wu C, Huang J (2016d) A combined application of biochar and phosphorus alleviates heat-induced adversities on physiological, agronomical and quality attributes of rice. Plant Physiol Biochem 103:191–198.

Fahad S, Hussain S, Saud S, Khan F, Hassan S, Jr A, Nasim W, Arif M, Wang F, Huang J (2016b) Exogenously applied plant growth regulators affect heat-stressed rice pollens. J Agron Crop Sci 202:139–150.

Fahad S, Hussain S, Saud S, Tanveer M, Bajwa AA, Hassan S, Shah AN, Ullah A,Wu C, Khan FA, Shah F, Ullah S, Chen Y, Huang J (2015a) A biochar application protects rice pollen from high-temperature stress. Plant Physiol Biochem 96:281–287.

Fahad S, Muhammad ZI, Abdul K, Ihsanullah D, Saud S, Saleh A, Wajid N, Muhammad A, Imtiaz AK, Chao W, Depeng W, Jianliang H (2018) Consequences of high temperature under changing climate optima for rice pollen characteristics-concepts and perspectives. Arch Agron Soil Sci 64:1479–1488. https://doi.org/10.1080/03650340.2018.1443213.

Fahad S, Nie L, Chen Y, Wu C, Xiong D, Saud S, Hongyan L, Cui K, Huang J (2015b) Crop plant hormones and environmental stress. Sustain Agric Rev 15:371–400.

Fahad S, Rehman A, Shahzad B, Tanveer M, Saud S, Kamran M, Ihtisham M, Khan SU, Turan V, Rahman MHU (2019a) Rice responses and tolerance to metal/metalloid toxicity. In: Hasanuzzaman M, Fujita M, Nahar K, Biswas JK (eds) Advances in rice research for abiotic stress tolerance. Woodhead Publ Ltd, Cambridge, England, pp 299–312.

Farooq M, Wahid A, Kobayashi N, Fujita D, Basra SMA (2009) Plant drought stress: effects, mechanisms and management. Agron Sustain Dev 29:185–212.

Flexas J, Diaz-Espejo A, Galmés J, Kaldenhoff R, Medrano H, Ribas-Carbo M (2007) Rapid variations of mesophyll conductance in response to changes in CO_2 concentration around leaves. Plant Cell Environ 30:1284–1298.

Flowers TJ, Colmer TD (2015) Plant salt tolerance: adaptations in halophytes. Ann Bot 115:327–331.

Garg BC, Gupta IC (1998) Physiology of salt tolerance of arid-zone crops IV. Rapeseed and Indian mustard, Central Arid Zone Research Institute, Jodhpur, India. Curr Agric 22:1–2.

Guy CL (1990) Cold acclimation and freezing stress tolerance: role of protein metabolism. Annu Rev Plant Phys 41:187–223.

Hasanuzzaman M, Nahar K, Fujita M (2013a) Extreme temperatures, oxidative stress and antioxidant defense in plants. In: Vahdati K, Leslie C (eds) Abiotic stress-plant responses and applications in agriculture. InTech, Rijeka, Croatia, pp 169–205.

Hasanuzzaman M, Nahar K, Mahabub Alam M, Roychowdhury R, Fujita M (2013b) Physiological, biochemical, and molecular mechanisms of heat stress tolerance in plants. Int J Mol Sci 14:9643–9684.

Hernández JA Corpas FJ, Gómez M, del Río LA, Sevilla F (1993) Salt induced oxidative stress mediated by activated oxygen species in pea leaf mitochondria. Physiol Plant 89:103–110.

Hesham FA, Fahad S (2020) Melatonin application enhances biochar efficiency for drought tolerance in maize varieties: Modifications in physio-biochemical machinery. Agron J 112(4):1–22.

Hoffland E, Findenegg GR, Nelemans JA (1989) Solubilisation of rock phosphate by rape II. Local root exudation of organic acids as a response to P starvation. Plant Soil 113:161–165.

Holford ICR (1997) Soil phosphorus, its measurements and its uptake by plants. Aust J Soil Res 35:227–239.

Johnson M (2016) Photosynthesis. Essays Biochem 60:255–273.

Intergovernmental panel on climate change (IPCC). Climate change (2007) The physical science basis. In contribution of working group I to the fourth assessment report of the intergovernmental panel on climate change; Cambridge University Press, Cambridge

Khan A, Shah F, Muhammad N, Khan MS, Ahmad MS, Farooq M, Adnan M, Jawad MS, Ullah H, Yousafzai AM (2016) Quantitative determination of lethal concentration Lc50 of atrazine on biochemical parameters; total protein and serum albumin of freshwater fish grass carp (*Ctenopharyngodonidella*). Pol J Environ Stud 25:1–7.

Kotak S, Larkindale J, Lee U, von Koskull-Doring P, Vierling E, Scharf KD (2007) Complexity of the heat stress response in plants. Curr Opin Plant Biol 10:310–316.

Krasensky J, Jonak C (2012) Drought, salt and temperature stress-induced metabolic rearrangements and regulatory networks. J Exp Bot 63:1593–1608.

Marschner H (1995) Mineral nutrition of higher plants, 2nd edn. Academic Press, London.

Mohanpuria P, Rana NK, Yadav SK (2007) Cadmium induced oxidative stress influence on glutathione metabolic genes of *Camellia sinensis* (L.) O. Kuntze. Environ Toxicol 22:368–374.

Munns R, Termaat A (1986) Whole plant response to salinity. Aust J Plant Physiol 13:143–160.

Naumburg E, Ellsworth DS (2000) Photosynthetic sunfleck utilization potential of understory saplings growing under elevated CO2 in FACE. Oecologia 122:163–174.

Nayyar H, Chander K, Kumar S, Bains T (2005) Glycine betaine mitigates cold stress damage in Chickpea. Agron Sustain Dev 25:381–388.

Parida AK, Das AB (2005) Salt tolerance and salinity effects on plants: a review. Ecotoxicol Environ Saf 60:324–349.

Passariello B, Giuliano V, Quaresima S, Barbaro M, Caroli S, Forte G, Garelli G, Iavicoli I (2002) Evaluation of the environmental contamination at an abandoned mining site. Microchem J 73:245–250.

Pinheiro C, Chaves MM (2011) Photosynthesis and drought: can we make metabolic connections from available data? J Exp Bot 62:869–882.
Ros R, Cooke DT, Martínez-Cortina C, Picazo I (1992) Nickel and cadmium-related changes in growth, plasma membrane lipid composition, ATPase hydrolytic activity and proton-pumping of rice (*Oryza sativa* L. cv. Bahia) shoots. J Exp Bot 43:1475–1481.
Römheld V, Kirkby E (2010) Research on potassium in agriculture: needs and prospects. Plant Soil 335:155–180.
Salt DE, Rauser WE (1995) Mg ATP-dependent transport of phytochelatins across the tonoplast of oat roots. Plant Physiol 107:1293–1301.
Saud S, Chen Y, Fahad S, Hussain S, Na L, Xin L, Alhussien SA (2016) Silicate application increases the photosynthesis and its associated metabolic activities in Kentucky bluegrass under drought stress and post-drought recovery. Environ Sci Pollut Res 23:17647–17655. https://doi.org/10.1007/s11356-016-6957-x.
Saud S, Chen Y, Long B, Fahad S, Sadiq A (2013) The different impact on the growth of cool season turf grass under the various conditions on salinity and drought stress. Int J Agric Sci Res 3:77–84.
Saud S, Fahad S, Cui G, Chen Y, Anwar S (2020) Determining nitrogen isotopes discrimination under drought stress on enzymatic activities, nitrogen isotope abundance and water contents of Kentucky bluegrass. Sci Rep 10:6415. https://doi.org/10.1038/s41598-020-63548-w.
Saud S, Fahad S, Yajun C, Ihsan MZ, Hammad HM, Nasim W, Amanullah Jr, Arif M, Alharby H (2017) Effects of nitrogen supply on water stress and recovery mechanisms in kentucky bluegrass plants. Front. Plant Sci 8:983. https://doi.org/10.3389/fpls.2017.00983.
Saud S, Li X, Chen Y, Zhang L, Fahad S, Hussain S, Sadiq A, Chen Y (2014) Silicon application increases drought tolerance of Kentucky bluegrass by improving plant water relations and morph physiological functions. SciWorld J 2014:1–10. https://doi.org/10.1155/2014/368694.
Schwartz C, Gerard E, Perronnet K, Morel JL (2001) Measurement of in situ phytoextraction of zinc by spontaneous metallophytes growing on a former smelter site. Sci Total Environ 279:215–221.
Shah N, Khan A, Ali R, Marimuthu K, Uddin MN, Rizwan M, Rahman KU, Alam M, Adnan M, Jawad SM, Hussain S (2020) Monitoring bioaccumulation (in gills and muscle tissues), hematology, and genotoxic alteration in *Ctenopharyngodon idella* exposed to selected heavy metals. BioMed Res Int 15:2020.
Sharma P, Dubey RS (2005) Lead toxicity in plants. Braz J Plant Physiol 17:35–52.
Sharma P, Jha AB, Dubey RS, Pessarakli M (2012) Reactive oxygen species, oxidative damage, and anti-oxidative defense mechanism in plants under stressful conditions. J Bot 2012:1–26.
Skeffington RA, Shewry PR, Perterson PJ (1976) Chromium uptake and transport in barley seedlings (*Hordeum vulgare* L.) Planta 19:807–810.
Somerville C (1995) Direct tests of the role of membrane lipid composition in low temperature-induce photoinhibition and chilling sensitivity in plants and cyanobacteria. Proc Natl Acad Sci 92:6215–6218.
Stadtman ER, Oliver CN (1991) Metal-catalyzed oxidation of proteins. J Biol Chem 266:2005–2008.
Tang X, Mu X, Shao H, Wang H, Brestic M (2015) Global plant-responding mechanisms to salt stress: physiological and molecular levels and implications in biotechnology. Crit Rev Biotechnol 35:425–437.
Thomashow MF (1999) Plant cold acclimation: freezing tolerance genes and regulatory mechanisms. Annu Rev Plant Phys 50:571–599.
Valliyodan B, Nguyen HT (2006) Understanding regulatory networks and engineering for enhanced drought tolerance in plants. Curr Opin Plant Biol 9:189–195.
Warne MS, Heemsbergen D, Stevens D, McLaughlin M, Cozens G, Whatmuff M, Broos K, Barry G, Bell M, Nash D, Pritchard D, Penney N (2008) Modeling the toxicity of copper and zinc salts to wheat in 14 soils. Environ Toxicol Chem 27:786–792.
Wery J, Silim SN, Knights EJ, Malhotra RS, Cousin R (1994) Screening techniques and sources and tolerance to extremes of moisture and air temperature in cool season food legumes. Euphytica 73:73–83.
Wójcik M, Tukiendorf A (2004) Phytochelatin synthesis and cadmium localization in wild type of Arabidopsis thaliana. Plant Growth Regul 44:71–80.
Zhang WH, Tyerman SD (1999) Inhibition of water channels by $HgCl_2$ in intact wheat root cells. Plant Physiol 120:849–857.
Zhu J, Dong CH, Zhu JK (2007) Interplay between cold-responsive gene regulation, metabolism and RNA processing during plant cold acclimation. Curr Opin Plant Biol 10:290–295.

3

Biological Control

Abdel Rahman M. Al-Tawaha[1], Harpreet Kaur Cheema[2], Marwa M. El-Deriny[3], Dina S. S. Ibrahim[3], Mazen A. Ateyyat[4], Huma Naz[5], Abdel Razzaq Al-Tawaha[6], Ali M. Qaisi[7], Amanullah[8], Imran[8], Abdur Rauf[9], Shah Khalid[8], Mohd Abas Shah[10], Ekaterina Kozuharova[11], Devarajan Thangadurai[12], Jeyabalan Sangeetha[13], Shah Fahad[14], Moh'd Ali Wedyan[15], Sonia Sheoran[16], Pradeep Sharma[16], Garima Singroh[16], Wafa'a A. Al-Taisan[17], and Yousef M Abu-Zaitoon[1]

[1]*Department of Biological Sciences, Al Hussein Bin Talal University, Ma'an, Jordan.*
[2]*Department of Plant Breeding and Genetics, Punjab Agricultural University, Ludhiana, India.*
[3]*Department of Nematodes diseases and central Lab of biotechnology, Plant Pathology Institute, Agriculture Research Center (ARC), Egypt.*
[4]*Faculty of Agricultural Technology, Al Balqa Applied University, As-Salt, Jordan.*
[5]*Mohammad Ali Nazeer Fatima Degree College, Hardoi, U.P., India*
[6]*Department of Crop Science, Faculty of Agriculture, Universiti Putra Malaysia, Serdang, Selangor, Malaysia*
[7]*Department of Pharmaceutical Sciences, School of Pharmacy, University Of Jordan Amman, Jordan.*
[8]*Agronomy Department, The University of Agriculture Peshawar, Pakistan*
[9]*Department of Chemistry, University of Swabi, Anbar, Khyber Pakhtunkhwa, Pakistan*
[10]*ICAR-Central (10) (10)Potato Research Station, Jalandhar, Punjab, India*
[11]*Medical University of Sofia, Faculty of Pharmacy, Department of Pharmacognosy, Sofia, Bulgaria*
[12]*Department of Botany, Karnatak University, Dharwad, Karnataka, India*
[13]*Department of Environmental Science, Central University of Kerala, Kasaragod, Kerala, India*
[14]*Department of Agronomy, University of Haripur, Haripur, Khyber Pakhtunkhwa, Pakistan*
[15]*Department of Biology and Biotechnology The Hashemite University-Zarqa, Jordan*
[16]*ICAR-Indian Institute of Wheat and Barley Research, Karnal, India*
[17]*Department of Biology, College of Science, Imam Abdulrahman Bin Faisal University, Dammam, Saudi Arabia*

3.1 Introduction

The use of natural enemies for control of phytophagous insects has a long historical background. Earliest examples are from China where nests of Pharoah's ants *Monomorium pharaonis* (Linn.) were used to control stored grains and *Oecophylla* spp. against foliage feeding caterpillars. The era of modern biological control (BC) was started in the 1880s with introduction of *Rodalia* beetles from Australia for the successful control of cotton cushiony scale in California. The term biological control (Biologische Bekampfung) was first used in 1914 by the German Plant pathologist Tubuef for the control of plant pathogens and by Harry Smith for insect control in 1919 (Baker 1987). The allelopathic properties of plants can be exploited successfully as a tool for pathogen and weed reduction.

Nearly 230 natural enemies are available globally with a market value of around $200 million. A total registered species of 77 bacteria, 68 fungi and yeasts, 36 viruses, and 2 protozoa accounting for 2300 commercial products are available for plant protection (Ravensberg 2015). By 2010, these 2384 different bio control agents including parasitoids and predators were introduced as 6158 cases in 148 countries against 588 pest species, of which 32.6% led to establishment and 10.1% satisfactorily controlled 172 different pest species. More than 170 biopesticide products based on entomopathogenic fungi are available against insects and mites under glasshouse, field crops, and vegetables (Faria and Wraight 2007).

Allelopathy is defined as the direct or indirect damaging or beneficial effects of one crop or plant on another through the release of chemical compounds into the environment. Growing extremely allelopathic plants has the possibility to considerably decrease our reliance on synthetic bioherbicides for pest management. Specific phytotoxins have been found in allelopathic black mustard, ricc, wheat, and barley varieties. On the other hand, the potential for use of allelopathic crops to reduce synthetic herbicide inputs has been discussed in many review (Turk and Tawaha 2002a,b; Turk and Tawaha 2003a,b; Turk et al. 2003c; Turk et al. 2005; Al-Tawaha and Odat 2010; Al-Zoubi and Tawaha 2019).

BC assumes major importance in the era of sustainable agriculture and increasing preference towards organic farming. Its increasing popularity can be attributed to safety in the era of modern BC (Waage and Greathead 1988). BC may not provide rapid insect pest control as achieved by insecticides, especially as desired in commercial crops. However, it may be more stable and lasting (Baker and Cook 1974). Successful use of BCAs depends on judicious management to form an integral component of biointensive pest management. Combining BCAs, botanicals, and cultivars with moderate level of resistance can adequately manage insect pests. This would surely minimize dependence on pesticides as and when required. Thus, successful combination of entomopathogens, predators, and parasitoids with soft and selective insecticides, semiochemicals, and habitat modification can produce profitable and sustainable pest management systems (Lacey and Shapiro-Ilan 2008).

3.2 Organisms Employed as Biological Control Agents (BCAs)

According to Cook (1988), the agents of BC can include (1) the pest or disease agent used against itself, (2) antagonists or natural enemies of the pest or disease agent, or (3) the host plant or animal managed or manipulated to defend itself. This includes the release of sterile males or the use of behaviour-modifying pheromones produced by insects to regulate the population of certain insect pests over wide geographic areas or within specific orchards or other agroecosystems and use of trap/cover crops (Cook 1988). Arthropods are mainly controlled by predators, parasitoids, and pathogens, while plant pathogens are usually managed through antagonistic microbes.

3.2.1 Biological Control Agents against Arthropods

The antagonists and natural enemies of pests include the classical BCAs imported, mass-reared, and released to control a pest population of insects and weeds (Cook 1988). Singh (2004) reviewed the success stories in classical BC of agricultural pests in India as well as success achieved in other countries using Indian BCAs. Worldwide, there are many documented success stories including control of *Nezara viridula* (Linnaeus) in Australia by *Trissolcus basalis* (Woll.) (Clarke 1992), parasitoids *Bathyplectus* spp. and *Microtonus aethiopoides* (Loan) against alfalfa weevil *Hypera postica* (Gyllenhal) in the United States, and *Apoanagyrus lopezi* (De Santis) against cassava mealy bug in Africa (IITA 1984). Numerous surveys on natural enemies (predators, parasites, parasitoids, and entomopathogens) have been conducted for various crops and regions and have been well summarized by Kogan et al. (1999a) and recently by Toepfer et al. and Mansour et al. Hajek et al. (2007) reviewed classical BC of insects and mites using pathogens and nematodes especially targeting the invasive species. Recently, Gupta and Kumar (2018) have reviewed the status of research on predatory and parasitic mites belonging to 27 families under 3 orders.

3.2.2 Biological Control Agents against Pathogens

BCAs supprcss thc pathogens by various mechanisms as direct effect (interaction between BCAs and pathogens), indirect effect (enhanced plant defence), and ecological effects (occupation of ecological niche). In the case of direct effect, these agents directly suppress pathogens by producing antibiotic, secreting lytic enzymes, competition for nutrients or indirectly by induction of plant resistance response. Simionato et al. have reviewed several micro-organisms and their bioactive metabolites as potential antagonists for pathogen BC. These micro-organisms include bacteria like *Bacillus, Pseudomonas, Agrobacterium, Pantoea,* and *Burkholderia*; fungi *Trichoderma, Aspergillus, Gliocladium, Paecilomyces,* and *Streptomyces*; and yeasts *Aureobasidium, Metschnikowia fructicola, Candida,* and *Arbuscular mycorrhiza* (Trias et al. 2008; Abbasi et al. 2015).

3.2.3 Biological Control Agents against Plant Parasitic Nematodes

Several fungi and bacteria have been reported as successful BCAs in suppression of plant parasitic nematodes in many monocropping cultivation; however, various factors including ecological, intrinsic, technological, commercial, and regulatory affect their efficacy (Moosavi and Zare 2015). Several fungi such as *Paecilomyces lilacinus, Verticillium lecanii, Verticillium chlamydosporium, Trichoderma harzianum, Trichoderma viride*, and *Gliocladium virens* have been reported to parasitize eggs of plant parasitic nematodes (Parvathy Reddy and Nagesh 1998). Plants like *Crotolaria* spp. can suppress phytonematodes including *Meloidogyne* spp., *Heterodera glycines* Ichinohe, and *Rotylenchus reniformis* Linford and Oliveira (Wang et al. 2002).

3.3 Biological Control of Nematodes

Plant-parasitic nematodes (PPN) constitute a major constraint to agriculture. Estimates of crop losses (12.6%), due to PPN for the 20 life-sustaining crops worldwide responsible for crop loss, equalled $215.77 billion annual yield losses. Additional 20 crops significant for food or export value have also (14.45%) annual yield loss of $142.47 billion (Abd-Elgawad 2016). Losses for the total 40 crops represented 13.5%, but the losses on 37 crops are estimated to be $358.24 billion annually (Abd-Elgawad and Askary 2015). Development of control measures of PPN is a great challenge. Because most PPN spend their lives in the soil or plant roots, delivery of a chemical to the immediate surroundings of a nematode is difficult (Chitwood 2003). Moreover, due to health and environmental problems related to the use of chemical nematicides, a resurgence of biocontrol alternatives as much safer and highly practicable for PPN management is apparent. The most effective biocontrol agent should antagonize the pathogen using multiple mechanisms as in *Pseudomonas* which uses both antibiosis and induction of host resistance to suppress the disease-causing microorganisms. These BCAs may originally include organisms such as nematophagous fungi, nematophagous bacteria, mites, plant growth-promoting rhizobacteria (PGPR), arbuscular mycorrhizal fungi (AMF), and predatory nematodes (PN) (Askary and Martinelli 2015).

Most research efforts on predatory nematodes (PN) have more focus on their biological attributes, predation ability, and prey-searching capability than their usability as BCAs in terms of mass production, stability (storage), longevity, and easy application. Control efficacy of PN is defined by the product of predation ability and prey-searching capability, which is similar to the real number of prey removed from the habitat via their predation (Khan and Kim 2007). Prey predation comprises the predation phases: encounter with prey, attack response, and feeding (composed of attack, extra-corporeal digestion, and ingestion) (Bilgrami 2008). Mononchid predators of the families *Anatonchidae, Lotonchidae,* and *Mononchidae* meet prey nematodes through accidental contact, indicating a low probability of prey capture. However, other predatory nematodes such ad dorylaimids, diplogasterids, and aphelenchids make intentional contact with prey nematodes in chemosensory responses to kairomones emitted by the prey, enhancing the probability of predator–prey contacts (Kim 2015). So, overall predation abilities of mononchids are reduced below those of the other PN. *Seinura* species belonging to aphelenchid predators attack their prey by injecting toxic substances for the immobilization of their prey, which is another feature that enhances the PN ability as a successful BCA (Bilgrami 2008). Diplogasterids and aphelenchids have higher searching capabilities because of their high colonizing abilities, short generation time, and high reproduction rates compared to mononchids or dorylaimids which have long generation time and low fecundity (Kim 2015). Under food-rich conditions, diplogasterids such as *Mononchoides* species may reproduce rapidly and can be practically applied for pest control by boosting population numbers using soil amendments of compost and manure (Bilgrami et al. 1983). The massive introduction of mononchids can be applied for intensive short-term control of PPN in heavily infested crops growing in protected areas, for example, glasshouses, due to their aggressive and vigorous feeding behaviours that make them suitable for nematode control (Bilgrami et al. 1983). In contrast, dorylaimids can be maintained via agricultural practices for the long-term control of chronic nematode diseases (Bilgrami 2008). Due to their omnivorous strategy, feeding on nematodes and bacteria having high reproduction rates, the PN culturing for mass production is the most feasible for diplogastrids (Bilgrami 2008, 1990a,b). Such culturing is the least acceptable for mononchids since they have fastidious predaceous strategy and feeding mostly on prey nematodes with low reproduction rates. Cultured growth is moderately feasible for dorylaimids and aphelenchids with both omnivorous behaviour and easy culture *in vitro* (Khan 2007). Kim (2015) advocated that techniques and formulations used for EPN can be applied in the commercialization of predatory nematodes, suggesting them as a feasible BCA, especially diplogasterids that share a common biological feature of forming dauer larvae (a type of stasis enabling them to survive harsh conditions) as typical EPN. Mass production of predatory nematodes with a long shelf-life is prerequisite for their easy application in the biocontrol of PPN, indicating a merit feature of diplogasterids, a demerit of mononchids, and a neutral feature for dorylaimids and aphelenchids. Clearly, all PN can be used for pest control using practical methods designed for their biocontrol features to obtain as best they can for control efficacy. Diplogasterids are best suited for the biocontrol of PPN due to their prey-searching capability and high levels of predation ability. Mononchids are least suited due to their low levels of predation ability. Dorylaimids and aphelenchids each possess high and low degrees of these respective traits and consequently are medially suited for biocontrol of PPN. Eventually, the weakest point of PN for the biocontrol of PPN is materialized in the difficulties in their mass production and storage as compared to other BCAs. These hurdles can be fairly lessened to improve biocontrol efficacy by using the most feasible and practical techniques possibly similar to those of EPN.

3.4 Integrated Pest Management (IPM)

BC was given importance as early as 1959, when Stern et al. (1959) defined integrated control as 'applied pest control which combines and integrates biological and chemical control.' BC is well identified as a key ecosystem service in IPM (Kogan 1998). Earlier IPM programmes were operated with pesticide management objectives based on thresholds rather than pest management objectives (Lewis et al. 1997); however, pesticides should be taken up as the last choice for plant health management. IPM is a broad, ecosystem-centred strategy that focuses on suppressing pest populations by combining various techniques, such as BC, habitat manipulation, modified cultural practices, resistant varieties, and pesticides, which tend to be reduced in well-established IPM programmes (Torres and Bueno 2018).

3.4.1 Integrating Biocontrol Agents with Other Components of IPM

In the IPM pyramid, BC is the most sensitive element to actions at other levels. BC, especially conservation of natural enemies, can play a major role along with agroecosystem modification and other tactics in the evolving IPM.

3.4.1.1 Insecticides

Integrating BC with other pest management approaches including insecticides can help minimize the insecticide-based management problems such as pesticide resistance and secondary pest outbreak. In commercial crops, BCAs and chemicals together can manage multi pest problems adequately. New generations of pesticides that are more selective and less damaging to BCAs are likely to be much more compatible with IPM. Conservation BC using selective insecticides is a scientifically supported, economic, and management decision to be applied to IPM programmes that should provide benefit to agriculture and environment (Torres and Bueno 2018). There is varying level of selectivity for new insecticides relative to the target pests, and these include insect growth regulators, spinosyns, diamides, pyrimidine azomethine, and pyridine carboxamide. Insecticide selectivity may vary among and within species of natural enemies (Carvalho et al. 2000).

Using IPM approach, parasitoids can be released to control *Lepidopteran* pests in crops like sugarcane, maize alongside with the adjusted spray schedules of insecticides preferably insect growth regulators (IGRs) to minimize exposure of parasitoids and natural enemies to pesticide residues. Several arthropod natural enemies such as *Chrysoperla carnea* (Stephens) have the potential to become resistant to pesticides over time which can be used as a positive trait to enhance integration of chemical and biological tools in IPM programmes (Bielza 2016). Thus, through rational planning, integrated use of insecticides and natural enemies can be implemented.

3.4.1.2 Host Plant Resistance

Effective BCAs can increase durability of host plant resistance by weakening natural selection for counter-resistance in pests (Stenberg 2017), delaying the development of resistance in the pest towards its host plant (Gould et al. 1991). The potential is also great where GM crops can be grown followed by BC measures to manage pest populations. Genetically modified plants are highly selective to their targets that help lower insecticide usage and conserve natural enemies (Naranjo 2009). Delay in development of resistance was observed in diamond back moth *Plutella xylostella* in the presence of natural enemy *Coleomegilla maculata* (De Geer) on Bt broccoli.

3.4.1.3 Landscape/Habitat Management

Landis et al. (2000) have emphasized on SNAP (shelter, nectar, alternate food/prey, pollen) and structural resources (Diehl et al. 2012) as means of environment manipulation to enhance classical BC. Establishment and population of natural enemies can be enhanced by providing additional resources in the form of insectary plants and banker plants that provide alternative food, prey, hosts, mating, and oviposition sites or shelters, and habitat management can enhance natural enemy overwintering and colonization success of predaceous mites to manage spider mites in field crops and fruits (Kogan et al. 1999b). In an Australian citrus system, citrus orchards with dense ground covers harboured higher predatory mesostigmatid mites and lesser of insect pest *Pezothrips* spp. (Colloff et al. 2013).

Several methods of conservation BC such as food from plants, food sprays, alternate prey/host species, artificial rearing systems, pest-in-first techniques, mixed diet effects, oviposition sites and shelters, and vegetation diversity can be used to support establishment of commercially produced and released natural enemies under greenhouse conditions. More abundant and diverse arthropod natural enemies and less abundant pests are reported under diverse and complex landscape conditions and increase in non-crop habitat (Bianchi et al. 2006; Gardiner et al. 2009). Abundance of natural enemies in the farm is usually directly influenced by diversity in natural habitat (Chrisholm et al. 2014), but there may be increase in pest abundance due to various factors. Sometimes, increased parasitism in diverse habitat may not necessarily mean reduction in pest numbers as pest may also derive benefit from the resources offered (Wäckers et al. 2007). Hence, elaborate studies on effects of manipulation of inter and intra-specific diversity and its role in IPM are still to be explored.

Complex multitrophic interactions influence the success of BCA in field on a large scale. Interactions among natural enemies may be synergistic or negative, resulting in a variation of consequences for prey suppression. Moreover, intraguild predation if involved may disrupt or enhance BC (Rosenheim et al. 1995; Gardiner et al. 2009).

3.4.2 Biological Control as a Component of Integrated Pest Management (IPM) for Various Crops

BC has been an integral component of IPM in various crops. Stern et al. (1959) had managed to integrate the use of a comparatively broad-spectrum organophosphate with conservation BC in the spotted alfalfa aphid–alfalfa system. Thereafter, many successful cases of successful integration of various management components with BCAs have been reported worldwide on different crops. In Africa, resistant cassava varieties, cultural control methods, and BC of green mites using predaceous mites hold great potential in IPM (Bellotti et al. 1999).

There are numerous reports of high benefits when insecticides or biopesticides are integrated with augmentative BC in various crops like sweet corn, tomato, cotton, soybean, and mango that have been well documented by Naranjo et al. (2015). Releasing predatory mites using *Metaseiulus occidentalis* (Nesbitt) to manage *Tetranychus* spp. in almonds as part of IPM yielded 14:1 to 34:1 benefit:cost ratios and 280–370% annual rates of return (Headley and Hoy 1987). Although research and development investment in (augmentative) biocontrol is lower than that for insecticides (van Lenteren 2012), natural enemy release generally results in lower returns compared with insecticide alternatives owing to higher costs of control, differential efficacy, or both (Naranjo et al. 2015).

Green clover worm, insect pest of Crimson clover and other legumes strip tilled in cotton, is a good winter and spring reservoir for predators and parasitoids such as *Cotesia marginiventris* that limit subsequent outbreaks of armyworms and loopers in cotton (Fleischer and Gaylor 1987; Ruberson et al. 1994). In Australia, lucerne strips in cotton stabilize generalist predator populations, then a predator food spray (Envirofeast) is sprayed on crop to attract and retain beneficials, and Bt and NPV are sprayed against *Helicoverpa armigera* in this IPM complex (Mensah and Khan 1997; Mensah 1998; Fitt 2000; Uddin et al. 2016). Supplemental food sprays to attract arthropod natural enemies avoided a net value of \$276/ha in yield loss from multiple insect pests (*H. armigera*, *Diparopsis watersi* (Roths.), *Earias* spp, *Pectinophora scutigera* (Hold.), *Nezara viridula* (L.), *Dysdercus sidae* (Montr.)) of organic cotton in Benin, West Africa (Mensah et al. 2012). Arizona cotton growers saved nearly \$500 million in control costs and yield loss since 1996 with the introduction of Bt cotton with conservation BC being a large contributor to these financial gains (Ellsworth et al. 2018). The IPM approach (*Trichogramma chilonis* Ishii, neem seed extract, HaNPV, mechanical collection of larvae, and use of chemicals as a last resort based on economic threshold levels of a pest) helps gain higher advantage from Bt cotton in India under varying pest situations (Bambawale et al. 2004). Non Bt refuge also provided food and shelter to the natural enemies (Lu et al. 2012).

Use of *Trichogramma japonicum* Ashmead and *Trichogramma chilonis* is important part of IPM to control stem borers in rice (Upamanya et al. 2013). *Pseudomonas fluorescens'* application to soil for control of root knot nematode *Meloidogyne graminicola* in non-Basmati rice in Karnataka and Assam increased yield by 13% (Sehgal et al. 2016). Integrated use of Bt sweet corn and selective insecticide spinosad and *Trichogramma ostriniae* (Pengand Chen) is developed as a decision guide managing European corn borer, *Ostrinia nubilalis* (Hubner) (Musser et al. 2006). Tailored flower strips can be effective tool for conservation BC, by increasing natural enemies of cereal leaf beetles in adjacent winter wheat along with reduction in crop damage (Tschumi et al. 2015). In potato, such floral strips greatly enhance abundance of key natural enemies of aphids (hoverflies, lacewings, and lady birds) and hoverfly species richness (Tschumi et al. 2016).

Management of stem borers in Columbia focuses on periodic releases of the tachinids *Lydella minense* (Townsend) and *Billaea claripalpis* (Wulp), and the egg parasitoid *Trichogramma exigum* Pinto and Platner (Vargas et al. 2015). Another successful example is inundative application of larval parasitoid *Cotesia flavipes* (Cameron) in Brazil against sugarcane borer.

IPM in soybean encompasses BC along with selective insecticides and other sustainable management approaches (Zalucki et al. 2009). Although pest infestation rates and defoliation levels are higher in IPM and BC treatments, productivity was not significantly different from that obtained in the treatment involving prophylactic control measures (Bueno et al. 2011). Predictable natural enemy units have been used to develop action thresholds for incorporation into IPM decisions for aphids' management in the crop (Hallett et al. 2014). Common bean grown earlier than soybean served as breeding hosts and nurse crop for a larval pupal parasitoid *Pediobius foveolatus* (Crawford) against Mexican bean beetle, *Epilachna varivestis* Mulsant (Mellors et al. 1983)

Longevity of parasitoid wasps *Microplitis mediator* Haliday and *Diadegma fenestrale* (Holmgren) as well as parasitization rates of *M. mediator* against the cabbage moth *Mamestra brassicae* (Linn.) was significantly increased by the presence of floral and/or extrafloral nectar of selective plants (buck wheat, corn flower *Centaurea cyanus* L., and common vetch *Vicia sativa* L.) (Geneau et al. 2012).

In Indian citrus orchards, release of beetle *Rodolia cardinalis* (Mulsant) introduced against cotton cushiony scale *Icerya purchasi* Maskell provided excellent control, and *Cryptolaemus montrouzierri* Mulsant against mealy bug *Planococcus citri* (Risso) provided substantial control (Singh 2004). Release of parasitoid *Aphelinus mali* (Haldeman) against woolly aphid *Eriosoma lanigerum* (Hausmann) and *Encarsia perniciosi* (Tower) against San Jose scale *Quadraspidiotus perniciousus* (Comstock) was substantially successful in apple orchards in India.

Phytoseiid mites *Phytoseiulus persimilis* Athias-Henriot and *Neoseiulus californicus* (McGregor) are released to manage spider mites *Tetranychus* spp. in melons, watermelons, and cantaloupes in Israel (Kogan et al. 1999b), green house vegetables (Palevsky et al. 2008), and strawberry grown under high plastic tunnels (Fraulo and Liburd 2007). Albendin et al. (2015) found no advantage in releasing multiple natural enemies (Phytoseiid mites) for the control of spider mites and Western flower thrips in strawberry. There is a varying level of satisfaction among farmers along with certain problems associated with the use of the BCAs in IPM such as influence of weather on their efficacy (Moser et al. 2008).

3.5 Climate Change and Beneficial Insects

Climate change has both direct and indirect impacts on agricultural sector. Climate change poses direct changes on plant communities that will indirectly cause both positive and negative impacts on agricultural insects and their natural enemies. Increasing temperature in temperate regions may lead to increase populations of some insect pests that will avoid control made by natural enemies, which will be reflected negatively on agricultural crops. Also, increasing temperature to extinct other insects and their specialist natural enemies, for example, increasing temperature up to 40 °C, may lead to destruction of pod-sucking bug that could not tolerate temperature rise.

Temperature increased 0.80 °C during the last 100 years, and the rise is expected to be 1.1–5.4 °C in the next 100 years. Another climate factor, CO_2 concentration, is expected to reach 740 ppm in 2100. These climate changes are direct or indirect attributes of human activities by misusing available natural resources which increased emissions of greenhouse gases leading to global warming. The agroecosystems have been drastically affected by climate changes as they represent an interaction between organisms (e.g. plants, insects, natural enemies) and climate components (e.g. temperature, precipitation).

An example of the agroecosystems that consists of a tri-trophic levels: plants, herbivores, and their natural enemies. The three levels act in mutualistic manner in which plants get benefit from natural enemies of herbivores by reducing the population of herbivores. On the other hand, plants provide food to herbivores and benefit their enemies by making them more available to the trophic level (Singh 2004). Due to climate change, cropping intensity and patterns changed, leading to release of new varieties of plants that may affect insect populations and dynamics either positively or negatively which is considered as indirect effect of climate change and would be reflected on natural enemies of these insects. Even though negative and positive impacts of climate changes were reported on insects and their natural enemies, neutral impacts were also recorded. Climate changes cause different impacts on insects and their natural enemies, especially their reproductive potential: pre-oviposition period, oviposition period, number of generations, fecundity, and fertility. Other recorded factors are biodiversity, distribution, hibernation, and aestivation. Forest natural enemies of insects are expected to be more affected by climate change due to changes recorded on disturbance of patterns of forest insects as a result of increasing temperature, high CO_2 concentrations, and increased drought areas due to changes in precipitations.

3.5.1 Temperature Impact on Insects and Natural Enemies

As a poikilothermic animal, insects (insect pests, predators, or parasitoids) are highly sensitive to changes in the temperature of their surroundings. Changes in temperature may lead to one or more of the followings:

- If temperatures become unsuitable to their life, migration will be a solution to overcome this problem.
- If temperatures become more optimal, outbreaks of insect pests will occur that will be reflected negatively on the vegetation as these insect pests will be out of control of their natural enemies and cause more damage to crops. So, this will increase the cost of crop protection tactics.
- Some insect pests will shift to other hosts which will lead to emergence of new pests.
- Temperature affects the developmental time of insect larvae, their survival, and reproduction potential of adult based upon magnitude of fluctuations and the mean temperature.
- Thomas et al. (2004) reported that global warming may lead to extinctions of some insects.
- Increasing temperature due to climate change results in decreasing the searching ability of predators that leads to increase in their prey populations with an indirect effect on population dynamics of prey insects.
- *Cyrtorhinus lividipennis* is an egg predator Brown Plant Hopper, and *Nilaparvata lugens* (Stål) had a decrease in handling times and an increase in instantaneous rates of attack due to increasing temperatures to 32 °C, which means that predator activity increases with raising temperature up to 32 °C.
- Increasing temperature affects aphids by reducing their response to alarm pheromones that make them more vulnerable to predation.
- Prey consumption of the female predator *Phytoseius plumifer* doubled by increasing temperature from 15 to 25 °C (Gorji et al. 2009).
- Uszko et al. (2017) recorded that warming temperature affects predator–prey dynamics in which predators tend to follow type III functional response more than type II functional response. However, type III functional response of predators to prey is a sigmoid relationship in which the density of consumed prey is positive density-dependent in some niches of prey density. Other predators showed different functional response to temperature, for example, the handling time and attack rate of dragonfly predator, *Celithemis fasciata* larvae has a linear relationship of temperature with a little relative change little between 15 and 20 °C when fed on variable densities of midge larvae.
- Distribution of parasitoids varies with variations in temperatures present in different altitudes (Granados-Echegoyen 2014). The parasitoids were available between 20 and 2900 m, but the highest number recorded at a height of 736 m where the genus *Heterospilus* was the most common. Also, different water temperatures resulted in different parasitic fauna of spined stickleback beetle, *Gasterosteus aculeatus* (Scharsack et al. 2016).

- Increasing temperature may lead to accelerating the development time of some insects making them less vulnerable to attack by predators and parasitoids (Rudolf and Lafferty 2011).
- Warming affects the structure of host–parasitoid interaction. Brose et al. (2012) reported that this relation will be stronger with warming but concluded that modifications of the host–parasitoid interactions depend on the size structure of these communities.
- Global warming reported to affect the reproductive potential of parasitoids, for example, *Campoletis chlorideae,* which is an endo larval parasitoid of *Helicoverpa armigera.* Successful development of this parasitoid with an inverse correlation with temperature within the range of 12 to 37 °C was reported. Pupal mortality increases either above 37 °C or below 12 °C. Increasing temperature leads to decrease in female longevity that affects its fecundity (Pandey and Tripathi 2008)
- High temperature also reported to increase parasitism in aphid as it has negative impact on the pheromone alarm system of aphids.
- Global warming decreases parasitism by producing a temporal asynchrony between parasitoids and their hosts. Asynchrony also occurs due differential response of parasitoids to any change in climatic conditions (Rudolf and Lafferty 2011; Heckmann et al. 2012).
- Warming may cause changes in performance of the host larvae of prey insect that will negatively affect predation (Emmerson and Raffaelli 2004).
- Temperature is mainly the most important environmental factor that affects the growth rate of the parasitoids and the activity of the immune system of their ecothermic insect hosts (Scharsack et al. 2016).

3.5.2 CO_2 Impact on Insects and Natural Enemies

Another factor that leads to climate change is changes in CO_2 levels that have an indirect effect on insects and their natural enemies through interrupting the plants–insects–natural enemies' tri-trophic relationships. Lee et al. expect a decline in BCAs and predators over the coming decades due to the rise in CO_2 levels. The following scenario is expected due to increase in CO_2 levels on caterpillars and their predatory wasps (Figure 3.1):

Increasing CO_2 levels have positive effects on photosynthesis, increasing plant growth (Shurin et al. 2012). Increasing plant growth provides herbivores with more food that increase their reproduction potential and their population, resulting in more food for predators and more hosts for parasitoids (Henri et al. 2012).

Increasing CO_2 levels means an increase in the carbon to nitrogen ratio in tissues of plants that may have negative impact on insects' development and may prolong stages vulnerable to attack by parasitoids (Coviella and Trumble 1999; Hunter 2001).

The sole effects of both CO_2 and temperature on insects and their natural may be different from that of the combined effects of both of them, in which one factor may balance the other factor. Increasing temperature may elevate the development of leaf miner moths, but increasing CO_2 may change the nutritional quality in the leaves of plants (de Sassi et al. 2012; Ives and Cardinale 2004), which results in lowering adults' survival and reduction in their size. This will result in changes in the parasitoid fauna of these adults.

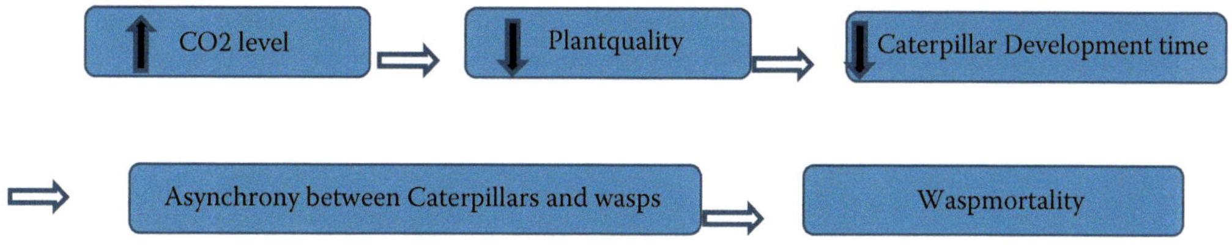

FIGURE 3.1 Effects of CO_2 Increase on the Tri-Trophic Level.

3.6 Entomovector Technology

Entomovector technology has a considerable potential in practical use for BC (Mommaerts and Smagghe 2011). Although entomovectoring is broadly defined as the vectoring or dissemination of items by insects, for example, pollen to malaria in number of publications, it is regarded sensu stricto as a useful method to fight pests and diseases of plant (Hokkanen and Menzler-Hokkanen 2009; Mommaerts and Smagghe 2011; Smagghe et al. 2012; Hokkanen et al. 2015; Menzler-Hokkanen and Hokkanen 2017). It is applicable for pests and diseases that attack the plants via the flowers. An entomovector is pollinating insect used as a vector to spread a substance used in the biocontrol. Potentially, the insect/vectors might be honey bees, bumble bees, solitary bees, some flies, butterflies, and moths. A powdered substance containing a virus, bacterium, or fungus is used to protect the host plant from a given disease or pest. The insect/vector, which normally spreads pollen among plants, is exposed to this substance. The choice of vector species is decided by a combination of native species in the area to be pollinated, the plant species to be treated, and the ease of care of the vector species (Hokkanen and Menzler-Hokkanen 2009; Mommaerts and Smagghe 2011; Smagghe et al. 2012; Hokkanen et al. 2015; Van Delm et al. 2015; Kay et al. 2017; Menzler-Hokkanen and Hokkanen 2017; Remy et al. 2017). Entomovectoring as a plant protection strategy is a vast field for research where all details should not be neglected. For instance, it demands the design of an appropriate bioassay to assess the risks of potential side effects of the powder formulations (e.g. carriers/diluents cellulose and kaolin) in the dispenser towards the vectoring insect (Mommaerts et al. 2012).

Honey bees are most convenient from the practical point of view as it is possible to place a tray containing the powder at a hive exit or by using fans to blow it into the hive. When honey bees are usually not the eligible pollinators, it is necessary to find a specific solution. The importance of honey bees, *Apis mellifera*, compared to the other species providing pollination has been debated for more than 20 years, and the final conclusion is that the role of wild bees in the pollination process should not be neglected (Greenleaf and Kremen 2006; Aebi et al. 2012; Ollerton et al. 2012; Brittain et al. 2013; Garibaldi et al. 2013). Honey bees are crucial for the pollination of crops (Southwick and Southwick 1992; Delaplane et al. 2000; Morse and Calderone 2000). However, a number of other bee and insect pollinators, such as orchard bees, bumblebees, and squash bees, which are not affected by either the Varroa mites or the Africanized bees, are considered as likely candidates for management and use in commercial agriculture (Kevan et al. 1990). An essential pollination service is performed as by domesticated honeybees (*Apis mellifera*) so by wild bees – feral honeybees, bumblebees, and hundreds of species of solitary bees (Sihag 1986; Corbet et al. 1991). Crops are rare to specialist pollinators like the squash bee *Peponapis pruinosa* (Tepedino 1981; Canto-Aguilar and Parra-Tabla 2000; López-Uribe et al. 2016), and even in these cases, with the domestication of the crops, other pollinators can play the role more or less adequately (Tepedino 1981; Nepi and Pacini 1993; Vidal et al. 2010). Bumblebees are among the most important wild pollinators of crops (Free 1993; Goulson 2006; Fandiño 2007). The main agricultural crop that bumblebee pollinate is the greenhouse tomato, *Lycopersicon esculentum* (Abrol 2011), and honeybees are found non-effective (Banda and Paxton 1990), although in Brazil, this crop is pollinated by *Exomalopsis analis* together with other local vibrating bees and not vibrating bees such as *Apis mellifera* and *Trigona spinipes* (Santos et al. 2014; Vinícius-Silva et al. 2017). Bumblebees are also superior pollinators of alfalfa and other forage crops such as red clover (Kearns and Thomson 2001). Additionally, bumblebees fly at low temperatures and in inclement weather, increasing their value as crop pollinators in the field (Heinrich 2004).

Medicinal plants should not be omitted regarding BC. The tendency for their cultivation increases (Schippmann et al. 2002). Same is valid for the aromatic plants like for instance lavender (Sihag 1986; Lis-Balchin 2003; Stanev et al. 2016). There is a predominant dependence of medicinal plants on insects for pollination (Kozuharova et al. 2017; Kozuharova 2018), and it also presents a potential for using these pollen vectors in the context of BC of pests and diseases, for example, entomovectoring (Hokkanen and Menzler-Hokkanen 2009; Smagghe et al. 2012; Hokkanen et al. 2015; Menzler-Hokkanen and Hokkanen 2017).

The pollinators' decline recorded during last decades is a serious hazard for both agricultural productivity and possibility to apply entomovectoring. Habitat destruction, loss of flower resources, and increased use of pesticides (neonicotinoids and others) are causing declines in pollinator abundance and diversity (Potts et al. 2010; Ollerton et al. 2014; Goulson et al. 2015; Potts et al. 2015). Decline of pollinators in North West Europe is well documented (Biesmeijer et al. 2006; Carvalheiro et al. 2013) as well as in North America (Colla and Packer 2008; Grixti et al. 2009; Cameron et al. 2011; Meeus et al. 2011) and South America (Plischuk et al. 2009), or it is referred as a global decline (Kevan and Viana 2003; Dafni et al. 2005; Potts et al. 2010; Goulson et al. 2015). Ollerton et al. (2014) reveal that 23 species of bees and flower-visiting wasps have gone extinct in the UK. They also found that the rate of extinction was highly variable and raised the big question whether these extinctions have stopped or whether they will continue in the future. The study demonstrates the importance of maintaining the year-on-year effort of recording natural history data (Ollerton et al. 2014). The relationship between use of pesticides and the decline of pollinators is an important and controversial topic that is unlikely to go away any time soon. Neonicotinoids are often applied as systemic seed treatments to crops and have reported negative impact on pollinators when they appear in floral nectar and pollen (Wright et al. 2015; Stanley et al. 2015a,b). There is strong evidence that this scales up to cause losses of bee biodiversity (Goulson et al. 2015; Woodcock et al. 2016).

3.7 Conclusion

This work reviewed the *status quo* of BC and suggests future developments aimed at BC under climate change. On the other hand, a substantial advantage of BC is its comparative protection for environment and human health compared to wide use of broad-spectrum pesticides. For the success of biocontrol technologies, aspects like establishment of production, formulation, delivery systems, and timing of field applications need to be considered for each cropping system. Further, quality control assays should be implemented to determine viability, stability, and efficacy. Coordination among universities, industry, and research institutes is also needed for development and adoption of appropriate biocontrol agents. Finally, there is a need for strong extension mechanism for educating farmers to reduce insecticide usage.

REFERENCES

Abbasi H, Ambreen A, Rushda S (2015) Vesicular *Arbuscular mycorrhizal* (VAM) fungi: a tool for sustainable agriculture. Am J Plant Nutr Fertil Tech 5:40–49.

Abd-Elgawad MM (2016) Biological control agents of plant-parasitic nematodes. Egypt J Biol Pest Co 26: 423–429.

Abd-Elgawad MM, Askary TH (2015) Impact of phytonematodes on agriculture economy. In: Askary TH, Martinelli PRP (eds) Biocontrol agents of phytonematodes. CAB International, UK, pp 3–49.

Abrol DP (2011) Pollination biology: biodiversity conservation and agricultural production. Springer Science and Business Media, Germany.

Aebi A, Vaissière BE, Delaplane KS, et al. (2012) Back to the future: Apis versus non-Apis pollination—a response to Ollerton et al. Trends Ecol Evol 27:142–143.

Albendin G, Garcia MDC, Molina JM (2015) Multiple natural enemies do not improve two spotted spider mite and flower western thrips control in strawberry tunnels. Chil J Agric Res 75. https://scielo.conicyt.cl/scielo.php?script=sci_arttextandpid=S0718–58392015000100009.

Al-Tawaha AM, Odat N (2010) Use of sorghum and maize allelopathic properties to inhibit germation and growth of wild barley (*Hordeum spontaneum*). Notulae Botanicae Horti Agrobotanici Cluj-Napoca 38: 124–127.

Al-Zoubi OM, Tawaha AM (2019) Allelopathic effect of beetroot (*Beta vulga`ris* L) on germination and growth *Zea mays* and *Vigna umbellate*. Int J Botany Stud 4:47–51.

Baker KF (1987) Evolving concepts of biological control of plant pathogens. Ann Rev Phytopathol 25:67–85.

Baker KF, Cook RJ (1974) Biological control of plant pathogens. W.H. Freeman and Co., San Francisco, California, pp 433.

Bambawale OM, Singh A, Sharma OP et al. (2004) Performance of Bt cotton (MECH-162) under integrated

pest management in farmers' participatory field trial in Nanded district, Central India. Curr Sci 86:1628–1633.
Banda HJ, Paxton RJ (1990) Pollination of greenhouse tomatoes by bees. In: VI International Symposium on Pollination 194–198.
Bellotti AC, Smith L, Lapointe SL (1999) Recent advances in cassava pest management. Annu Rev Entomol 44:343–370.
Bellows TS, Headrick DH (1999) Arthropods and vertebrates in biological control of plants. In: Principles and applications of biological control. Handbook of biological control. pp 505–516. https://doi.org/10.1016/B978-012257305-7/50064-3.
Bianchi FJJA, Booij CJH, Tscharntke T (2006) Sustainable pest regulation in agricultural landscapes: a review on landscape composition, biodiversity and natural pest control. Proc R Soc B 273:1715–1727.
Bielza P (2016) Insecticide resistance in natural enemies. In: Horowitz A, Ishaaya I (eds) Advances in insect control and resistance management. Springer, Cham, pp 313–329. https://doi.org/10.1007/978-3-319-31800-4_16.
Biesmeijer JC, Roberts SPM, Reemer M et al. (2006) Parallel declines in pollinators and insect-pollinated plants in Britain and the Netherlands. Science 313:351–354.
Bilgrami AL (1990b) Predaceous dorylaims. In: Jairajpuri MS, Alam MM, Ahmad I (eds) Nematode bio-control (aspects and prospects). CBS Publishers and Distributors, Delhi, India, pp 133–138.
Bilgrami AL (1990a) Diplogasterid predators. In: Jairajpuri MS, Alam MM, Ahmad I (eds) Nematode bio-control (aspects and prospects). CBS Publishers and Distributors, Delhi, India, pp 125–132.
Bilgrami AL (2008) Biological control potentials of predatory nematodes. In: Ciancio A, Mukergi KG (eds) Integrated management and biocontrol of vegetable and grain crops nematodes. Springer, New York, pp 3–28.
Bilgrami AL, Ahmad I, Jairajpuri MS (1983) Some factors influencing predation by *Mononchus aquaticus*. Revue de Nematologie 6:325–326.
Brittain C, Williams N, Kremen C, Klein AM (2013). Synergistic effects of non-Apis bees and honey bees for pollination services. Proc R Soc B 280: 20122767.
Brose U, Jennifer A, Dunne JM et al. (2012) Climate change in size-structured ecosystems. Philos Trans R Soc of Lond 367:2903–2912.
Bueno AF, Batistela MJ, Bueno RCO et al. (2011) Effects of integrated pest management, biological control and prophylactic use of insecticides on the management and sustainability of soybean. J Crop Prot 30:937–945. https://doi.org/10.1016/j.cropro.2011.02.021.
Cameron SA, Lozier JD, Strange JP et al. (2011). Patterns of widespread decline in North American bumble bees. Proc Natl Acad Sci 108:662–667.
Canto-Aguilar MA, Parra-Tabla V (2000) Importance of conserving alternative pollinators: assessing the pollination efficiency of the squash bee, *Peponapis limitaris* in *Cucurbita moschata* (Cucurbitaceae). J Insect Conserv 4:201–208.
Carvalheiro LG, Kunin WE, Keil P et al. (2013) Species richness declines and biotic homogenisation have slowed down for NW-European pollinators and plants. Ecol Lett 16:870–878.
Carvalho GA, Parra JRP, Baptista GC (2000) Seletividade de alguns produtos fitossanitários a duas linhagens de *Trichogramma pretiosum* Riley, 1879 (Hymenoptera: Trichogrammatidae). Ciência e Agrotecnologia 25:583–591.
Chitwood DJ (2003) Nematicides. In: Plimmer JR (ed) Encyclopedia of agrochemicals, Vol. 3. John Wiley and Sons, New York, pp 1104–1115.
Chrisholm PJ, Gardiner MM, Moon EG et al. (2014) Tools and techniques for investigating impacts of habitat complexity on biological control. Biol Control 75:48–57.
Clarke AR (1992) Current distribution and pest status of *Nezara viridula* (L.) (Hemiptera: *Pentatomidae*) in Australia. Aust J Entomol 31:289–297.
Colla SR, Packer L (2008) Evidence for decline in eastern North American bumblebees (Hymenoptera: Apidae), with special focus on *Bombus affinis* Cresson. Biodivers Conserv 17:1379.
Colloff M, Lindsay E, Cook D (2013) Natural pest control in citrus as an ecosystem service: integrating ecology, economics and management at the farm scale. Biol Control 67:170–177.
Cook RJ (1988) Biological control and holistic plant-health care in agriculture. AJAA 3:51–62.
Corbet SA, Williams IH, Osborne JL (1991) Bees and the pollination of crops and wild flowers in the European community. Bee World 72:47–59.

Coviella C, Trumble J (1999) Effects of elevated atmospheric carbon dioxide on insect-plant interactions. Conserv Biol 13:700–712.
Dafni A, Kevan PG, Husband BC (2005) Practical pollination biology. Enviroquest Ltd., Cambridge, Ontario, Canada.
IITA (1984) Dissemination, dispersal, and impact of *E. lopezi*—a natural enemy of the cassava mealybug. In: Research highlights for 1984. International Institute of Tropical Agriculture, Ibadan, Nigeria, pp 35–39.
Delaplane KS, Mayer DR, Mayer DF (2000) Crop pollination by bees. CABI, UK.
Diehl E, Wolters V, Birkhofer K (2012) Arable weeds in organically managed wheat fields foster carabid beetles by resource- and structure-mediated effects. Arthropod Plant Interact 6:75–82.
Ellsworth PC, Fournier A, Frisvold G, et al. (2018) Chronicling the socio-economic impact of integrating biological control, technology, and knowledge over 25 years of IPM in Arizona. In: Mason PG, Gillespie DR, Vincent C (eds) Proceedings of the 5th international symposium on biological control of arthropods. CAB International, UK, pp 214–216.
Emmerson MC, Raffaelli D (2004) Predator–prey body size, interaction strength and the stability of a real food web. J Anim Ecol 73:399–409.
Fandiño MTA (2007) Management of *Bombus atratus* bumblebees to pollinate lulo (*Solanum quitoense* L), a native fruit from the Andes of Colombia, Vol. 50. Cuvillier Verlag, Germany.
Faria MR, Wraight SP (2007) Mycoinsecticides and mycoacaricides: a comprehensive list with worldwide coverage and international classification of formulation types. Biol Control 43:237–256.
Fitt GP (2000) An Australian approach to IPM in cotton: integrating new technologies to minimize insecticide dependence. J Crop Prot 19:793–800.
Fleischer SJ, Gaylor MJ (1987) Seasonal abundance of *Lygus lineolaris* (*Heteroptera: Miridae*) and selected predators in early season uncultivated hosts: implications for managing movement into cotton. Environ Entomol 16:379–389.
Fraulo AB, Liburd OE (2007) Biological control of two spotted spider mite, *Tetranychus urticae*, with predatory mite, *Neoseiulus californicus*, in strawberries. Exp Appl Acarol 43:109–119.
Fravel DR (2005) Commercialisation and implementation of biocontrol. Annu Rev Phytopathol 43:337–359.
Free JB (1993) Insect pollination of crops, 2nd edn. Academic Press, London.
Gardiner MM, Fiedler AK, Costamagna AC, Landis DA (2009) Integrating conservation biological control into IPM systems. In: Radcliffe EB, Hutchinson WD, Cancelado RE (eds) Integrated pest management. Cambridge University Press, Cambridge, UK.
Garibaldi LA, Steffan-Dewenter I, Winfree R et al. (2013) Wild pollinators enhance fruit set of crops regardless of honey bee abundance. Science 339: 1608–1611 https://doi.org/10.1126/science.1230200.
Geneau CE, Wackers FL, Luka H et al. (2012) Selective flowers to enhance biological control of cabbage pests by parasitoids. Basic Appl Ecol 13:85–93.
Gorji M, Fathipour Y, Kamali K (2009) The effect of temperature on the functional response and prey consumption of *Phytuseius plumifer* (Acari: Phytoseiidae) on the two-spotted spider mite. Acarina 17:231–237.
Gould F, Kennedy GG, Johnson MT (1991) Effects of natural enemies on the rate of herbivore adaptation to resistant host plants. Entomol Exp Appl 58:1–14.
Goulson D (2006) Bumblebees their behaviour and ecology. Oxford University Press Inc., New York.
Goulson D, Nicholls E, Botías C et al. (2015) Bee declines driven by combined stress from parasites, pesticides, and lack of flowers. Science 347:1255957.
Granados-Echegoyen C (2014) Spatial distribution of braconids (Hymenoptera) records from the state of Oaxaca. Nat Prod Bioprospect 30:564–594.
Lis-Balchin M (ed) (2003) Lavender: the genus *Lavandula*. CRC Press, London and New York.
Greenleaf SS, Kremen C (2006) Wild bees enhance honey bees' pollination of hybrid sunflower. Proc Natl Acad Sci 103:13890–13895.
Grixti JC, Wong LT, Cameron SA, et al. (2009) Decline of bumble bees (Bombus) in the North American Midwest. Biol Conserv 142:75–84.
Gupta SK, Kumar PS (2018) The underestimated worth of predatory and parasitic mites in India: does it really have to import exotic species for biological control. CAB Rev 13:031. https://doi.org/10.1079/PAVSNNR201813031.
Hajek AE, McManus ML, Delalibera I (2007) A review of introductions of pathogens and nematodes for classical biological control of insects and mites. Biol Control 41:1–13.

Hallett RH, Bahlai C, Xue Y, Schaafsma AW (2014) Incorporating natural enemy units into a dynamic action threshold for the soybean aphid, *Aphis glycines* (*Homoptera*: *Aphididae*). Pest Manag Sci 70:879–888.
Headley JC, Hoy MA (1987) Benefit/cost analysis of an integrated mite management program for almonds. J Econ Entomol 80:555–559.
Heckmann L, Drossel B, Brose U et al. (2012) Interactive effects of body-size structure and adaptive foraging on food-web stability. Ecol Lett 15:243–250.
Henri DC, Seager D, Weller T, et al. (2012) Potential for climate effects on the size-structure of host–parasitoid indirect interaction networks. Philos Trans R Soc B367:3018–3024.
Hokkanen HM, Menzler-Hokkanen I (2009) Successful use of honey bees for grey mould biocontrol on strawberries and raspberries in Finland. Apidologie 40:659.
Hokkanen HM, Menzler-Hokkanen I, Lahdenpera ML (2015) Managing bees for delivering biological control agents and improved pollination in berry and fruit cultivation. SAR 4:89.
Hunter MD (2001) Effects of elevated atmospheric carbon dioxide on insect-plant interactions. Agric For Entomol 3:153–159.
Ives AR, Cardinale BJ (2004) Food-web interactions govern the resistance of communities after non-random extinctions. Nature 429:174–177.
Kay C, Wackers F, Pozo MI et al. (2017, September) Entomovectoring technology in kiwifruit pollination. In: IX International Symposium on Kiwifruit, pp 381–390.
Kearns CA, Thomson, JD (2001) Natural history of bumblebees. University Press of Colorado, Colorado, USA.
Kevan PG, Viana BF (2003) The global decline of pollination services. Biodiversity 4:3–8.
Kevan PG, Clark EA, Thomas VG (1990) Insect pollinators and sustainable agriculture. Am J Altern Agric 5:13–22.
Khan Z, Kim YH (2007) A review on the role of predatory soil nematodes in the biological control of plant parasitic nematodes. Appl Soil Ecol 35:370–379.
Kim YH (2015) Predatory nematodes as biocontrol agents of phytonematodes. In: Askary TH, Martinelli PRP (eds) Biocontrol agents of phytonematodes. CAB International, UK, pp 393–420.
Kogan M (1998) Integrated pest management: historical perspectives and contemporary developments. Annu Rev Entomol 43:243–270.
Kogan M, Gerling D, Maddox JV (1999b) Enhancement of biological control of annual agricultural environments. In: Bellows TS, Fisher TW (eds) Handbook of biological control. Academic Press, London, pp 789–818. https://doi.org/10.1016/B978-012257305-7/50077-1.
Kogan M, Croft BA, Suthurst RF (1999a) Applications of ecology for integrated pest management In: Huffaker CB, Gutierrez AP (eds) Ecological entomology. Wiley, New York.
Kozuharova E (2018) Functional flower morphology and entomophilous pollination syndromes in Cape Kaliakra Nature Reserve (North Black Sea coast, Bulgaria). Acta Zool Bulg 11:87–90.
Kozuharova E, Aneva I, Milkoteva K, et al. (2017) No bees no medicinal plants? 3rd International Conference on Natural Products Utilization From Plants to Pharmacy Shelf, 18–21 October 2017, Bansko, Bulgaria
Askary TH, Martinelli PRP (eds) (2015) Biocontrol agents of phytonematodes. CAB International, UK, pp 470.
Lacey LA, Shapiro-Ilan DI (2008) Microbial control of insect pests in temperate orchard systems: potential for incorporation into IPM. Annu Rev Entomol 53:121–144.
Landis DA, Wratten SD, Gurr GM (2000) Habitat management to conserve natural enemies of arthropod pests in agriculture. Annu Rev Entomol 45:175–201.
Lewis WJ, van Lenteren JC, Phatak SC, Tumlinson JH (1997) A total system approach to sustainable pest management. Proc Natl Acad Sci 94:12243–12248.
López-Uribe MM, Cane JH, Minckley RL, Danforth BN (2016) Crop domestication facilitated rapid geographical expansion of a specialist pollinator, the squash bee *Peponapis pruinosa*. Proc R Soc B 283: 20160443.
Lu YH, Wu KM, Jiang YY et al. (2012) Widespread adoption of Bt cotton and insecticide decrease promotes biocontrol services. Nature 487:362–365.
Meeus I, Brown MJ, De Graaf DC et al. (2011) Effects of invasive parasites on bumble bee declines. Conserv Biol 25:662–671.
Mellors WK, Forrester OT, Schwalbe CP (1983) Parasite production in nurse plots after inoculative release of

Pediobius foveolatus (*Hymenoptera*: *Eulophidae*) against Mexican bean beetle larvae (*Coleoptera*: *Cocinellidae*). J Econ Entomol 76:1452–1455.
Mensah RK (1998) Habitat diversity: implications for conservation and use of predatory insects in cotton systems in Australia. Int J Pest Manag 45:91–100.
Mensah RK, Khan M (1997) Use of *Medicago sativa* (L.) interplantings/trap crops in the management of the green mirid, *Creontiades dilutus* (Stal) in commercial cotton in Australia. Int J Pest Manag 43:197–202.
Mensah R, Vodouhe D, Sanfillippo D et al. (2012) Increasing organic cotton production in Benin West Africa with a supplementary food spray product to manage pests and beneficial insects. Int J Pest Manag 58:53–64.
Menzler-Hokkanen I, Hokkanen HM (2017) Entomovectoring: an agroecological practice of using bees for biocontrol. In: Wesel A (ed) Agroecological practices for sustainable agriculture: principles, applications, and making the transition. World Scientific Publishing Company, Singapore.
Mommaerts V, Smagghe G (2011) Entomovectoring in plant protection. Arthropod Plant Interact 5:81.
Mommaerts V, Put K, Vandeven J, et al. (2012) Miniature-dispenser-based bioassay to evaluate the compatibility of powder formulations used in an entomovectoring approach. Pest Manag Sci 68:922–927.
Moosavi MR, Zare R (2015) Factors affecting commercial success of biocontrol agents of phytonematodes. In: Askary TH, Martinelli RRP (eds) Biocontrol agents of phytonematodes. CAB International, Croydon, pp 423–445.
Morse RA, Calderone NW (2000) The value of honey bees as pollinators of US crops. Bee Culture 128:1–15.
Moser R, Pertot I, Elad Y, Raffaelli R (2008) Farmers' attitudes toward the use of biocontrol in IPM strawberry production in three countries. Biol Control 47:125–132.
Musser FR, Nyrop JP, Shelton AM (2006) Integrating biological and chemical controls in decision making: European corn borer (*Lepidoptera*: *Crambidae*) control in sweet corn as an example. J Econ Entomol 99:1538–1549.
Naranjo SE (2009) Impacts of BT crops on non-target organisms and insecticide use patterns. CAB Rev 4:1–23.
Naranjo SE, Ellsworth PC, Frisvold GB (2015) Economic value of biological control in integrated pest management of managed plant systems. Annu Rev Entomol 60:621–645.
Nepi M, Pacini E (1993) Pollination, pollen viability and pistil receptivity in *Cucurbita pepo*. Ann Bot 72:527–536.
Ollerton J, Price V, Armbruster WS, Memmott J, et al. (2012) Overplaying the role of honey bees as pollinators: a comment on Aebi and Neumann (2011) Trends Ecol Evol 27:141–142.
Ollerton J, Erenler H, Edwards M, Crockett R (2014) Extinctions of aculeate pollinators in Britain and the role of large-scale agricultural changes. Science 6215:1360–1362.
Palevsky E, Walzer A, Gal S, Schausberger P (2008) Evaluation of dry-adapted strains of the predatory mite *Neoseiulus californicus* for spider mite control on cucumber, strawberry and pepper. Exp Appl Acarol 45:15–27.
Pandey AK, Tripathi CPM (2008) Effect of temperature on the development, fecundity, progeny sex ratio and life-table of *Campoletis chlorideae*, an endo-larval parasitoid of the pod borer, *Helicoverpa armigera*. Biol Control 53:461–471.
Parvathy Reddy P, Nagesh M (1998) Fungal and bacterial antagonists for biological suppression of plant parasitic nematodes on horticultural crops. In: Singh SP, Hussaini SS (eds) Biological suppression of plant diseases, phytoparasitic nematodes and weeds-present scenario and future thrusts. PDBC, Bangalore, pp 138–153.
Plischuk S, Martín-Hernández R, Prieto L et al. (2009) South American native bumblebees (Hymenoptera: Apidae) infected by *Nosema ceranae* (Microsporidia), an emerging pathogen of honeybees (*Apis mellifera*). Environ Microbiol Rep 1:131–135.
Potts SG, Biesmeijer JC, Kremen C et al. (2010) Global pollinator declines: trends, impacts and drivers. Trends Ecol Evol 25:345–353.
Potts S, Biesmeijer K, Bommarco R et al. (2015) Status and trends of European pollinators. Key findings of the STEP project. Pensoft Publishers, Bulgaria.
Ravensberg JW (2015) Commercialisation of microbes: present situation and future prospects. In: Lugtenberg B (ed) Principles of plant-microbe interactions. Springer, Leiden, pp 309–317.
Remy S, Smessaert J, Claes M et al. (2017) Biocontrol to protect apple and pear flowers against fire blight using bumblebees. Commun Agric Appl Biol Sci 82:311–320.

Rosenheim JA, Kaya HK, Ehler LE, et al. (1995) Intraguild predation among biological-control agents: theory and evidence. Biol Control 5:303–335.

Ruberson JR, Herzog GA, Lambert WR, Lewis WJ (1994) Management of the beet armyworm (*Lepidoptera: Noctuidae*) in cotton: role of natural enemies. Florida Entomol 77:440–453.

Rudolf VHW, Lafferty KD (2011) Stage structure alters how complexity affects stability of ecological networks. Ecol Lett 14:75–79.

Santos AOR, Bartelli BF, Nogueira-Ferreira FH (2014) Potential pollinators of tomato, *Lycopersicon esculentum* (Solanaceae), in open crops and the effect of a solitary bee in fruit set and quality. J Econ Entomol 107:987–994.

de Sassi C, Staniczenko PP, Tylianakis JM (2012) Warming and nitrogen affect size structuring and density dependence in a host–parasitoid food web. Philos Trans R Soc 367:3033–3041.

Scharsack JP, Franke F, Erin NI et al. (2016) Effects of environmental variation on host–parasite interaction in three-spined sticklebacks (*Gasterosteus aculeatus*). Zoology 119:375–383.

Schippmann U, Leaman DJ, Cunningham AB (2002) Impact of cultivation and gathering of medicinal plants on biodiversity: global trends and issues. FAO, Biodiversity and the ecosystem approach in agriculture, forestry and fisheries, Rome, 12–13 October 2002. Inter-Departmental Working Group on Biological Diversity for Food and Agriculture, Rome.

Sehgal M, Ravindra H, Somasekhara Y et al. (2016) IPM for traditional rice. In: Vennila S, Birah A, et al. (eds) Success stories of integrated pest management in India. ICAR-National Research Centre for Integrated Pest Management, New Delhi, pp 18–22.

Shurin JB, Clasen JL, Greig HS et al. (2012) Warming shifts top-down and bottom-up control of pond food web structure and function. Philos Trans R Soc B367:3008–3017.

Sihag RC (1986) Insect pollination increases seed production in cruciferous and umbelliferous crops. J Apic Res 25:121–126.

Singh SP (2004) Some success stories in classical biological control of agricultural pests in India. Asia-Pacific Association of Agricultural Research Institutions, FAO Regional Office for Asia and the Pacific, Bangkok, pp 73.

Smagghe G, Mommaerts V, Hokkanen H, Menzler-Hokkanen I (2012) Multitrophic interactions: the entomovector technology. Arthropod Plant Interact 127–157.

Southwick EE, Southwick Jr L (1992) Estimating the economic value of honey bees (Hymenoptera: Apidae) as agricultural pollinators in the United States. J Econ Entomol 85:621–633.

Stanev S, Zagorcheva T, Atanassov I (2016) Lavender cultivation in Bulgaria-21st century developments, breeding challenges and opportunities. Bulg J Agric Sci 22:584–590.

Stanley DA, Smith KE, Raine NE (2015a) Bumblebee learning and memory is impaired by chronic exposure to a neonicotinoid pesticide. Sci Rep 5:16508.

Stanley DA, Garratt MP, Wickens JB, et al. (2015b) Neonicotinoid pesticide exposure impairs crop pollination services provided by bumblebees. Nature 7583:548–550.

Stenberg JA (2017) A conceptual framework for integrated pest management. Trends Plant Sci 22:759–769.

Stern VM, Smith RF, van den Bosch R, Hagen KS (1959) The integrated control concept. Hilgardia 29:81–101.

Tepedino VJ (1981) The pollination efficiency of the squash bee (*Peponapis pruinosa*) and the honey bee (*Apis mellifera*) on summer squash (*Cucurbita pepo*). J Kans Entomol Soc 359–377.

Thomas CD, Cameran A, Green RE (2004) Extinction risk from climate change. Nature 427:145–148.

Torres JB, Bueno AF (2018) Conservation biological control using selective insecticides- a valuable tool for IPM. Biol Control 126:53–64.

Trias R, Bañeras L, Montesinos E, Badosa E (2008) Lactic acid bacteria from fresh fruit and vegetables as biocontrol agents of phytopathogenic bacteria and fungi. Int Microbiol 11:231–236.

Tschumi M, Albrecht M, Collatz J et al. (2016) Tailored flower strips promote natural enemy biodiversity and pest control in potato crops. J Appl Ecol 53:1169–1176. https://doi.org/10.1111/1365-2664.12653.

Tschumi M, Albrecht M, Entling MH, Jacot K (2015) High effectiveness of tailored flower strips in reducing pests and crop plant damage. Proc R Soc B 282

Turk MA, Tawaha AM (2002b) Inhibitory effects of aqueous extracts from black mustard (*Brassica nigra* L.) on germination and growth of wheat. Pak J Biol Sci 5:278–280.

Turk MA, Tawaha AM (2002a) Inhibitory effects of aqueous extracts of black mustard on germination and growth of lentil. Pak J Agron 1:28–30.

Turk MA, Tawaha AM (2003a) Allelopathic effect of black mustard (*Brassica nigra* L.) on germination and growth of wild oat (*Avena fatua* L.). Crop Prot 22:673.
Turk MA, Tawaha AM (2003b) Allelopathic effects of black mustard (*Brassica nigra*) on germination and growth of wild barley (*Hordeum spontaneum*). J Agron Crop Sci 189:298–303.
Turk MA, Shatnawi MK, Tawaha AM (2003c) Inhibitory effects of aqueous extracts of black mustard on germination and growth of alfalfa. Weed Biol Manag 3:37–40.
Turk MA, Lee KD, Tawaha AM (2005) Inhibitory effects of aqueous extracts of black mustard on germination and growth of radish. Res J Agric Biol Sci 1:227–231.
Uddin MN, Ali M, Muhammad, Farooq M, Ahmad N, Jamil J, Kalsoom, Adnan M, Shah N, Khan A (2016) Characterizing microbial populations in petroleum-contaminated soils of Swat District, Pakistan. Pol J Environ Stud 25:11691–11697.
Upamanya GK, Dutta P, Sarma R et al. (2013) Biological management of rice stem borer in the farmers field of Assam. Insect Environ 19:73–76.
Uszko W, Diehl S, Englund G (2017) Effects of warming on predator–prey interactions – a resource-based approach and a theoretical synthesis. Ecol Lett 20:513–523.
Van Delm T, Van Beneden S, Mommaerts V et al. (2015) Control of *Botrytis cinerea* in strawberries with *Gliocladium catenulatum* vectored by bumblebees. J Berry Res 5 23–28.
Vargas G, Gomez LA, Michaud JP (2015) Sugarcane stem borers of the Colombian Cauca River Valley: current pest status, biology, and control. Florida Entomol 98:728–735.
Vidal MDG, Jong DD, Wien HC, Morse RA (2010) Pollination and fruit set in pumpkin (*Cucurbita pepo*) by honey bees. Braz J Bot 33:106–113.
Vinícius-Silva R, Parma DDF, Tostes RB et al. (2017) Importance of bees in pollination of *Solanum lycopersicum* L.(Solanaceae) in open-field of the Southeast of Minas Gerais State Brazil. Hoehnea 44: 349–360.
Waage J, Greathead DJ (1988) Biological control: challenges and opportunities. In: Wood RKS, Way MJ (eds) Biological control of pests, pathogens, and weeds: developments and prospects. The Royal Society, London, pp 1–18.
Wäckers FL, Romeis J, van Rijn P (2007) Nectar and pollen feeding by insect herbivores and implications for multitrophic interactions. Annu Rev Entomol 52:301–323. https://doi.org/10.1146/annurev.ento.52.110405.091352.
Wang KH, Sipes BS, Schmitt DP (2002) Crotolaria as a cover crop for nematode management: a review. Nematropica 32:35–57.
Woodcock BA, Isaac NJ, Bullock JM, et al. (2016) Impacts of neonicotinoid use on long-term population changes in wild bees in England. Nat Commun 7:12459. https://doi.org/10.1038/ncomms12459.
Wright GA, Softley S, Earnshaw H (2015) Low doses of neonicotinoid pesticides in food rewards impair short-term olfactory memory in foraging-age honeybees. Sci Rep 5:15322. https://doi.org/10.1038/srep15322.
Zalucki MP, Adamson D, Furlong MJ (2009) The future of IPM: whither or wither. Aust J Entomol 48:85–96.

4

Future Perspectives of Oxidative Stress and Antioxidant Defence System in Plants

Sibgha Noreen[1],
Muhammad Salim Akhter[1],
Ali Noman[2],
Ume Ummara[1],
Seema Mahmood[1],
Mohamed Hashim[3]
[1]Institute of Pure and Applied Biology, Bahauddin Zakariya University, Multan, Pakistan
[2]Department of Botany, Government College University, Faisalabad, Pakistan
[3]Botany Department, Assiut University, Egypt

4.1 Introduction

The agrarian farming community faces considerable environmental threats in raising their crop productivity under different ecologies. The agrarians ought to adopt certain interventions to thwart the deleterious impact of influencing factors. The external threats consist of both biotic and abiotic stresses that limit the crop productivity to a greater proportion and thereby lessen farmers' livelihood. It is estimated that global population will rise to 10 billion by 2050, ultimately increasing the demand of more food, feed, fibre, and biofuels, which is nearly higher by 70% over the current level of agricultural and livestock production (Saud et al. 2013; Shah et al. 2013; Saud et al. 2014; Hafiz et al. 2016; Saud et al. 2016; Aziz et al. 2017a,b; Habib ur Rahman et al. 2017; Kamaran et al. 2017; Saud et al. 2017; uz Zaman et al. 2017; Wajid et al. 2017; Yang et al. 2017; Zahida et al. 2017; Adnan et al. 2018a,b; Akram et al. 2018a,b; Depeng et al. 2018; Hafiz et al. 2018; Tariq et al. 2018; Adnan et al. 2019; Ahmad et al. 2019; Baseer et al. 2019; Hafiz et al. 2019; Jan et al. 2019; Muhammad et al. 2019; Sajjad et al. 2019; Wu et al. 2019; Adnan et al. 2020; Farhana et al. 2020; Farhat et al. 2020; Hafiz et al. 2020a,b; Hussain et al. 2020; Ilyas et al. 2020; Mubeen et al. 2020; Rehman 2020; Saleem et al. 2020a,b,c; Saud et al. 2020; Shafi et al. 2020; Subhan et al. 2020; Wahid et al. 2020; Wu et al. 2020; Zafar-ul-Hye et al. 2020a,b).

The external threats could be fought by developing stress-tolerant and highly productive crop varieties especially under stressful conditions. The development of stress-tolerant crops varieties could be carried out through genetic manipulation, which is not only time-consuming, but requires heavy investment in human and capital resources. On the other hand, enhancing the intrinsic tolerance level in plant system through exogenous application of some chemicals is a valuable tool in the subsistence farming system, which is highly important to raise crops in stressful conditions (Fahad and Bano 2012; Fahad et al. 2013; Fahad et al. 2014a,b; Fahad et al. 2015a,b; Fahad et al. 2016a,b,c,d; Fahad et al. 2017; Fahad et al. 2018; Fahad et al. 2019a,b; Hesham and Fahad 2020). The plant hormones, vitamins, and phenolic compounds help in various physiological and biochemical processes in plant species under different environmental stresses (Ahmad et al. 2010; Ashraf et al. 2010). The natural occurrence of phenolic compounds and secondary metabolites is involved in plant allelopathy in the agroecosystem (John and Sarada 2012), as well as signalling in plant–insect, plant–microorganism, and growth regulators' interactions.

These solutes are organic substances having low molecular weight and do not produce toxic effects at their higher concentration in plant system. These compatible solutes protect the plant by bringing stability in enzymes, detoxifying reactive oxygen species (ROS), and maintaining cell membrane integrity and alleviation in oxidative stress (Ashraf and Foolad 2007; Mafakheri et al. 2010). The exogenous application of some osmoprotectants would be a potential option to ensure food security through adequate food supply from the farms. This chapter introduces the basic phenomena that how to mitigate harshness of environmental factors through foliar application of certain chemicals for enhancing the internal tolerance level in crop plant species, as well to harvest economic returns by usage of certain chemicals on crop plants. The beneficial effects about spraying of some chemicals are mentioned below:

4.1.1 Proline

Proline, an amino acid, is fairly distributed in most crop species which accumulates in plants in response to higher proportions under stressful conditions. Moreover, activities of anti-oxidative enzymes are improved (Hayat et al. 2013).

4.1.1.1 Application of Proline

The reactive oxygen species (ROS) are continuously produced during different metabolic pathways in plant system. However, ROS is also considered beneficial in protecting the plants from biotic stress, treachery element formation, lignification, and other developmental processes (Fath et al. 2002). On the other hand, oxidative stress results in greater accumulation of ROS-damaged nucleic acid, peroxidation of lipids, degradation of proteins, outflow of K^+ from cells, and destruction of chlorophyll (Schutzendubel and Polle 2002).

The increased ROS generation and its damage under stressful condition can be controlled by proline application (Chen and Dickman 2005). Proline caused reduction in level of ROS in roots of Arabidopsis and reduced ROS-induced K^+ efflux through roots (Cuin and Shabala 2007). The activities of catalase (CAT), peroxidase (POX), and superoxide dismutase (SOD) are enhanced by external application of proline in tobacco suspension cultures under salt stress (Hoque et al. 2007). Furthermore, proline application up-regulates the activities of ascorbate-glutathione (ASC-GSH), ascorbate peroxidase (APX), monohydro-ascorbate reductase (MDHAR), and dihydro-ascorbate reductase (DHAR) for reduction in oxidative stress in saline environment (Hoque et al. 2007; Hossain et al. 2010). There are evidences that growth of maize was improved by foliar spray of 100 mM proline under stress conditions (Ali and Ashraf 2011). Spray of proline under copper stress causes increase in SOD, CAT, POX, proline, and Ca^{2+} contents to detoxify the harshness of copper stress in wheat crop (Noreen et al. 2018). The foliar application of 30 mM proline caused improvement in photosynthesis rate as wells as chlorophyll constituents in maize plant under water stress (Ali et al. 2008a,b). The growth of bean (*Vigna radiata* L.) was improved by proline application through better accumulation of carbohydrates and chlorophyll pigments and accelerating the rate of CAT and POD under water stress (Ali et al. 2013). The accumulation of proline by plant species is being used as a tool for screening purpose (Ashraf and Foolad 2007). However, care may be taken to avoid spraying higher concentrations of proline which may cause negative effects on growth and development (Garde-Cerdán et al. 2014). The beneficial effect of exogenous supply of proline on different antioxidants is summarized in Table 4.1.

4.2 Salicylic Acid

Salicylic acid (SA) is an endogenous growth regulator, belonging to diverse group of plant phenols, and its content is about 1.0 μg g^{-1} fresh weight in most plant species (Hayat et al. 2010). It acts as an important signalling molecule during plant defence responses (Raskin et al. 1987; Fahad and Bano 2012). The application of SA up-regulates plant pathogenesis-related (PR) genes and induces disease resistance. The infection of tobacco mosaic virus triggers increases in SA and develops systemic acquired resistance (SAR) (Yalpani et al.

TABLE 4.1

Activities of Antioxidant Enzymes as Affected by Proline Under Abiotic Stresses

Types of abiotic stresses	Types of exogenous application	Dose	Crop	Effects on antioxidants	References
Salinity	Rooting medium	0 and 10 mM	*Cucumis melo* L.	Increases the activity of SOD, POD, and CAT	Kaya et al. (2007)
Salinity	Rooting medium	0 and 20 mM	*Nicotiana tabacum* L.	Increases activity of glutathione of glutathione-S-transferase, glyoxalase II, glutathione peroxidase, and glyoxalase I	Hoque et al. (2008)
Heavy metal stress (Cd)	Seed priming	0, 1, and 10 mM	*Nicotiana tabacum* L.	Increases SOD and CAT activities without reducing Cd contents under Cd stress	Islam et al. (2009)
Oxidative stress	Leaves' submergence in solutions	0 and 20 mM	*Vitis vinifera* L.	Increases SOD, POD, APX, and CAT activities and decreases MDA and H_2O_2 contents	Ozden et al. (2009)
Cold stress	Buds priming	0 and 20 mM	*Camellia sinensis* L.	Enhances glutathione-S-transferase and glutathione reductase (GR), glyoxalase I, and glyoxalase II enzymes	Kumar and Yadav (2009)
Oxidative stress	Pulse application	0, 1, 2, 4, 5, 8, and 10 mM	*Rosa hybrida* L.	Increases SOD (Mn-SOD), proline dehydrogenase (PDH) content, and GSH/GSSG ratio	Kumar et al. (2010)
Salinity	Rooting medium	0 and 15 mM	*Vigna radiata* L.	Increases glutathione peroxidase, glutathione S-transferase, and glyoxalase II enzyme activities	Hossain and Fujita (2010)
Salinity	Rooting medium	0, 25, and 50 mM	*Olea europaea* L.	Increases SOD, APX, and CAT activities for 1.1, 1.3, and 1.4 times in young leaves	Ben Ahmed et al. (2010)
Chilling stress	Foliar application	0 and 10 μM	*Cicer arietinum* L.	Decreases H_2O_2, while increases in SOD, POD, CAT, GSH, and APX contents	Kaur et al. (2011)
Drought stress	Foliar application	0 and 10 mM	*Arabidopsis thaliana* L.	Decreases MDA content and increases activities of antioxidants	Moustakas et al. (2011)
Salinity	Rooting medium	0 and 10 mM	*Oryza sativa* L.	Decreases hydrogen peroxide (H_2O_2) content and increases the activity of antioxidative enzymes like (SOD, POD, APX, CAT) and transcript up-regulation of genes encoding antioxidant enzymes (Cu/ZnSOD, MnSOD, CytAPX, and CatC)	Nounjan et al. (2012)

(*Continued*)

TABLE 4.1 (Continued)

Types of abiotic stresses	Types of exogenous application	Dose	Crop	Effects on antioxidants	References
Drought stress	Foliar application	0 and 30 mM	*Zea mays* L.	Increases oil antioxidants like α-, γ-, and δ-tocopherols' activity	Ali et al. (2013)
Salinity	Foliar application	0, 25, and 50 mM	*Vicia faba* L.	Increases antioxidants' enzymes (peroxidase and polyphenol oxidase)	Dawood et al. (2014)
Salinity	Seed priming	0 and 1.0 mM	*Hordeum vulgare* L.	Increases activities of antioxidant enzymes (SOD, CAT, and POD)	Agami (2014a)
Heavy metal stress (Cd)	Rooting medium	0 and 20 mM	*Phoenix dactylifera* L.	Decreases H_2O_2 content while increases APX and POD activity	Zouari et al. (2016)
Heavy metal stress (Cu)	Foliar application	0 and 80 mM	*Triticum aestivum* L.	Reduces germination of ROS, increase in proline content, accumulation of proline	Noreen et al. (2018)
Nitrogen nutrition	Foliar application	40 and 50 mM	Grape vines	Reduces plant growth under higher concentration	Garde-Cerdán et al. (2014)
Salinity	Foliar spray	20 to 30 mM	Sugar beet	Reduces water potential due to increase in proline concentration	Farkhondeh et al. (2012)
Drought	Foliar spray	100 mM	*Zea mays* L.	Improves growth and development	Ali and Ashraf (2011)
Drought	Foliar spray	100 mg L^{-1}	*Gossypium hirsutum* L.	Causes increase in seed cotton yield	Noreen et al. (2013)
Drought	Foliar spray	30 mM	*Zea mays* L.	Increases photosynthesis rate, chlorophyll pigments	Ali et al. (2008)
Salinity	Foliar spray	30 mM	Crop species	Accumulation of proline by plant is a tool for screening under salinity	Ashraf and Foolad (2007)

1993; Fahad and Bano 2012). The application of SA causes increase in plant disease resistance in Arabidopsis, cucumber, and potato plants (Gunes et al. 2007). It also induces resistance in response to extreme temperatures, salinity, and oxidative stress (Larkindale et al. 2005). The plants having a little locomotion strive to withstand the vagaries of environment through adaption of certain intrinsic measures, that is, up-regulation of phytohormones (Farooq et al. 2010; Fahad et al. 2014a,b), inducing antioxidant defence network (Gill and Tuteja 2010; Fahad and Bano 2012).

4.2.1 Application of Salicylic Acid

Under abiotic stress, SA induces antioxidant enzyme system and increases synthesis of SOD, POD, CAT, APX, ASC, and DHA, which protect the plant from oxidative damage (Hayat et al. 2008); however, some reports showed that SA application inhibits the activities of CAT (Janda et al. 2003). The spray of SA at low concentration leverages the work of non-enzymatic antioxidant activities (Arfan et al. 2007) and increases anti-oxidative capacity in sunflower (Noreen et al. 2009), mitigating the effect of drought stress in wheat (Waseem et al. 2006) and salt stress in wheat (Arfan et al. 2007) by scavenging ROS radicals (Rivas-San Vicente and Plasencia 2011). SA application improves biological and grain yield of wheat under drought stress (Singh and Usha 2003; Noreen et al. 2017) through increased nutrients uptake, improved water relations, maintained stomatal conductance (Arfan et al. 2007), and enhanced photosynthesis and growth-related parameters (Khan et al. 2014). Table 4.2 illustrates the antioxidant enzymatic response of exogenous SA application.

4.3 α-Tocopherol

Vitamin E, like tocotrienols and tocopherols, is constituted of compounds essential for animal cells. Vitamin E is exclusively synthesized by photosynthetic eukaryotes and other oxygenic photosynthetic organisms such as cyanobacteria. In order to prevent lipid oxidation, the plants mainly accumulate tocochromanols in oily seeds and fruits or in young tissues undergoing active cell divisions (Colombo 2010).Vitamin E [chroman-6-ols (tocochromanols; tocopherols + tocotrienols)] is a fat soluble vitamin and ingested along with fat-containing food products. The human body has warehouses to store Vitamin E; thereby, it is not necessary to consume it on daily basis (Zingg 2007). The vegetable oils (olive, sunflower, corn, and soybean) contain significant quantities of tocopherols, while palm oil and oils extracted from barley and rice bran are major sources of tocotrienols (Sookwong et al. 2007; Colombo 2010).

The leaves and seed of various crop species contain tocopherols, ranging from 5 to 200 mg kg^{-1}, and its quantum is enhanced to significant proportion in response to stress, while mitigating the harshness of ROS (Munné-Bosch and Alegre 2002). Among the members of tocopherols, α-tocopherol is the most active isomer with relation to β-, γ-, and δ-tocopherols (Hyun et al. 2011; Szarka et al. 2012). It is potentiated with other hormones, viz., jasmonic acid, salicylic acid, and ascorbic acid. Resultantly, the tocopherols participate in stress signalling mechanisms (Falk et al. 2004) and adaptation mechanism against photosynthesis- induced ROS.

4.3.1 Application of α-Tocopherol

Different physiological and biochemical parameters of plants are improved to a greater extent due to exogenous application of tocopherols under salt stress environment (Semida et al. 2014). The external application of tocopherol enhances growth and development of *Vicia faba* (Orabi and Abdelhamid 2016) and germination (Ayad et al. 2009) under salinity stress. The improvement in growth structure results due to better water potential, greater storage of antioxidants, reduced oxidative stress, and better networking among various growth regulators in plant system (El-Bassiouny et al. 2005). The externally applied α-tocopherols increases storage of phenols in sunflower (Rivero et al. 2001) and controls metabolic reaction during growth period (Rady et al. 2011) under abiotic stress conditions. However, a little response has been observed in mung bean cultivars under drought stress conditions (Kostopoulou et al. 2014). The

TABLE 4.2

Activities of Antioxidant Enzymes as Affected by Salicylic Acid Under Abiotic Stresses

Types of abiotic stresses	Types of exogenous application	Dose	Crop	Effects on antioxidants	References
Heat stress	Foliar application and rooting medium	0, 0.1, 0.25, 0.5, 1, and 1.5 mmol	*Poa pratensis* L.	Increases antioxidant enzymes, superoxide dismutase (SOD), and catalase (CAT) and decreases H_2O_2 and superoxide anion	He et al. (2005)
Heat stress	Foliar application	0 and 1 mM	*Cucumis sativa* L.	SOD, CAT, GPX, APX, and GR activities	Shi et al. (2006)
Salinity	Seed priming	0 and 50 ppm	*Triticum aestivum* L.	Increases antioxidants activity	Afzal et al. (2006)
Salinity	Foliar application	0, 0.25, 0.50, and 1.00 mM	*Cucumis sativus* L.	Increases antioxidants activity	Yildirim et al. (2008)
Salinity	Seed priming	0 and 0.2 mM	*Vicia faba* L.	Enhances accumulation of antioxidant enzymes (CAT, POD, APX, and GR)	Azooz (2009)
Drought stress	Foliar application	100 $mg\ L^{-1}$	*Oryza sativa* L.	Reduces H_2O_2 and MDA production, increases synthesis of metabolites, and enhances capacity of antioxidant system	Farooq et al. (2010)
Cold stress	Seed priming	0 and 10^{-4} M	*Phaseolus vulgaris* L.	Increases antioxidants activity, viz., CAT, polyphenol oxidase, and POD	Gharib and Hegazi (2010)
Salinity	Foliar application	0, 0.1, 0.5, and 1.0 mM	*Brassica juncea* L.	Increases APX, CAT, GR, and glutathione content	Syeed et al. (2011)
Salinity	Foliar application	0, 0.5, and 1.0 mM	*Vigna radiata* L.	Increases APX, GR, NR, and ATPS and reduces glutathione (GSH) content	Nazar et al. (2011)
Salinity	Seed priming	0, 50, 100, and 150 $mg\ L^{-1}$	*Lycopersicum esculentum* L.	Increases antioxidants level	Ghoohestani et al. (2012)
Heavy metal stress (Cd)	Seed priming	0 and 500 μM	*Triticum aestivum* L.	Increases activities of SOD, CAT, and POD and proline content	Agami and Mohamed (2013)
Drought stress	Seed priming	0, 25, and 50 ppm	*Secale montanum* L.	Increases CAT and APX	Ansari et al. (2013)
Salinity	Foliar application	0 and 0.5 mmol	*Torreya grandis* L.	Decreases MDA and increases SOD, POD, and CAT	Li et al. (2014)

phenols are accumulated to a greater quantum in response to spray of tocopherols in wheat crop under drought conditions (Farouk 2011). The exogenously applied α-tocopherols cause increase in different enzymatic antioxidants (POD, CAT), but have a little effect on SOD in mung bean under stress conditions. The antioxidant enzymatic activities of SOD and POD are not affected by treating the *Leymus chinensis* seedlings with tocopherols under drought environment (Jie et al. 2008), also by external application of tocopherols in citrus seedlings (Kostopoulou et al. 2014) and *Vicia faba* (Orabi and Abdelhamid 2016) under saline conditions. The effects of exogenous application of tocopherols on different antioxidants are summarized in Table 4.3.

4.4 Trehalose

The trehalose is categorized as non-reducing disaccharide sugar and widely distributed in nature (bacteria, fungi, plants, and invertebrates such as insects and nematodes). The sugar was named by Berthelot in 1858 who found it in Trehala, a desert manna produced by the weevil *Larinus nidificans* (Berthelot 1858). Prior to Berthelot, Wiggers discovered trehalose in 1832 while studying solutions of ergot of rye, a plant disease caused by the fungus *Claviceps purpurea*. More recently, trehalose was reported to protect the integrity of cells in many organisms against environmental stresses due to its physical, colligative, and chemical properties (Crowe et al. 1984; Argüelles 2000).

4.4.1 Application of Trehalose

Trehalose in instantly absorbed by leaves and roots and then transported to other organs of plants (Gao et al. 2013). The various biological and structural proteins and membrane liquids are stabilized by external application of trehalose (Luo et al. 2010; Ma et al. 2013). Some of other attributes, viz., biological yield, photosynthetic capacity, water relations, and accumulation of enzymatic and non-enzymatic antioxidant compound, are increased by external application of 30 mM trehalose in maize plants (Ali and Ashraf 2011). It has usefulness in rehydrating water binding as well as antioxidant properties (Becker et al. 1996). During occurrence of hydrolysis, early germination of plants meets source of carbon for synthesis and glucose for energy (Thevelein 1984). The anhydrobiotic organs contain higher amounts of trehalose (disaccharides and oligosaccharides) for their survival over a longer period. It also acts as a protein stabilizer and other compatible solutes, consisting of polyols, sugars, and amino acids (De Virgilio et al. 1994). The damages caused by higher temperatures to cell are protected by stabilizing proteins in the presence of sufficient quantity of trehalose, preventing denaturation and aggregation of specific proteins at cold temperatures (Singer and Lindquist 1998) and protecting cells against ROS radicals (Ali and Ashraf 2011). The beneficial effects of trehalose on different crop species are summarized in Table 4.4.

4.5 Brassinosteroids

Brassinosteroids (BRs) are a class of plant poly hydroxyl steroids that have been recognized as a new kind of phytohormones that play an essential role in plant development. Brassinosteroids are plant growth-promoting molecules found at low concentrations throughout the plant kingdom and are widely distributed in lower and higher plants. Brassinosteroids (BRs) belong to a class of poly hydroxyl steroids and were isolated in 1979. Since then, 70 BRs compounds have been discovered. In the plant system, the vegetative growth and development, initiation of flowers, and development of fruits are mainstreamed by presence of BRs (Sasse 2003). In brassinosteroids, brassinolide, the active constituent, takes part in cell elongation and light regulation. The absence of BRs results in development of dwarf phenotypes with curled dark-green rosette leaf, shorter internodes, male sterility, short hypocotyls, and expanded cotyledons in darkness (Bishop and Koncz 2002). The increased salt tolerance level is induced through various mechanisms in response to environmental factors (Bajguz and Hayat 2009).

TABLE 4.3

Activities of Antioxidant Enzymes as Affected by α-Tocopherols Under Abiotic Stresses

Types of abiotic stresses	Types of exogenous application	Dose	Crop	Effects on antioxidants	References
Cryopreservation-induced oxidative stress	Rooting medium	11–15 mM	Rubus species	Increases SOD, POD, and CAT activities	Uchendu et al. (2009)
Salinity	Seed priming	0, 25, and 50 mg L^{-1}	*Helianthus annuus* L.	Increases APX, CAT, and POD activities	Sadak et al. (2010)
Salinity	Foliar application	0, 200, and 400 mg L^{-1}	*Linum usitatissimum* L.	Increases APX and CAT activities	El Hariri et al. (2010)
Oxidative damage due to various abiotic stresses	Rooting medium	2.3 kg ha^{-1}	*Oryza sativa* L.	Decreases production of lipid peroxide radical and increases activities of antioxidant enzymes	Mohammed and Tarpley (2011)
Salinity	Foliar application	100 mg L^{-1}	*Triticum aestivum* L.	Decreases MDA content and increase activities of CAT and POD	Farouk (2011)
Heat stress	Rooting medium	0 and 5 μM	*Triticum aestivum* L.	Increases activities of enzymatic antioxidants, viz. SOD, CAT, APX, and GSH, and non-enzymatic antioxidants, viz., ascorbate, glutathione, and proline	Kumar et al. (2013)
Salinity	Rooting medium	0.46 and 0.93 mM	*Linum usitatissimum* L.	Maintains the catalytic activities of antioxidants	Sh (2014)
Salinity	Foliar application	0, 0.25, 0.50, and 1.0 mM	*Vicia faba* L.	Increases SOD, GSH, POD, and CAT activities and decreases MDA content	Semida et al. (2014)
Salinity	Foliar application	0, 0.5, and 1.0 mM	*Allium cepa* L.	Increases enzymatic (superoxide dismutase, catalase, ascorbate peroxidase, and glutathione reductase) and non-enzymatic (ascorbic acid and glutathione) antioxidant activities	Semida et al. (2016)

Salinity	Foliar application	0, 50, or 100 mg L^{-1}	*Vicia faba* L.	Increases activity of SOD, POD, APX, CAT, phenylalanine ammonia lyase (PAL), and proline content	Orabi and Abdelhamid (2016)
Drought stress	Foliar application	0, 100, 200, and 300 mg L^{-1}	*Vigna radiata* L.	Reduces MDA and H_2O_2 content, while increases antioxidant activity of SOD, POD, and CAT	Sadiq et al. (2016)
Drought stress	Foliar application	0, 100, 200, and 300 mg L^{-1}	*Vigna radiata* L.	Enhances the activities of antioxidant enzymes, viz., SOD, POD, and CAT	Sadiq et al. (2017)
Salinity	Foliar application	0 and 1.0 mM	*Phaseolus vulgaris* L.	Increases osmoprotectants, viz., proline, non-enzymatic (ascorbic acid, glutathione, and TOC) and enzymatic (SOD, CAT, and GPOX) antioxidants	Hemida et al. (2017)
Salinity	Foliar application	0 and 1 mM	*Capsicum annum* L.	Boosts the activities of SOD, CAT, APX, and GR	Taha et al. (2017)
Salinity	Foliar application	0, 250, and 500 ppm	*Glycine max* L.	Enhances activities of enzymatic and non-enzymatic antioxidants	Rahmawati and Damanik (2018)

TABLE 4.4

Activities of Antioxidant Enzymes as Affected by Trehalose Under Abiotic Stresses

Type of stress	Type of application	Dose	Crop	Effects on antioxidants	References
Drought Stress	Foliar application	0 and 30 mM	Maize	Increases activities of POD and CAT and decreases the SOD activity	Ali and Ashraf (2011)
Heavy metal stress	Foliar application	0.5, 1, 2, and 5 mM	Duckweed (*Lemna gibba* L.)	Increases SOD, CAT, and APX activities and decreases proline content.	Duman et al. (2011)
Drought stress	Foliar application	0 and 30 mM	Maize	Increases oil antioxidant activity of oil DPPH (2,2-diphenyl-1-picrylhydrazyl), free radical scavenging activity and oil tocopherols, total flavonoids, and total phenolics contents	Ali et al. (2012)
Salinity	Rooting medium	10 mM	Rice	Enhances activities of SOD, POX, APX, and CAT	Nounjan et al. (2012)
Drought Stress	Culture medium	50 mM	Wheat	Enhances activities of enzymatic antioxidants including SOD, and decreases GR, AsA, GSH, POD, and CAT	Ma et al. (2013)
Drought Stress	Rooting medium	5 mM	Brassica species	Improves ascorbate (AsA), glutathione (GSH) contents, AsA/DHA (ratio of AsA and dehydroascorbate), and GSH/GSSG (reduced to oxidized GSH) ratios; enhances ascorbate peroxidase (APX), dehydroascorbate reductase (DHAR), glutathione reductase (GR), glutathione peroxidase (GPX) activities. Up-regulates catalase (CAT), glutathione S-transferase (GST), glyoxalase I (Gly I), AsA/DHA ratio, and MDHAR activity	Alam et al. (2014a)
Salinity	Rooting medium	10 mM	Rice	Counteracts levels SOD and MDHAR; and elevates CAT, GST, APX, and DHAR activities; and maintains GPX and GR activities	Mostofa et al. (2015)

4.5.1 Application of Brassinosteroids

The application of 10 μM 24-epibrassinolide enhanced biological yield, photosynthetic rate, stomatal conductance, chlorophyll constituents, proline content and activities of SOD, POD, CAT, and nitrate reductase in *Pisum aestivum.* Treating the seed of *Cajanus cajan* with 10^{-7} M and 0.5×10^{-9} M epibrassinolide causes higher accumulation of free amino acids and soluble proteins in roots under saline environment (Sharma et al. 2013), while grain yield of phenolic acid in pods of *Pisum vulgaris* is increased by spray of 50 mg L^{-1} under high temperature stress (El-Bassiouny et al. 2012). The role of external application of BRs on antioxidant defence system is listed in Table 4.5.

4.6 Jasmonic Acid

Jasmonic Acid (JA) is an essential oil and extracted from Jasmin plant (Demole et al. 1962). The endogenous JA helps to maintain various physiological and biochemical processes through enhancing antioxidant activities in plants under stressful condition. The foliar-applied JA stimulates growth and yield parameters by inducing scavenging enzymes and absorption of greater amount of K^+ from rhizosphere (Qiu et al. 2014) and increases water relations in soybean plants (Anjum et al. 2011). It improves photosynthetic process, chlorophyll constituents, growth, and proline content by spray of 30 μM methyl jasmonate in soybean (Yoon et al. 2009) and *Pisum sativum* (Fedina and Tsonev 1997). The stressful effect of copper stress can be mitigated by spray of JA (Poonam et al. 2013). Moreover, *Brassica juncea* species perform well by spray of 0.5 mM JA under drought in field conditions (Alam et al. 2014). The foliar application of JA along with plant growth regulators and salicylic acid is a useful strategy to mitigate the harmful effects of environmental factors. The role of exogenous application of Jasmonic acid on antioxidant activities of different crop species under different abiotic stresses is listed in Table 4.6.

4.7 Ethylene

Ethylene, having both antagonistic and synergistic effects, has its beneficial influences on improving the quality of fruit and various physiological process in crop plants (Lee and LaRue 1992). The treating seeds of mung bean (*Vicia radiata* L.) and pea (*Pisum sativum* L.) with ethylene capacitate the plants to improve the tolerance level to molecular oxygen (O_3) with concurrent reduction in hydrogen peroxide (H_2O_2) (Mehlhorn 1990). The biological yield and grain yield of *Pisum vulgaris* L. are improved by treating with ethylene diurea under molecular oxygen (O_3) (Brunschön-Harti et al. 1995). The external application of ethylene did not decrease the number of rhizobial infectious in root, while it was restricted in the basal epidermal cell (Lee and LaRue 1992).

4.8 Indole Acetic Acid

Indole acetic acid (IAA) is a hormone that belongs to 'auxin group.' It induces the growth and development of plants through cell enlargement and root proliferation processes. Numerous research-based evidence show that foliar spray of IAA has a key role in the accumulation of nutrients from rhizosphere, physiological processes, and alleviating environmental stress in various crop plant (Egamberdieva 2009; Egamberdieva et al. 2015; Fahad et al. 2014a,b). The exogenous application of 15 mg L^{-1} IAA on *Phaseolus mungo* L. plants enhanced the antioxidant defence system in response to salt stress (100 mM NaCl) environment. The biological yield, chlorophyll content, nitrate reductase, catalase, and seed yield were improved, while the reduction occurred in quantity of proline, free amino acids, and total soluble sugars (Gravel et al. 2007). The seedlings of *Pisum sativum* L. treated with 100 μM IAA led to alleviation of manganese (250 μM) through increased quantum of SOD, APX, GR, and DHAR activities while enhancing oxidative stress in response to Mn toxicity. Furthermore, the germination process of *Cicer arietinum* has hampered in response to cadmium stress. However, the

TABLE 4.5

Activities of Antioxidant Enzymes as Affected by Brassinosteroids Under Abiotic Stresses

Type of stress	Type of application	Dose	Crop	Effects on antioxidant	References
Temperature stress	Foliar application	0.01 to 1.0 mg L^{-1}	*Lycopersicon esculentum* M.	Increases activities of SOD, APX, GPOD, and CAT	
Drought stress	Foliar application	0.01 and 1 μM	*Lycopersicon esculentum* L	Increases activities of POD, CAT, APX, GR, and SOD	Behnamnia et al. (2009)
Drought stress	Foliar application	0.01 μM	*Oryza sativa* L.	Increases activities of SOD, APX, and CAT	Farooq et al. (2009)
Cadmium stress	Foliar application	0, 10^{-11}, 10^{-9}, and 10^{-7}	*Raphanus sativus* L.	Increases activities of APOX, CAT, GR, POD, and SOD	Sharma et al. (2010)
Salinity stress	Foliar application	5 and 10 μM (EBL)	*Pisum sativum* L.	Increases activities of SOD, POD, CAT, NRA, and NiRA	
Drought stress	Foliar application	0.1 mg^{-1}	*Zea mays* L.	Increases activities of SOD and CAT	Anjum et al. (2011)
Chilling stress	Foliar application	10^{-8} and 10^{-6} M	*Cucumis sativus* L.	Increases activities of CAT, POX, and SOD	Fariduddin et al. (2011)
Salinity Stress	Foliar application	10^{-10} and 10^{-8} M (EBL)	*Lycopersicon esculentum* L.	Increases activities of GPOX, SOD, and CAT	Slathia et al. (2012)
Salinity stress	Foliar application	10^{-7} to 10^{-9} M (EBL)	*Oryza sativa* L.	Increases activities of SOD, APX, CAT, and GR	Sharma et al. (2013)
High temperature stress	Foliar application	0.05 to 0.2 μM	*Solanum melanogena* L.	Increases activities of SOD, GSH, POD, CATT, and APX	Wu et al. (2014)
Zinc stress	Foliar application	0.5, 1.0, and 2.0 μM	*Raphanus sativus* L.	Increases activities of CAT, SOD, APX, GR, POD, and GSH	Ramakrishna and Rao (2015)

TABLE 4.6

Activities of Antioxidant Enzymes as Affected by Jasmonic Acid Under Abiotic Stresses

Type of stress	Type of application	Dose	Crop	Effects on antioxidants	References
Drought stress	Foliar application	50 mM	Pear	Increases glycine betaine, betaine aldehyde, and dehydrogenase.	Gao et al. (2004)
Drought stress	Foliar application	0.5 mM	*Brassica campestris* L.	Increases activities of GR (glutathione reductase) and Glycine L.	Alam et al. (2014)
Salinity stress	Foliar application	20 and 30 µM,	Soybean (*Glycine max* L.)	Increases proline content	Yoon et al. (2009)
Heavy metal (Cd, Cu) stress	Foliar application	0.5 µg L^{-1}	*Arabidopsis thaliana* L.	Enhances activity of glutathione reductase	Maksymiec et al. (2005)
Heavy metal stress	Foliar application	0.5 µg L^{-1}	*Phaseolus coccineus* L.	Enhances activities of glutathione reductase	Maksymiec et al. (2005)
Heavy metal(Cd) stress	Foliar application	–	*Vicia faba* L.	Enhances activities of glycine betaine, ascorbate peroxidase, superoxide dismutase, and catalase.	Ahmad et al. (2017)
Drought stress	Foliar application	10 µM	*Brassica oleraceae* L.	Enhances activities of catalase, ascorbate peroxidase, and superoxide dismutase.	Wu et al. (2012)
Heavy metal (Ni) stress	Seed priming	1 mM	Soybean (*Glycine max* L.)	Enhances amount of glycine betaine	Sirhindi et al. (2016)
Heavy metal (Cd) stress	Jasmonic acid solution	5 µM	Rice (*Oryza sativa* L.)	Increases activities of GSH, catalase, glutathione reductase, and superoxide dismutase.	Singh and Shah (2014)

soaking of seed with 4-chloroindole-3-acetic acid led to improvement in relative water content, and nitrate reductase and peroxidase having beneficial effects in escalating the rate of germination in *Vigna radiate* and in wheat seedlings (Majid et al. 2013).

4.9 Gibberellic Acid (GA_3)

Gibberellic acid works as phytohormone and maintains various developmental processes, since the inception of seed germination to maturity (Sun and Gubler 2004; Fahad et al. 2014a,b) and strengthening the adaptive strategy by plants under external threats. The mitigation efforts taken by soybean (*Glycine max* L.) treated with GA_3 result in regulating bioactive GAs (GA_1 and GA_3), increasing amount of jasmonic acid with simultaneously reducing the level of abscisic acid and salicylic acid contents (Hamayun et al. 2010; Fahad et al. 2014a,b).

The deleterious effects of salinity (200 µM NaCl) can be ameliorated by presoaking seed of mung bean with 200 mg $L^{-1}GA_3$, which is mainly due to improvement in growth and physiological parameters and reduction in Na^+ and Cl^- contents in various plant organs. The productivity of groundnut in terms of pod, kernel, and haulm yield could be activated by foliar spray of 40 mg L^{-1} GA under normal and drought stress condition. Treating the mung bean plant with 200 mg L^{-1} GA causes increase in biological yield, number of pods, leaf area index, and seed yield under drought stress (Abdel and Al-Rawi 2011). The tolerance level in *Vicia radiate* plants could be enhanced by 100 µM GA_3 foliar spray through increased quantum of α-amylase under temperature differential levels ranging from 30 to 40 °C. The accelerated growth of seedling is due to greater amounts of α-amylase in roots, shoots, and cotyledons of plants (Mansoor and Naqvi 2012). The heavy metals stress impacted by the cadmium and lead could be averted by treating the plants of *Vicia faba* and *Vicia albus* with spray of 100 ppm GA. Treating the plants with GA causes increase in CAT, POD, and amylase activities.

4.10 Ascorbic Acid

Ascorbic acid (AsA) is abundantly present in leaves of crop plants, and its proportionate quantity is significantly higher than glutathione (Smirnoff and Wheeler 2000; Ibrahim et al. 2012; Fahad et al. 2016a,b,c,d). It is involved in works of scavenging ROS and balancing stomatal conductance for enhancing photosynthetic capacity and networking with certain enzymatic activities (Gallie 2012; Hussein and Alva 2014). The stabilization of photosynthetic machinery is manifested through series of activities of signal transduction pathways in response to AsA (Karuppanapandian et al. 2011). The amount of AsA in plant system varies greatly in response to stressful conditions and helps in adaptation of plants through readjusting certain processes, by closure of stomata for avoiding loss of water through transpiration (Idrees et al. 2011; Fahad et al. 2016a,b,c,d).

Ascorbic acid (AsA) is known to protect organelles and cells from the adversaries of ROS, which overaccumulate due to stress-induced oxidative damages (Mukhtar et al. 2016; Naz et al. 2016). It controls cell division and cell expansion, acts as a cofactor of many enzymes, modulates plant sense, and is involved in photosynthesis, hormone biosynthesis, and regeneration of antioxidants (Gallie 2012). Although AsA is found in different plant tissues, its levels are abundant in meristems, photosynthetic cells, and different fruits (HongBo et al. 2005; Mazid et al. 2011). Higher concentration of AsA has been reported in fully developed chloroplasts of mature leaves. Under normal conditions, AsA is available mostly in reduced form (Khan et al. 2011). In most plant species, the levels of AsA are not adequate to mitigate effectively the adverse effects of a stress (Shafiq et al. 2014). Thus, under such circumstances, exotic application of AsA is considered advisable (Mukhtar et al. 2016; Naz et al. 2016).

4.10.1 Application of Ascorbic Acid

Exogenous application of AsA through different modes has been widely reported to improve salinity tolerance of many plants (Hussein and Alva 2014; Barus et al. 2015; Fahad et al. 2016a,b,c,d). For example, in an earlier study, the effects of various levels of AsA as a seed treatment (0.25, 0.5, and 1 mM) were assessed in callus culture/plants of sugarcane grown under saline conditions (Munir et al. 2013). Of all AsA pretreatments used, generally 0.25 and 0.5 mM had a positive effect on plant growth, activities of antioxidant enzymes, and soluble proteins in both plants and callus culture. Under drought stress, Azooz et al. (2013) found that seed soaking in AsA improved leaf chlorophyll, plant water status, and soluble proteins in *Vicia faba* plants. Similarly, Reiahi and Farahbakhsh (2013) found that seed priming with AsA (0.5, 1.0, and 1.5 mM) improved seed germination and seedling resistance of sorghum (Sorghum bicolour) to drought stress.

The role of exogenous application of AsA has been studied widely for salinity and drought tolerance capabilities, but very few studies have been carried out to assess the influence of AsA on altering the defence potential of different plants against temperature stress. In a study, foliar-applied AsA has been found very effective in improving the growth of straw berry plants under high temperature (44°C) (Ergin et al. 2014). For example, exogenously applied AsA (3 mM) increased cell turgidity of the strawberry plants and hence their growth. Foliar or seed treatment with 20 and 40 mg L^{-1} AsA improved the seedling growth and yield production of maize under low temperature stress (Ahmad et al. 2014). A summary of the results of different studies on exogenous application of AsA is presented in Table 4.7.

4.11 Abscisic Acid

Abscisic acid (ABA) being an intrinsic plant hormone has a greater role in defence system and adaptive mechanism in response to abiotic stress (Fahad et al. 2014a,b). Various growth and developmental processes are regulated by signals generated by ABA under normal and stressful conditions (Fahad et al. 2014a,b). Under

TABLE 4.7

Role of AsA Under Abiotic Stress

Stress	Mode of application	Dose of AsA	Crop	Results	References
Drought	Pre sowing	1 and 2 mM	Sunflower	Enhanced germination rate, germination percentage, seed stamina index, and fresh and dry weights	Fatemi (2014)
Salinity	Pre sowing	1 mM	Barley	Improved proline, RWC, chlorophyll, enzymatic antioxidants, and leaf anatomy	Agami (2014)
Salinity	Pre sowing	55, 110, and 165 μM	Sunflower	An improved germination percentage, seedling fresh and dry weights, shoot and root lengths, and vigour index	Razaji et al. (2012)
Drought	Spray	500, 1000, and 2000 mg/L	Olive	Enhanced plant height, leaf number, leaf area, and lateral shoot number	Mayi et al. (2014)
Drought	Spray	50, 100, and 150 mg/L	Canola	Improved shoot and root fresh weights, root dry weight, qN, NPQ, shoot and root P, and AsA contents	Shafiq et al. (2014)
Salinity	Spray	0.1, 0.5, and 1 mM	*Saccharum* spp.	POD, SOD proline contents, and growth improved, while protein contents were decreased	Ejaz et al. (2012)
Salinity	Spray	0.7 mM	Wheat	Improved leaf area, chlorophyll, and carotenoid contents and proline, while decreased H_2O_2 contents	Azzedine et al. (2011)
Salinity	Spray	0.7 mM	Wheat	A significantly improved chlorophyll and carotenoid contents, leaf area, and proline, while decreased H_2O_2 contents	Fercha et al. (2011)

extreme water shortage, generation of signals regulates stomatal opening and closure (Lim et al. 2015). Mung bean plants treated with 10 μM ABA showed a greater tolerance to drought stress through increase in biological yield and quantum of nodules on the roots; chlorophyll contents enhanced accumulation of proline and protein, augmenting antioxidants' enzymatic network (Farooq and Bano 2006). The exogenous application of 100 μM ABA regulated the quantum of phyto hormones and polyamines which led to tolerance to drought stress in soybean plants (Hassanein et al. 2009). Seedlings of *Vicia faba* treated with 0.1 mM ABA showed greater improvement in biological yield and antioxidant defence system under drought stress environment (Ali et al. 2013). The foliar spray of 2.5 μM ABA on *Cicer arietinum* plants increased amount of ABA and osmolytes with simultaneous reduction in MDA and H_2O_2 under high temperature stress (35–45°C) (Kumar et al. 2012). The chilling stress could also mitigate by foliar application of 1 mM ABA in *Vigna radiate* seedlings due to reduction in lipid peroxidation, membrane leakage, and hydrogen peroxide with concurrent increase in chlorophyll contents, carbohydrates, and proline contents (Saleh 2007).

4.12 Thiamine

Thiamine, also known as vitamin B1, was the first vitamin type B identified. Free thiamine, thiamine monophosphate (TMP), and thiamine pyrophosphate (TPP) are the three most predominant forms of B1 that exist in the cells. Vitamin B1is a colourless, water soluble vitamin made solely by plants and microorganisms and acts as essential micronutrient in the human diet (Funk 1975; Pourcel et al. 213). Thiamine occurrence in plants is widely distributed across organs, namely, leaves, flowers, fruits, seeds, roots, tubers, and bulb (Asensi-Fabado and Munné-Bosch 2010). In plants, thiamine is known to have its role as a cofactor for important metabolic activities (Colinas and Fitzpatrick 2015). Thiamine is known to be an essential regulator that plays an important role in plant's primary regulatory system (Bocobza and Aharoni 2014). Living organisms require the active form of thiamine which is known as thiamine pyrophosphate (TPP) in order to play the role as an important cofactor. TPP is a crucial component required in many metabolic activities such as acetyl-CoA biosynthesis, amino acid biosynthesis, Krebs cycle, and Calvin cycle (Du et al. 2011). In plants, thiamine plays a role as a response molecule towards abiotic and biotic stresses, and data from the literature suggest that boosting thiamine content could increase resistance to stresses (Dong et al. 2016). Comprehensive studies on the effect of abiotic stresses on the regulation of thiamine in oil palm were also done where various types of stresses, namely, oxidative, salinity, and osmotic stresses, have been induced in oil palm where an increase in gene expression and also total thiamine content was observed post stress inductions (Abidin et al. 2016). The greater accumulation of thiamine invokes plant system to avoid biotic stress.

4.13 Naphthalene Acetic Acid (NAA)

Naphthalene acetic acid (NAA) comes under the umbrella of 'auxin' and is extensively used in *in vitro* plant regeneration and vegetative propagation (Navalón et al. 1997). It is also being applied foliarly to enhance quantity and yield of fruits in orchards (Abbasi et al. 2013). The recent advances have shown that NAA has potential capacity to ameliorate the harshness of external factors in different crop species. The foliar application of NAA (4.5% active ingredient) resulted in increased biological yield of pods and 100 seed weight and seed yield of *Vigna unguiculata* (Bhalse 2012).

4.14 Triacontanol (TRIA)

Triacontanol (TRIA) is a potential antioxidant enzyme and regulates synthesis of enzymatic and non-enzymatic peroxidation of lipids (Verma et al. 2011). It is categorized secondary plant growth and works in expressing genetic potential of crops through enhancing physiological efficacy of cells in plant system. It is involved in upregulating photosynthetic machinery, even at a very little content in system. TRIA is fairly present in some plants, viz., runner bean (*Phaseolus multiflorus* L.), white clovers (*Trifolium repens* L.), alfalfa (*Medicago saliva* L.), blue berry (*Vaccinium ashei* L.), radish, and barely seedlings. The biological yield, photosynthetic capacity, and chlorophyll contents in some crop plants are greatly influenced by application of TRIA (Verma et al. 2011). The productivity capacity of groundnut, chickpea, pigeon pea, rice, maize, and wheat was proportionately improved by foliar application of TRIA under saline environment. The improvement in productivity of plants is due to enhancement in photosynthesis, mainstreaming various enzymes, and greater uptake of nutrients and water from rhizosphere under enhanced accumulation of TRIA by plants (Verma et al. 2011; Shahbaz et al. 2013). The external application of TRIA could prove beneficial in raising the productivity of cereals vegetable and fruit crops.

4.15 Polyamines

Polyamines belong to group of low molecular weight aliphatic amines and have three isomers, viz., diamine putrescine, triamine spermidine, and tetraamine spermine. These compounds work in growth and development, accelerating seed germination, seedling growth, and flower initiation. The growth and development of *Pisum vulgaris* is enhanced by foliar spray of 10 μM diamine putrescine under saline (1% NaCl) environment. The increase in growth results due to upregulation of amylase and protease enzyme during germination stage. The adverse effects of sodium chloride are mitigated with concurrent activating defence-response genes in response to foliar spray of diamine putrescine. The *Pisum vulgaris* treated with polyamines accumulates greater amount of proline, thereby reducing the harshness of abiotic stresses (Jiménez-Bremont et al. 2006). The exogenous application of 0.1 mM of tetra amine spermine enhances antioxidant defence and ABA in soybean plants by reducing osmotic stress, which eventually causes increase in protein content and grain yield attributes (Radhakrishnan and Lee 2013).

4.16 Glutathione

Glutathione (GSH) belongs to thiol group and potentiates in the work biochemical function. The harshness of external factors is minimized due to its active involvement in bonding with metals and encouraging water contents and phytochelatin synthesis. Treating *Vigna cutjang* with reduced glutathione (10 to 3 mM) led to reduction in glycolate oxidase activity and proline content, enhancing relative water content under water stress at −1.0 MPa. The adverse effects of water stress could be alleviated by treating *Phaseolus vulgaris* seedlings with reduced glutathione (Mukherjee and Choudhuri 1983), because of increased quantity in glutathione and ascorbic acid, and stimulating β-amylase, acid phosphate, and guaiacol peroxidase (Nagesh Babu and Devoraj 2008). The adverse effects of salt stress (200 mM NaCl) on mung bean (*Vicia radiate* L.) were mitigated by treating seedling with reduced glutathione with onset reduction in oxidative stress, with simultaneous increase in antioxidant defence battery (Nahar et al. 2015).

4.17 Glycine Betaine

Glycine betaine (N,N,N-trimethyl glycine) is a zwitterionic quaternary amine and exists in neutral pH. Glycine betaine (GB) was isolated from sugar beet (*Beta vulgaris*) (Scheibler 1869) and is fairly distributed in plants (Chen and Murata 2011; Karabudak et al. 2014). It is water soluble amino acid that does not interfere with organic osmolytes (Iqbal et al. 2011). GB is involved in a number of processes, that is, photosynthetic machinery, preventing photo inhibition, net photosystem II, macro molecule stabilization (Chen and Murata 2011), and cytoprotector as a chemical chaperone (Figueroa-Soto and Valenzuela-Soto 2018), and reduces the intensity of reactive oxygen species (Ashraf and Foolad 2007).

4.17.1 Application of Glycine Betaine

Its accumulation works as organic osmolyte, enhances osmotic adjustment and oxidative defence system (Wang et al. 2010), and reduces quantity of irrigation water due to its anti-transpiration properties (Agboma et al. 1997; Reddy et al. 2013). The addition of GB enhances the activities of AsA, GSH, APX, DHAR, GST, GPX, and CAT under cadmium stress (Raza et al. 2007; Hossain et al. 2010). The various crop species, viz., sugar beet (Yamada et al. 2009), spinach (Chen and Murata et al. 2011), and wheat (Aldesuquy 2014), accumulate the greatest quantity of GB especially under stressful conditions. The exogenous application of GB results in increase in economic yield of tobacco, wheat, barely, sorghum, soybean, and cotton (Ashraf and Foolad 2007; Hoque et al. 2007; Noreen et al. 2013). The response of different plant species to exogenous application of glycine betaine under different types of stresses is listed in Table 4.8.

4.18 Hydrogen Peroxide

Hydrogen peroxide (H_2O_2) is a potent signalling molecule that is actively engaged in a number of physiological functions of plants. Hydrogen peroxide (H_2O_2) is generally present in apoplastic space and works in proper functioning of growth and development and also cell death. The effects of H_2O_2 on biological processes vary greatly in response to its concentration, site of production, stage of plant growth, and development and cumulative severity of earlier stress conditions. The concentration of H_2O_2 in a small quantity results in signalling molecules, while higher amount induces death of cell. The death of cell is also an essential phenomenon for continuation of developmental processes, which include aleurone cell death, leaf senescence, responses to pathogenesis, and allelopathic plant–plant interactions. The effects of H_2O_2 on various biological processes are outlined in Table 4.9.

4.19 Future Perspectives

In the days to come, climate change is expected to increase the vulnerability of agricultural systems. The projected changes in increased temperature, changes in rainfall patterns, and increased frequency of extreme weather events are expected to have negative impacts on crops, especially under stressful environments. During the discourse of evolution, plant species have adapted to accrue beneficial effects of excessive amounts of ROS production for signalling purpose. The endeavours to quench the deleterious effects of ROS are of a paramount importance for raising the crops. Without adaptation of strategies, environmental stresses would have far-reaching effects on limiting the agricultural production. Under the

TABLE 4.8

Activities of Antioxidant Enzymes as Affected by Glycine Betaine Under Abiotic Stresses

Type of stress	Type of application	Dose	Crop	Effects on antioxidant	References
Drought stress	Foliar application	50, 100, and 150 mg L^{-1}	*Oryza sativa* L.	Increases the activities of SOD, CAT, and APX	Farooq et al. (2008)
Cadmium stress	Foliar application	1 and 10 mM	Cultured tobacco cells	Increases the activities of SOD and CAT	Islam et al. (2009)
Salt stress	Foliar application	0, 50, and 100 mM	*Zea mays* L.	Increases the activities of SOD, CAT, and POD	Nawaz and Ashraf (2010)
Cadmium stress	Foliar application	5 mM	Mung bean	Increases the activities of GSH, GST, CAT, and APX	Hossain et al. (2010)
Water stress	Foliar application	25 and 50 mM	*Carapa guianesis*	Increases the activities of APX and CAT	Cruz et al. (2013)

TABLE 4.9

Activities of Antioxidant Enzymes as Affected by H_2O_2 Under Abiotic Stresses

Type of abiotic stress	Type of exogenous application	Dose	Crop	Effects on antioxidants	References
Chilling	Foliar application	200 mM	Mung bean	Increases H_2O_2-induced chilling tolerance by elevation of glutathione content.	Yu et al. (2003)
Heat stress	Foliar application	9 mmol L^{-1}	*Agrostis stolonifera*	Increases APX activity, reduces POX activity, and lowers CAT activity.	Larkindale and Huang (2004)
Salinity	Rooting medium	1 mM	Maize	Increases activities of antioxidative enzymes (superoxide dismutase, ascorbate peroxidase, guaiacol peroxidase, glutathione reductase, and catalase)	de Azevedo Neto et al. (2005)
Osmotic and heat stress	Rooting medium	0.05 mM	Wheat	Transduces signal and increases the activities of SOD, APOX, CAT, and NADPH oxidase	Agarwal et al. (2005)
Salinity	Seed priming	1, 40, 80, and 120 mM	Wheat	Activates antioxidants in seeds, seedlings, and physiological phenomena and prevents oxidative damage	Wahid et al. (2007)
Salinity	Rooting medium	10 mM	Citrus	Increases the antioxidant enzymes, i.e., SOD, CAT, APX, and CR	Tanou et al. (2009)
Salinity	Rooting medium	0.05 M	Wheat	Increases the activities of superoxide dismutase, peroxidase, catalase, ascorbate peroxidase, glutathione, and carotenoids.	Li et al. (2011)

stressful conditions, the exogenous application of osmoprotectants on crop plants is a potential option to mitigate eventual stress and to maintain productivity of crops without economic loss to the farmers. The germplasm improvement for stress tolerance and natural resource management lays basic infrastructure to reduce susceptibility of agricultural systems to climate change. The maximum benefits could be realized, when crop technology options are combined with advanced adaptation tools at the farm level.

REFERENCES

Abbasi NA, Zafar L, Khan HA, Qureshi AA (2013) Effects of naphthalene acetic acid and calcium chloride application on nutrient uptake, growth, yield and post harvest performance of tomato fruit. Pak J Bot 45:1581–1587.

Abdel CG, Al-Rawi IMT (2011) Response of mungbean (*Vigna radiata* L., Wilczek) to gibberellic acid (GA3) rates and varying irrigation frequencies. Int J Biosci 1:85–92.

Abidin AAZ, Wong S, Rahman NSA, Idris ZHC, Balia Yusof Z (2016) Osmotic, oxidative and salinity stresses upregulate the expressions of thiamine (Vitamin B1) biosynthesis genes (THIC and THI1/THI4) in oil palm (*Elaies guineensis*). J Oil Palm Res 28:308–319.

Adnan M, Fahad S, Khan IA, Saeed M, Ihsan MZ, Saud S, Riaz M, Wang D, Wu C (2019) Integration of poultry manure and phosphate solubilizing bacteria improved availability of Ca bound P in calcareous soils. 3 Biotech 9:368.

Adnan M, Fahad S, Muhammad Z, Shahen S, Ishaq AM, Subhan D, Zafar-ul-Hye M, Martin LB, Raja MMN, Beena S, Saud S, Imran A, Zhen Y, Martin B, Jiri H, Rahul D (2020) Coupling phosphate-solubilizing

bacteria with phosphorus supplements improve maize phosphorus acquisition and growth under lime induced salinity stress. Plants 9:900. https://doi.org/10.3390/plants9070900.
Adnan M, Shah Z, Sharif M, Rahman H (2018b) Liming induces carbon dioxide (CO_2) emission in PSB inoculated alkaline soil supplemented with different phosphorus sources. Environ Sci Poll Res 25:9501–9509.
Adnan M, Zahir S, Fahad S, Arif M, Mukhtar A, Imtiaz AK, Ishaq AM, Abdul B, Hidayat U, Muhammad A, Inayat-Ur R, Saud S, Muhammad ZI, Yousaf J, Amanullah, Hafiz MH, Wajid N (2018a) Phosphate-solubilizing bacteria nullify the antagonistic effect of soil calcification on bioavailability of phosphorus in alkaline soils. Sci Rep 8:4339. https://doi.org/10.1038/s41598-018-22653-7.
Afzal I, Basra SM, Farooq M, Nawaz A (2006) Alleviation of salinity stress in spring wheat by hormonal priming with ABA, salicylic acid and ascorbic acid. Int J Agric Biol 8:23–28.
Agami R (2014a) Applications of ascorbic acid or proline increase resistance to salt stress in barley seedlings. Biol Plant 58:341–347.
Agami RA (2014b) Applications of ascorbic acid or proline increase resistance to salt stress in barley seedlings. Biol Plant 58:341–347. https://doi.org/10.1007/s10535-014-0392-y.
Agami RA, Mohamed GF (2013) Exogenous treatment with indole-3-acetic acid and salicylic acid alleviates cadmium toxicity in wheat seedlings. Ecotoxicol Environ Saf 94:164–171. https://doi.org/10.1016/j.ecoenv.2013.04.013.
Agarwal S, Sairam R, Srivastava G, Tyagi A, Meena R (2005) Role of ABA, salicylic acid, calcium and hydrogen peroxide on antioxidant enzymes induction in wheat seedlings. Plant Sci 169:559–570.
Agboma P, Jones M, Peltonen-Sainio P, Rita H, Pehu E (1997) Exogenous glycinebetaine enhances grain yield of maize, sorghum and wheat grown under two supplementary watering regimes. J Agron Crop Sci 178:29–37.
Ahmad P, Alyemeni MN, Wijaya L, Alam P, Ahanger MA, Alamri SA (2017) Jasmonic acid alleviates negative impacts of cadmium stress by modifying osmolytes and antioxidants in faba bean (*Vicia faba* L.). Arch Agron Soil Sci 63:1889–1899.
Ahmad I, Basra SMA, Wahid A (2014) Exogenous application of ascorbic acid, salicylic acid and hydrogen peroxide improves the productivity of hybrid maize at low temperature stress. Int J Agric Biol 16:825–830.
Ahmad P, Jaleel CA, Salem MA, Nabi G, Sharma S (2010) Roles of enzymatic and nonenzymatic antioxidants in plants during abiotic stress. Crit Rev Biotechnol 30:161–175.
Ahmad S, Kamran M, Ding R, Meng X, Wang H, Ahmad I, Fahad S, Han Q (2019) Exogenous melatonin confers drought stress by promoting plant growth, photosynthetic capacity and antioxidant defense system of maize seedlings. PeerJ 7:e7793. http://doi.org/10.7717/peerj.7793.
Akram R, Turan V, Hammad HM, Ahmad S, Hussain S, Hasnain A, Maqbool MM, Rehmani MIA, Rasool A, Masood N, Mahmood F, Mubeen M, Sultana SR, Fahad S, Amanet K, Saleem M, Abbas Y, Akhtar HM, Waseem F, Murtaza R, Amin A, Zahoor SA, ul Din MS, Nasim W (2018a) Fate of organic and inorganic pollutants in paddy soils. In: Hashmi MZ, Varma A (eds) Environmental pollution of paddy soils, soil biology. Springer International Publishing Ag, Switzerland, pp 197–214.
Akram R, Turan V, Wahid A, Ijaz M, Shahid MA, Kaleem S, Hafeez A, Maqbool MM, Chaudhary HJ, Munis, MFH, Mubeen M, Sadiq N, Murtaza R, Kazmi DH, Ali S, Khan N, Sultana SR, Fahad S, Amin A, Nasim W (2018b) Paddy land pollutants and their role in climate change. In: Hashmi MZ, Varma A (ed) Environmental pollution of paddy soils, soil biology. Springer International Publishing Ag Switzerland, pp 113–124.
Alam MM, Nahar K, Hasanuzzaman M, Fujita M (2014a) Exogenous jasmonic acid modulates the physiology, antioxidant defense and glyoxalase systems in imparting drought stress tolerance in different Brassica species. Plant Biotechnol Rep 8:279–293.
Alam MM, Nahar K, Hasanuzzaman M, Fujita M (2014b) Trehalose-induced drought stress tolerance: a comparative study among different Brassica species. Plant Omics 7:271.
Aldesuquy HS (2014) Glycine betaine and salicylic acid induced modification in water relations and productivity of drought wheat plants. J Stress Physiol Biochem 10:55–73.
Ali Q, Anwar F, Ashraf M, Saari N, Perveen R (2013) Ameliorating effects of exogenously applied proline on seed composition, seed oil quality and oil antioxidant activity of maize (*Zea mays* L.) under drought stress. Int J Mol Sci 14:818–835.
Ali Q, Ashraf M (2011) Induction of drought tolerance in maize (*Zea mays* L.) due to exogenous application of trehalose: growth, photosynthesis, water relations and oxidative defence mechanism. J Agron Crop Sci197:258–271.

Ali Q, Ashraf M, Anwar F, Al-Qurainy F (2012) Trehalose-induced changes in seed oil composition and antioxidant potential of maize grown under drought stress. J Am Oil Chem Soc 89:1485–1493.

Ali Q, Ashraf M, Shahbaz M, Humera H (2008b) Ameliorating effect of foliar applied proline on nutrient uptake in water stressed maize (*Zea mays* L.) plants. Pak J Bot 40:211–219.

Ali B, Hayat S, Hasan SA, Ahmad A (2008a) A comparative effect of IAA and 4-Cl-IAA on growth, nodulation and nitrogen fixation in *Vigna radiata* (L.) Wilczek. Acta Physiol Plant 30:35–41.

Anjum S, Wang L, Farooq M, Hussain M, Xue L, Zou C (2011) Brassinolide application improves the drought tolerance in maize through modulation of enzymatic antioxidants and leaf gas exchange. J Agron Crop Sci 197:177–185.

Ansari O, Azadi M, Sharif-Zadeh F, Younesi E (2013) Effect of hormone priming on germination characteristics and enzyme activity of mountain rye (Secale montanum) seeds under drought stress conditions. J Stress Physiol Biochem 9.

Arfan M, Athar HR, Ashraf M (2007) Does exogenous application of salicylic acid through the rooting medium modulate growth and photosynthetic capacity in two differently adapted spring wheat cultivars under salt stress? J Plant Physiol 164:685–694.

Argüelles JC (2000) Physiological roles of trehalose in bacteria and yeasts: a comparative analysis. Arch Microbiol 174:217–224.

Asensi-Fabado MA, Munné-Bosch S (2010) Vitamins in plants: occurrence, biosynthesis and antioxidant function. Trends Plant Sci 15:582–592.

Ashraf M, Akram N, Arteca RN, Foolad MR (2010) The physiological, biochemical and molecular roles of brassinosteroids and salicylic acid in plant processes and salt tolerance. Crit Rev Plant Sci 29:162–190.

Ashraf M, Foolad M (2007) Roles of glycine betaine and proline in improving plant abiotic stress resistance. Environ Exp Bot 59:206–216.

Ayad HS, El-Din K, Reda F (2009) Efficiency of stigmasterol and α-tocopherol application on vegetative growth, essential oil pattern, protein and lipid peroxidation of geranium (*Pelargonium graveolens* L.). J Appl Sci Res 5:887–892.

Aziz K, Daniel KYT, Fazal M, Muhammad ZA, Farooq S, Fan W, Fahad S, Ruiyang Z (2017a) Nitrogen nutrition in cotton and control strategies for greenhouse gas emissions: a review. Environ Sci Pollut Res 24:23471–23487. https://doi.org/10.1007/s11356-017-0131-y.

Aziz K, Daniel KYT, Muhammad ZA, Honghai L, Shahbaz AT, Mir A, Fahad S (2017b) Nitrogen fertility and abiotic stresses management in cotton crop: a review. Environ Sci Pollut Res 24:14551–14566. https://doi.org/10.1007/s11356-017-8920-x.

Azooz M (2009) Salt stress mitigation by seed priming with salicylic acid in two faba bean genotypes differing in salt tolerance. Int J Agric Biol 11:343–350.

Azooz MM, Alzahrani AM, Youssef MM (2013) The potential role of seed priming with ascorbic acid and nicotinamide and their interactions to enhance salt tolerance in broad bean (‘*Vicia faba*’ L.). Aust J Crop Sci 7:2091.

Azzedine F, Gherroucha H, Baka M (2011) Improvement of salt tolerance in durum wheat by ascorbic acid application. Журнал стресс-физиологии и биохимии 7:27–37.

Bajguz A, Hayat S (2009) Effects of brassinosteroids on the plant responses to environmental stresses. Plant Physiol Biochem 47:1–8.

Barus WA, Rauf A, Rosmayati CH (2015) Improvement of salt tolerance in some varieties of rice by ascorbic acid application. Int J Sci Technol Res 4:2277–8616.

Baseer M, Adnan M, Fazal M, Fahad S, Muhammad S, Fazli W, Muhammad A, Jr. Amanullah, Depeng W, Saud S, Muhammad N, Muhammad Z, Fazli S, Beena S, Mian AR, Ishaq AM (2019) Substituting urea by organic wastes for improving maize yield in alkaline soil. J Plant Nutr. https://doi.org/10.1080/01904167.2019.1659344.

Becker A, Schlöder P, Steele J, Wegener G (1996) The regulation of trehalose metabolism in insects. Experientia 52:433–439.

Behnamnia M, Kalantari KM, Ziaie J (2009) The effects of brassinosteroid on the induction of biochemical changes in *Lycopersicon esculentum* under drought stress. Turk J Bot 33:417–428.

Ben Ahmed C, Ben Rouina B, Sensoy S, Boukhriss M, Ben Abdullah F (2010) Exogenous proline effects on photosynthetic performance and antioxidant defense system of young olive tree. J Agric Food Chem 58:4216–4222. https://doi.org/10.1021/jf9041479.

Berthelot M (1858) Sur le trehalose, nouvelle espece de sucre. Compt Rend Hebd Seanc Acad Sci Paris 46:1276–1279.

Bhalse S (2012) Effect of naphthalene acetic acid (NAA) and maleic hydrazide (MH) on plant growth and green pod yield of Cowpea (*Vigna unguiculata* L. Walp.) cv. Pusa Komal. RVSKVV, Gwalior, India

Bishop GJ, Koncz C (2002) Brassinosteroids and plant steroid hormone signaling. Plant Cell 14:S97–S110.

Bocobza SE, Aharoni A (2014) Small molecules that interact with RNA: riboswitch-based gene control and its involvement in metabolic regulation in plants and algae. Plant J 79:693–703.

Brunschön-Harti S, Fangmeier A, Jäger H-J (1995) Influence of ozone and ethylenediurea (EDU) on growth and yield of bean (*Phaseolus vulgaris* L.) in open-top field chambers. Environ Pollut 90:89–94.

Chen C, Dickman MB (2005) Proline suppresses apoptosis in the fungal pathogen Colletotrichum trifolii. Proc Natl Acad Sci 102:3459–3464.

Chen TH, Murata N (2011) Glycinebetaine protects plants against abiotic stress: mechanisms and biotechnological applications. Plant Cell Environ 34:1–20.

Colinas M, Fitzpatrick TB (2015) Natures balancing act: examining biosynthesis de novo, recycling and processing damaged vitamin B metabolites. Curr Opin Plant Biol 25:98–106.

Colombo ML (2010) An update on vitamin E, tocopherol and tocotrienol—perspectives. Molecules 15:2103–2113.

Crowe JH, Crowe LM, Chapman D (1984) Preservation of membranes in anhydrobiotic organisms: the role of trehalose. Science 223:701–703.

Cruz F, Castro G, Júnior DS, Festucci-Buselli R, Pinheiro H (2013) Exogenous glycine betaine modulates ascorbate peroxidase and catalase activities and prevent lipid peroxidation in mild water-stressed *Carapa guianensis* plants. Photosynthetica 51:102–108.

Cuin TA, Shabala S (2007) Compatible solutes reduce ROS-induced potassium efflux in Arabidopsis roots. Plant Cell Environ. 30:875–885.

de Azevedo Neto AD, Prisco JT, Enéas-Filho J, Medeiros J-VR, Gomes-Filho E (2005) Hydrogen peroxide pre-treatment induces salt-stress acclimation in maize plants. J Plant Physiol 162:1114–1122.

De Virgilio C, Hottiger T, Dominguez J, Boller T, Wiemken A (1994) The role of trehalose synthesis for the acquisition of thermotolerance in yeast: I. Genetic evidence that trehalose is a thermoprotectant. Eur J Biochem 219:179–186.

Dawood MG, Taie HAA, Nassar RMA, Abdelhamid MT, Schmidhalter U (2014) The changes induced in the physiological, biochemical and anatomical characteristics of Vicia faba by the exogenous application of proline under seawater stress. S Afr J Bot 93:54–63. https://doi.org/10.1016/j.sajb.2014.03.002.

Demole E, Lederer E, Mercier D (1962) Isolement et détermination de la structure du jasmonate de méthyle, constituant odorant caractéristique de l'essence de jasmin. Helv Chim Acta 45:675–685.

Depeng W, Fahad S, Saud S, Muhammad K, Aziz K, Mohammad NK, Hafiz MH, Wajid N (2018) Morphological acclimation to agronomic manipulation in leaf dispersion and orientation to promote "Ideotype" breeding: evidence from 3D visual modeling of "super" rice (*Oryza sativa* L.). Plant Physiol Biochem 135. https://doi.org/10.1016/j.plaphy.2018.11.010.

Dong W, Thomas N, Ronald PC, Goyer A (2016) Overexpression of thiamin biosynthesis genes in rice increases leaf and unpolished grain thiamin content but not resistance to *Xanthomonas oryzae* pv. oryzae. Front Plant Sci 7:616.

Du Q, Wang H, Xie J (2011) Thiamin (vitamin B1) biosynthesis and regulation: a rich source of antimicrobial drug targets? Int J Biol Sci 7:41.

Duman F, Aksoy A, Aydin Z, Temizgul R (2011) Effects of exogenous glycinebetaine and trehalose on cadmium accumulation and biological responses of an aquatic plant (*Lemna gibba* L.). Water Air Soil Pollut 217:545–556.

Egamberdieva D (2009) Alleviation of salt stress by plant growth regulators and IAA producing bacteria in wheat. Acta Physiol Plant 31:861–864.

Egamberdieva D, Jabborova D, Hashem A (2015) Pseudomonas induces salinity tolerance in cotton (*Gossypium hirsutum*) and resistance to Fusarium root rot through the modulation of indole-3-acetic acid. Saudi J Biol Sci 22:773–779.

Ejaz B, Sajid ZA, Aftab F (2012) Effect of exogenous application of ascorbic acid on antioxidant enzyme activities, proline contents, and growth parameters of *Saccharum* spp. hybrid cv. HSF-240 under salt stress. Turk J Biol 36:630–640.

El-Bassiouny AM, Ghoname AA, El-Awadi M, Fawzy Z, Gruda N (2012) Ameliorative effects of brassinosteroids on growth and productivity of snap beans grown under high temperature. Gesunde Pflanzen 64:175–182.

El-Bassiouny HM, Gobarah ME, Ramadan AA (2005) Effect of antioxidants on growth, yield and favism causative agents in seeds of *Vicia faba* L. plants grown under reclaimed sandy soil. J Agron 4:281–287.

El Hariri D, Sadak MS, El-Bassiouny H (2010) Response of flax cultivars to ascorbic acid and?--tocopherol under salinity stress conditions. Int J Acad Res 2

Ergin S, Aydogan C, Ozturk N, Turhan E (2014) Effects of ascorbic acid application in strawberry plants during heat stress. Türk Tarım ve Doğa Bilimleri Dergisi 1:1486–1491.

Fahad S, Adnan M, Hassan S, Saud S, Hussain S, Wu C, Wang D, Hakeem KR, Alharby HF, Turan V, Khan MA, Huang J (2019b) Rice responses and tolerance to high temperature, In: Hasanuzzaman M, Fujita M, Nahar K, Biswas JK (eds) Advances in rice research for abiotic stress tolerance. Woodhead Publ Ltd, Cambridge, England, pp 201–224.

Fahad S, Bajwa AA, Nazir U, Anjum SA, Farooq A, Zohaib A, Sadia S, NasimW, Adkins S, Saud S, Ihsan MZ, Alharby H,Wu C,Wang D, Huang J (2017) Crop production under drought and heat stress: plant responses and management options. Front Plant Sci 8:1147. https://doi.org/10.3389/fpls.2017.01147.

Fahad S, Bano A (2012) Effect of salicylic acid on physiological and biochemical characterization of maize grown in saline area. Pak J Bot 44:1433–1438.

Fahad S, Chen Y, Saud S,Wang K, Xiong D, Chen C,Wu C, Shah F, Nie L, Huang J (2013) Ultraviolet radiation effect on photosynthetic pigments, biochemical attributes, antioxidant enzyme activity and hormonal contents of wheat. J Food Agri Environ 11:1635–1641.

Fahad S, Hussain S, Bano A, Saud S, Hassan S, Shan D, Khan FA, Khan F, Chen Y, Wu C, Tabassum MA, Chun MX, Afzal M, Jan A, Jan MT, Huang J (2014a) Potential role of phytohormones and plant growth-promoting rhizobacteria in abiotic stresses: consequences for changing environment. Environ Sci Pollut Res 22:4907–4921. https://doi.org/10.1007/s11356-014-3754-2.

Fahad S, Hussain S, Matloob A, Khan FA, Khaliq A, Saud S, Hassan S, Shan D, Khan F, Ullah N, Faiq M, Khan MR, Tareen AK, Khan A, Ullah A, Ullah N, Huang J (2014b) Phytohormones and plant responses to salinity stress: A review. Plant Growth Regul 75:391–404. https://doi.org/10.1007/s10725-014-0013-y.

Fahad S, Hussain S, Saud S, Hassan S, Chauhan BS, Khan F et al (2016a) Responses of rapid viscoanalyzer profile and other rice grain qualities to exogenously applied plant growth regulators under high day and high night temperatures. PLoS One 11:e0159590. https://doi.org/10.1371/journal.pone.0159590.

Fahad S, Hussain S, Saud S, Hassan S, Ihsan Z, Shah AN,Wu C, Yousaf M, Nasim W, Alharby H, Alghabari F, Huang J (2016c) Exogenously applied plant growth regulators enhance the morphophysiological growth and yield of rice under high temperature. Front Plant Sci 7:1250. https://doi.org/10.3389/fpls.2016.01250.

Fahad S, Hussain S, Saud S, Hassan S, Tanveer M, Ihsan MZ, Shah AN, Ullah A, Nasrullah KF, Ullah S, AlharbyH NW, Wu C, Huang J (2016d) A combined application of biochar and phosphorus alleviates heat-induced adversities on physiological, agronomical and quality attributes of rice. Plant Physiol Biochem 103:191–198.

Fahad S, Hussain S, Saud S, Khan F, Hassan S, Jr A, Nasim W, Arif M, Wang F, Huang J (2016b) Exogenously applied plant growth regulators affect heat-stressed rice pollens. J Agron Crop Sci 202:139–150.

Fahad S, Hussain S, Saud S, Tanveer M, Bajwa AA, Hassan S, Shah AN, Ullah A,Wu C, Khan FA, Shah F, Ullah S, Chen Y, Huang J (2015a) A biochar application protects rice pollen from high-temperature stress. Plant Physiol Biochem 96:281–287.

Fahad S, Muhammad ZI, Abdul K, Ihsanullah D, Saud S, Saleh A, Wajid N, Muhammad A, Imtiaz AK, Chao W, Depeng W, Jianliang H (2018) Consequences of high temperature under changing climate optima for rice pollen characteristics-concepts and perspectives. Arch Agron Soil Sci 64:1479–1488. https://doi.org/10.1080/03650340.2018.1443213.

Fahad S, Nie L, Chen Y, Wu C, Xiong D, Saud S, Hongyan L, Cui K, Huang J (2015b) Crop plant hormones and environmental stress. Sustain Agric Rev 15:371–400.

Fahad S, Rehman A, Shahzad B, Tanveer M, Saud S, Kamran M, Ihtisham M, Khan SU, Turan V, Rahman MHU (2019a) Rice responses and tolerance to metal/metalloid toxicity. In: Hasanuzzaman M, Fujita M, Nahar K, Biswas JK (eds) Advances in rice research for abiotic stress tolerance. Woodhead Publ Ltd, Cambridge, England, pp 299–312.

Falk J, Krahnstöver A, van der Kooij TA, Schlensog M, Krupinska K (2004) Tocopherol and tocotrienol accumulation during development of caryopses from barley (*Hordeum vulgare* L.). Phytochemistry 65:2977–2985.

Farhana G, Ishfaq A, Muhammad A, Dawood J, Fahad S, Xiuling L, Depeng W, Muhammad F, Muhammad F, Syed AS (2020) Use of crop growth model to simulate the impact of climate change on yield of various wheat cultivars under different agro-environmental conditions in Khyber Pakhtunkhwa, Pakistan. Arabian J Geosci 13:112. https://doi.org/10.1007/s12517-020-5118-1.

Farhat A, Hafiz MH, Wajid I, Aitazaz AF, Hafiz FB, Zahida Z, Fahad S, Wajid F, Artemi C (2020) A review of soil carbon dynamics resulting from agricultural practices. J Environ Manage 268:110319.

Fariduddin Q, Yusuf M, Chalkoo S, Hayat S, Ahmad A (2011) 28-Homobrassinolide improves growth and photosynthesis in *Cucumis sativus* L. through an enhanced antioxidant system in the presence of chilling stress. Photosynthetica 49:55–64.

Farkhondeh R, Nabizadeh E, Jalilnezhad N (2012) Effect of salinity stress on proline content, membrane stability and water relations in two sugar beet cultivars. Int J Agric Sci 2:385–392.

Farooq M, Aziz T, Basra SMA, Cheema MA, Rehman H (2008a) Chilling tolerance in hybrid maize induced by seed priming with salicylic acid. J Agron Crop Sci 194:161–168. https://doi.org/10.1111/j.1439-037X.2008.00300.x.

Farooq U, Bano A (2006) Effect of abscisic acid and chlorocholine chloride on nodulation and biochemical content of *Vigna radiata* L. under water stress. Pak J Bot 38:1511–1518.

Farooq M, Basra SMA, Wahid A, Ahmad N, Saleem BA (2009a) Improving the drought tolerance in rice (*Oryza sativa* L.) by exogenous application of salicylic acid. J Agron Crop Sci 195:237–246. https://doi.org/10.1111/j.1439-037X.2009.00365.x.

Farooq M, Basra S, Wahid A, Cheema Z, Cheema M, Khaliq A (2008b) Physiological role of exogenously applied glycinebetaine to improve drought tolerance in fine grain aromatic rice (*Oryza sativa* L.). J Agron Crop Sci 194:325–333.

Farooq M, Wahid A, Basra S (2009b) Improving water relations and gas exchange with brassinosteroids in rice under drought stress. J Agron Crop Sci 195:262–269.

Farooq M, Wahid A, Lee DJ, Cheema S, Aziz T (2010) Drought stress: comparative time course action of the foliar applied glycinebetaine, salicylic acid, nitrous oxide, brassinosteroids and spermine in improving drought resistance of rice. Journal of Agronomy and Crop Science 196:336–345.

Farouk S (2011) Ascorbic acid and α-tocopherol minimize salt-induced wheat leaf senescence. J Stress Physiol Biochem 7:58–79.

Fatemi SN (2014) Ascorbic acid and its effects on alleviation of salt stress in sunflower. Annu Res Rev Biol 3656–3665.

Fath A, Bethke P, Beligni V, Jones R (2002) Active oxygen and cell death in cereal aleurone cells. J Exp Bot 53:1273–1282.

Fedina I, Tsonev T (1997) Effect of pretreatment with methyl jasmonate on the response of *Pisum sativum* to salt stress. J Plant Physiol 151:735–740.

Fercha A, Hocine G, Mebarek B (2011) Improvement of salt tolerance in durum wheat by ascorbic acid application. J Stress Physiol Biochem 7

Figueroa-Soto CG, Valenzuela-Soto EM (2018) Glycine betaine rather than acting only as an osmolyte also plays a role as regulator in cellular metabolism. Biochimie 147:89–97.

Funk C (1975) The etiology of the deficiency. Anal Chim Acta 76:176–177.

Gallie DR (2012) The role of L-ascorbic acid recycling in responding to environmental stress and in promoting plant growth. J Exp Bot 64:433–443.

Gao XP, Wang XF, Lu YF, Zhang LY, Shen YY, Liang Z, Zhang DP (2004) Jasmonic acid is involved in the water-stress-induced betaine accumulation in pear leaves. Plant Cell Environ 27:497–507.

Gao J, Wang N, Xu S-S, Li Y, Wang Y, Wang G-X (2013) Exogenous application of trehalose induced $H_2 O_2$ production and stomatal closure in *Vicia faba*. Biol Plant 57:380–384.

Garde-Cerdán T, López R, Portu J, González-Arenzana L, López-Alfaro I, Santamaría P (2014) Study of the effects of proline, phenylalanine, and urea foliar application to Tempranillo vineyards on grape amino acid content. Comparison with commercial nitrogen fertilisers. Food Chem 163:136–141.

Gharib F, Hegazi A (2010) Salicylic acid ameliorates germination, seedling growth, phytohormone and enzymes activity in bean (*Phaseolus vulgaris* L.) under cold stress. J Am Sci 6:675–683.

Ghoohestani A, Gheisary H, Zahedi SM, Dolatkhahi A (2012) Effect of seed priming of tomato with salicylic acid, ascorbic acid and hydrogen peroxideon germination and plantlet growth in saline conditions. Int J Agron Plant Prod 3:700–704.

Gill SS, Tuteja N (2010) Reactive oxygen species and antioxidant machinery in abiotic stress tolerance in crop plants. Plant Physiol Biochem 48:909–930.

Gravel V, Antoun H, Tweddell RJ (2007) Growth stimulation and fruit yield improvement of greenhouse tomato plants by inoculation with *Pseudomonas putida* or *Trichoderma atroviride*: possible role of indole acetic acid (IAA). Soil Biol Biochem 39:1968–1977.

Gunes A, Inal A, Alpaslan M, Eraslan F, Bagci EG, Cicek N (2007) Salicylic acid induced changes on some physiological parameters symptomatic for oxidative stress and mineral nutrition in maize (*Zea mays* L.) grown under salinity. J Plant Physiol 164:728–736.

Habib ur Rahman M, Ashfaq A, Aftab W, Manzoor H, Fahd R, Wajid I, Md. Aminul I, Vakhtang S, Muhammad A, Asmat U, Abdul W, Syeda RS, Shah S, Shahbaz K, Fahad S, Manzoor H, Saddam H, Wajid N (2017) Application of CSM-CROPGRO-Cotton model for cultivars and optimum planting dates: evaluation in changing semi-arid climate. Field Crops Res 238:139–152. http://dx.doi.org/10.1016/j.fcr.2017.07.007.

Hafiz MH, Abdul K, Farhat A, Wajid F, Fahad S, Muhammad A, Ghulam MS, Wajid N, Muhammad M, Hafiz FB (2020b) Comparative effects of organic and inorganic fertilizers on soil organic carbon and wheat productivity under arid region. Commun Soil Sci Plant Anal 51:1406–1422. https://doi.org/10.1080/00103624.2020.1763385.

Hafiz MH, Farhat A, Ashfaq A, Hafiz FB, Wajid F, Carol Jo W, Fahad S, Gerrit H (2020a) Predicting kernel growth of maize under controlled water and nitrogen applications. Int J Plant Prod 14:609–620. https://doi.org/10.1007/s42106-020-00110-8.

Hafiz MH, Farhat A, Shafqat S, Fahad S, Artemi C, Wajid F, Chaves CB, Wajid N, Muhammad M, Hafiz FB (2018) Offsetting land degradation through nitrogen and water management during maize cultivation under arid conditions. Land Degrad Dev 1–10. https://doi.org/10.1002/ldr.2933.

Hafiz MH, Muhammad A, Farhat A, Hafiz FB, Saeed AQ, Muhammad M, Fahad S, Muhammad A (2019) Environmental factors affecting the frequency of road traffic accidents: a case study of sub-urban area of Pakistan. Environ Sci Pollut Res 26:11674–11685. https://doi.org/10.1007/s11356-019-04752-8.

Hafiz MH, Wajid F, Farhat A, Fahad S, Shafqat S, Wajid N, Hafiz FB (2016) Maize plant nitrogen uptake dynamics at limited irrigation water and nitrogen. Environ Sci Pollut Res 24:2549–2557. https://doi.org/10.1007/s11356-016-8031-0.

Hamayun M, Khan SA, Khan AL, Shin J-H, Ahmad B, Shin D-H, Lee I-J (2010) Exogenous gibberellic acid reprograms soybean to higher growth and salt stress tolerance. J Agric Food Chem 58:7226–7232.

Hassanein RA, Hassanein AA, El-din AB, Salama M, Hashem HA (2009) Role of jasmonic acid and abscisic acid treatments in alleviating the adverse effects of drought stress and regulating trypsin inhibitor production in soybean plant. Aust J Basic Appl Sci 3:904–919.

Hayat R, Ali S, Amara U, Khalid R, Ahmed I (2010) Soil beneficial bacteria and their role in plant growth promotion: a review. Annal Microbiol 60:579–598.

Hayat S, Hasan SA, Fariduddin Q, Ahmad A (2008) Growth of tomato (*Lycopersicon esculentum*) in response to salicylic acid under water stress. J Plant Interact 3:297–304.

Hayat S, Hayat Q, Alyemeni MN, Ahmad A (2013) Proline enhances antioxidative enzyme activity, photosynthesis and yield of *Cicer arietinum* L. exposed to cadmium stress. Acta Bot Croat 72:323–335.

He Y, Liu Y, Cao W, Huai M, Xu B, Huang B (2005) Effects of salicylic acid on heat tolerance associated with antioxidant metabolism in kentucky bluegrass. Crop Sci 45:988–995. https://doi.org/10.2135/cropsci2003.0678.

Hemida KA, Eloufey AZ, Seif El-Yazal MA, Rady MM (2017) Integrated effect of potassium humate and α-tocopherol applications on soil characteristics and performance of *Phaseolus vulgaris* plants grown on a saline soil. Arch Agron Soil Sci 63:1556–1571.

Hesham FA, Fahad S (2020) Melatonin application enhances biochar efficiency for drought tolerance in maize varieties: modifications in physio-biochemical machinery. Agron J 1–22.

HongBo S, ZongSuo L, MingAn S (2005) Changes of anti-oxidative enzymes and MDA content under soil water deficits among 10 wheat (*Triticum aestivum* L.) genotypes at maturation stage. Colloids Surf B 45:7–13.

Hoque MA, Banu MNA, Nakamura Y, Shimoishi Y, Murata Y (2008) Proline and glycinebetaine enhance antioxidant defense and methylglyoxal detoxification systems and reduce NaCl-induced damage in cultured tobacco cells. J Plant Physiol 165:813–824.

Hoque MA, Banu MNA, Okuma E, Amako K, Nakamura Y, Shimoishi Y, Murata Y (2007) Exogenous proline and glycinebetaine increase NaCl-induced ascorbate–glutathione cycle enzyme activities, and proline improves salt tolerance more than glycinebetaine in tobacco bright yellow-2 suspension-cultured cells. J Plant Physiol 164:1457–1468.

Hossain MA, Fujita M (2010) Evidence for a role of exogenous glycinebetaine and proline in antioxidant defense and methylglyoxal detoxification systems in mung bean seedlings under salt stress. Physiol Mol Biol Plant 16:19–29. https://doi.org/10.1007/s12298-010-0003-0.

Hossain MA, Hasanuzzaman M, Fujita M (2010) Up-regulation of antioxidant and glyoxalase systems by exogenous glycinebetaine and proline in mung bean confer tolerance to cadmium stress. Physiol Mol Biol Plants 16:259–272.

Hussain MA, Fahad S, Rahat S, Muhammad FJ, Muhammad M, Qasid A, Ali A, Husain A, Nooral A, Babatope SA, Changbao S, Liya G, Ibrar A, Zhanmei J, Juncai H (2020) Multifunctional role of brassinosteroid and its analogues in plants. Plant Growth Regul 92:141–156. https://doi.org/10.1007/s10725-020-00647-8.

Hussein M, Alva A (2014) Effects of zinc and ascorbic acid application on the growth and photosynthetic pigments of millet plants grown under different salinity. Agric Sci 5:1253.

Hyun TK, Kumar K, Rao KP, Sinha AK, Roitsch T (2011) Role of α-tocopherol in cellular signaling: α-tocopherol inhibits stress-induced mitogen-activated protein kinase activation. Plant Biotechnol Rep 5:19–25.

Ibrahim MH, Jaafar HZ, Rahmat A, Rahman ZA (2012) Involvement of nitrogen on flavonoids, glutathione, anthocyanin, ascorbic acid and antioxidant activities of Malaysian medicinal plant *Labisia pumila* Blume (Kacip Fatimah). Int J Mol Sci 13:393–408.

Idrees M, Naeem M, Aftab T, Khan MMA (2011) Salicylic acid mitigates salinity stress by improving antioxidant defence system and enhances vincristine and vinblastine alkaloids production in periwinkle [*Catharanthus roseus* (L.) G. Don]. Acta Physiol Plant 33:987–999.

Ilyas M, Mohammad N, Nadeem K, Ali H, Aamir HK, Kashif H, Fahad S, Aziz K, Abid U (2020) Drought tolerance strategies in plants: a mechanistic approach. J Plant Growth Regul. https://doi.org/10.1007/s00344-020-10174-5.

Iqbal N, Ashraf Y, Ashraf M (2011) Modulation of endogenous levels of some key organic metabolites by exogenous application of glycine betaine in drought stressed plants of sunflower (*Helianthus annuus* L.). Plant Growth Regul 63:7–12.

Islam MM, Hoque MA, Okuma E, Banu MNA, Shimoishi Y, Nakamura Y, Murata Y (2009) Exogenous proline and glycinebetaine increase antioxidant enzyme activities and confer tolerance to cadmium stress in cultured tobacco cells. J Plant Physiol 166:1587–1597.

Jan M, Anwar-ul-Haq M, Adnan NS, Muhammad Y, Javaid I, Xiuling L, Depeng W, Fahad S (2019) Modulation in growth, gas exchange, and antioxidant activities of salt-stressed rice (*Oryza sativa* L.) genotypes by zinc fertilization. Arabian J Geosci 12:775. https://doi.org/10.1007/s12517-019-4939-2.

Janda T, Szalai G, Rios-Gonzalez K, Veisz O, Páldi E (2003) Comparative study of frost tolerance and antioxidant activity in cereals. Plant Sci 164:301–306.

Jie G, Liu G-s, Juan G, Zhang J (2008) Effects of vitamin E on the activities of protective enzymes and membrane lipid peroxidation in *Leymus chinensis* under drought stress. Chem Res Chinese Univ 24:80–83.

Jiménez-Bremont J, Becerra-Flora A, Hernández-Lucero E, Rodríguez-Kessler M, Acosta-Gallegos JA, Ramírez-Pimentel J (2006) Proline accumulation in two bean cultivars under salt stress and the effect of polyamines and ornithine. Biol Plant 50:763–766.

John J, Sarada S (2012) Role of phenolics in allelopathic interactions. Allelopathy J 29.

Kamaran M, Wenwen C, Irshad A, Xiangping M, Xudong Z, Wennan S, Junzhi C, Shakeel A, Fahad S, Qingfang H, Tiening L (2017) Effect of paclobutrazol, a potential growth regulator on stalk mechanical strength, lignin accumulation and its relation with lodging resistance of maize. Plant Growth Regul 84:317–332. https://doi.org/10.1007/s10725-017-0342-8.

Karabudak T, Bor M, Özdemir F, Türkan İ (2014) Glycine betaine protects tomato (*Solanum lycopersicum*) plants at low temperature by inducing fatty acid desaturase7 and lipoxygenase gene expression. Mol Biol Rep 41:1401–1410.

Karuppanapandian T, Moon J-C, Kim C, Manoharan K, Kim W (2011) Reactive oxygen species in plants: their generation, signal transduction, and scavenging mechanisms. Aust J Crop Sci 5:709.

Kaur G, Kumar S, Thakur P, Malik JA, Bhandhari K, Sharma KD, Nayyar H (2011) Involvement of proline in response of chickpea (*Cicer arietinum* L.) to chilling stress at reproductive stage. Sci Hortic 128:174–181. https://doi.org/10.1016/j.scienta.2011.01.037.

Kaya C, Tuna AL, Ashraf M, Altunlu H (2007) Improved salt tolerance of melon (*Cucumis melo* L.) by the addition of proline and potassium nitrate. Environ Exp Bot 60:397–403. https://doi.org/10.1016/j.envexpbot.2006.12.008.

Khan MIR, Asgher M, Khan NA (2014) Alleviation of salt-induced photosynthesis and growth inhibition by salicylic acid involves glycinebetaine and ethylene in mungbean (*Vigna radiata* L.). Plant Physiol Biochem 80:67–74.

Khan T, Mazid M, Mohammad F (2011) A review of ascorbic acid potentialities against oxidative stress induced in plants. J Agrobiol 28:97–111.

Kostopoulou Z, Therios I, Molassiotis A (2014) Resveratrol and its combination with α-tocopherol mediate salt adaptation in citrus seedlings. Plant Physiol Biochem 78:1–9.

Kumar S, Kaushal N, Nayyar H, Gaur P (2012) Abscisic acid induces heat tolerance in chickpea (*Cicer arietinum* L.) seedlings by facilitated accumulation of osmoprotectants. Acta Physiol Plant 34:1651–1658.

Kumar N, Pal M, Singh A, SaiRam RK, Srivastava GC (2010) Exogenous proline alleviates oxidative stress and increase vase life in rose (*Rosa hybrida* L. 'Grand Gala'). Sci Hortic 127:79–85. https://doi.org/10.1016/j.scienta.2010.09.009.

Kumar S, Singh R, Nayyar H (2013) α-Tocopherol application modulates the response of wheat (*Triticum aestivum* L.) seedlings to elevated temperatures by mitigation of stress injury and enhancement of antioxidants. J Plant Growth Regul 32:307–314. https://doi.org/10.1007/s00344-012-9299-z.

Kumar V, Yadav SK (2009) Proline and betaine provide protection to antioxidant and methylglyoxal detoxification systems during cold stress in *Camellia sinensis* (L.) O. Kuntze. Acta Physiol Plant 31:261–269.

Larkindale J, Hall JD, Knight MR, Vierling E (2005) Heat stress phenotypes of Arabidopsis mutants implicate multiple signaling pathways in the acquisition of thermotolerance. Plant Physiol 138:882–897.

Larkindale J, Huang B (2004) Thermotolerance and antioxidant systems in *Agrostis stolonifera*: involvement of salicylic acid, abscisic acid, calcium, hydrogen peroxide, and ethylene. J Plant Physiol 161:405–413.

Lee KH, LaRue TA (1992) Exogenous ethylene inhibits nodulation of *Pisum sativum* L. cv sparkle. Plant Physiol 100:1759–1763.

Li T, Hu Y, Du X, Tang H, Shen C, Wu J (2014) Salicylic acid alleviates the adverse effects of salt stress in *Torreya grandis* cv. Merrillii seedlings by activating photosynthesis and enhancing antioxidant systems. PLOS One 9:e109492.

Li J-T, Qiu Z-B, Zhang X-W, Wang L-S (2011) Exogenous hydrogen peroxide can enhance tolerance of wheat seedlings to salt stress. Acta Physiol Plant 33:835–842.

Lim C, Baek W, Jung J, Kim J-H, Lee S (2015) Function of ABA in stomatal defense against biotic and drought stresses. Int J Mol Sci 16:15251–15270.

Luo Y, Li F, Wang G, Yang X, Wang W (2010) Exogenously-supplied trehalose protects thylakoid membranes of winter wheat from heat-induced damage. Biol Plant 54:495–501.

Ma C, Wang Z, Kong B, Lin T (2013) Exogenous trehalose differentially modulate antioxidant defense system in wheat callus during water deficit and subsequent recovery. Plant Growth Regul 70:275–285.

Mafakheri A, Siosemardeh A, Bahramnejad B, Struik P, Sohrabi Y (2010) Effect of drought stress on yield, proline and chlorophyll contents in three chickpea cultivars. Aust J Crop Sci 4:580.

Majid A, Mohsen S, Mandana A, Saeid J-H, Ezatollah E, Fariborz S (2013) The effects of different levels of salinity and indole-3-acetic acid (IAA) on early growth and germination of wheat seedling. J Stress Physiol Biochem 9.

Maksymiec W, Wianowska D, Dawidowicz AL, Radkiewicz S, Mardarowicz M, Krupa Z (2005) The level of jasmonic acid in *Arabidopsis thaliana* and *Phaseolus coccineus* plants under heavy metal stress. J Plant Physiol 162:1338–1346.

Mansoor S, Naqvi FN (2012) Effect of gibberrelic acid on á-amylase activity in heat stressed mung bean (*Vigna radiata* L.) seedlings. Afr J Biotechnol 11:11414–11419.

Mayi AA, Ibrahim ZR, Abdurrahman AS (2014) Effect of foliar spray of humic acid, ascorbic acid, cultivars and their interactions on growth of olive (*Olea european* L.) transplants cvs. Khithairy and Sorany. J Agric VetSci 7:18–30.

Mazid M, Khan TA, Khan ZH, Quddusi S, Mohammad F (2011) Occurrence, biosynthesis and potentialities of ascorbic acid in plants. Int J Plant Anim Environ Sci 1:167–184.

Mehlhorn H (1990) Ethylene-promoted ascorbate peroxidase activity protects plants against hydrogen peroxide, ozone and paraquat. Plant Cell Environ 13:971–976.

Mohammed AR, Tarpley L (2011) Characterization of rice (*Oryza sativa* L.) physiological responses to a-tocopherol, glycine betaine or salicylic acid application. J Agric Sci 3:3.

Mostofa MG, Hossain MA, Fujita M (2015) Trehalose pretreatment induces salt tolerance in rice (*Oryza sativa* L.) seedlings: oxidative damage and co-induction of antioxidant defense and glyoxalase systems. Protoplasma 252:461–475.

Moustakas M, Sperdouli I, Kouna T, Antonopoulou C-I, Therios I (2011) Exogenous proline induces soluble sugar accumulation and alleviates drought stress effects on photosystem II functioning of Arabidopsis thaliana leaves. Plant Growth Regul 65:315.

Mubeen M, Ashfaq A, Hafiz MH, Muhammad A, Hafiz UF, Mazhar S, Muhammad Sami ul Din, Asad A, Amjed A, Fahad S, Wajid N (2020) Evaluating the climate change impact on water use efficiency of cotton-wheat in semi-arid conditions using DSSAT model. J Water Climate Chang 11:1661–1675. https://doi.org/10.2166/wcc.2019.179/622035/jwc2019179.pdf.

Muhammad Z, Abdul MK, Abdul MS, Kenneth BM, Muhammad S, Shahen S, Ibadullah J, Fahad S (2019) Performance of *Aeluropus lagopoides* (mangrove grass) ecotypes, a potential turfgrass, under high saline conditions. Environ Sci Pollut Res 26:13410–13421. https://doi.org/10.1007/s11356-019-04838-3.

Mukherjee S, Choudhuri M (1983) Implications of water stress-induced changes in the levels of endogenous ascorbic acid and hydrogen peroxide in Vigna seedlings. Physiol Plant 58:166–170.

Mukhtar A, Akram N, Ashraf M, Shafiq S, Aisha R (2016) Foliar-applied ascorbic acid enhances anti-oxidative potential and drought tolerance in cauliflower (*Brassica oleracea* L. var. Botrytis). Agrochimica 60:100–113.

Munir N, Naz S, Aslam F, Shahzadi K, Javad S (2013) Effect of various levels of ascorbic acid pretreatment on alleviation of salt stress in salt sensitive sugarcane genotype SPF-213. J Agric Res 51.

Munné-Bosch S, Alegre L (2002) The function of tocopherols and tocotrienols in plants. Crit Rev Plant Sci 21:31–57.

Nagesh Babu R, Devaraj V (2008) High temperature and salt stress response in French bean (*Phaseolus vulgaris*). Aust J Crop Sci 2:40–48.

Nahar K, Hasanuzzaman M, Alam M, Fujita M (2015) Roles of exogenous glutathione in antioxidant defense system and methylglyoxal detoxification during salt stress in mung bean. Biol Plant. 59:745–756.

Navalón A, Blanc R, Vilchez JL (1997) Determination of 1-naphthylacetic acid in commercial formulations and natural waters by solid-phase spectrofluorimetry. Microchim Acta 126:33–38.

Nawaz K, Ashraf M (2010) Exogenous application of glycinebetaine modulates activities of antioxidants in maize plants subjected to salt stress. J Agron Crop Sci 196:28–37.

Naz H, Akram NA, Ashraf M (2016) Impact of ascorbic acid on growth and some physiological attributes of cucumber (*Cucumis sativus*) plants under water-deficit conditions. Pak J Bot 48:877–883.

Nazar R, Iqbal N, Syeed S, Khan NA (2011) Salicylic acid alleviates decreases in photosynthesis under salt stress by enhancing nitrogen and sulfur assimilation and antioxidant metabolism differentially in two mungbean cultivars. J Plant Physiol 168:807–815. https://doi.org/10.1016/j.jplph.2010.11.001.

Noreen S, Akhter MS, Yaamin T, Arfan M (2018) The ameliorative effects of exogenously applied proline on physiological and biochemical parameters of wheat (*Triticum aestivum* L.) crop under copper stress condition. J Plant Interact 13:221–230.

Noreen S, Ashraf M, Hussain M, Jamil A (2009) Exogenous application of salicylic acid enhances anti-oxidative capacity in salt stressed sunflower (*Helianthus annuus* L.) plants. Pak J Bot 41:473–479.

Noreen S, Athar HUR, Ashraf M (2013) Interactive effects of watering regimes and exogenously applied osmoprotectants on earliness indices and leaf area index in cotton (*Gossypium hirsutum* L.) crop. Pak J Bot 45:1873–1881.

Noreen S, Fatima K, Athar H, Ahmad S, Hussain K (2017) Enhancement of physio-biochemical parameters of wheat through exogenous application of salicylic acid under drought stress. J Anim Plant Sci 27:153–163.

Nounjan N, Nghia PT, Theerakulpisut P (2012) Exogenous proline and trehalose promote recovery of rice seedlings from salt-stress and differentially modulate antioxidant enzymes and expression of related genes. J Plant Physiol 169:596–604.

Orabi SA, Abdelhamid MT (2016) Protective role of α-tocopherol on two Vicia faba cultivars against

seawater-induced lipid peroxidation by enhancing capacity of anti-oxidative system. J Saudi Soc Agric Sci 15:145–154. https://doi.org/10.1016/j.jssas.2014.09.001.
Ozden M, Demirel U, Kahraman A (2009) Effects of proline on antioxidant system in leaves of grapevine (*Vitis vinifera* L.) exposed to oxidative stress by H_2O_2. Sci Hortic 119:163–168. https://doi.org/10.1016/j.scienta.2008.07.031.
Poonam S, Kaur H, Geetika S (2013) Effect of jasmonic acid on photosynthetic pigments and stress markers in *Cajanus cajan* (L.) Millsp. seedlings under copper stress. Am J Plant Sci 4:817.
Pourcel L, Moulin M, Fitzpatrick TB (2013) Examining strategies to facilitate vitamin B1 biofortification of plants by genetic engineering. Front Plant Sci 4:160.
Qiu Z, Guo J, Zhu A, Zhang L, Zhang M (2014) Exogenous jasmonic acid can enhance tolerance of wheat seedlings to salt stress. Ecotoxicol Environ Saf 104:202–208.
Radhakrishnan R, Lee I-J (2013) Spermine promotes acclimation to osmotic stress by modifying antioxidant, abscisic acid, and jasmonic acid signals in soybean. J Plant Growth Regul 32:22–30.
Rady M, Sadak MS, El-Bassiouny H, El-Monem AA (2011) Alleviation the adverse effects of salinity stress in sunflower cultivars using nicotinamide and α-tocopherol. Aust J Basic Appl Sci 5:342–355.
Rahmawati N, Damanik R (2018) Effect of foliar application of α-tocopherol on vegetative growth and some biochemical constituents of two soybean genotypes under salt stress. In: IOP conference series: earth and environmental science, Vol. 1. IOP Publishing, UK, pp 012049.
Ramakrishna B, Rao SSR (2015) Foliar application of brassinosteroids alleviates adverse effects of zinc toxicity in radish (*Raphanus sativus* L.) plants. Protoplasma 252:665–677.
Raskin I, Ehmann A, Melander WR, Meeuse BJ (1987) Salicylic acid: a natural inducer of heat production in Arum lilies. Science 237:1601–1602.
Raza SH, Athar HR, Ashraf M, Hameed A (2007) Glycinebetaine-induced modulation of antioxidant enzymes activities and ion accumulation in two wheat cultivars differing in salt tolerance. Environ Exp Bot 60:368–376.
Razaji A, Asli D, Farzanian M (2012) The effects of seed priming with ascorbic acid on drought tolerance and some morphological and physiological characteristics of safflower (*Carthamus tinctorius* L.). Ann Biol Res 3:3984–3989.
Reddy KR, Henry WB, Seepaul R, Lokhande S, Gajanayake B, Brand D (2013) Exogenous application of glycinebetaine facilitates maize (*Zea mays* L.) growth under water deficit conditions. Am J Exp Agric 3:1.
Rehman M, Fahad S, Saleem MH, Hafeez M, Habib ur Rahman M, Liu F, Deng G (2020) Red light optimized physiological traits and enhanced the growth of ramie (*Boehmeria nivea* L.). Photosynthetica 58:922–931.
Reiahi N, Farahbakhsh H (2013) Ascorbate and drought stress effects on germination and seedling growth of sorghum. Int J Agro Plant Prod 4:901–910.
Rivas-San Vicente M, Plasencia J (2011) Salicylic acid beyond defence: its role in plant growth and development. J Exp Bot 62:3321–3338.
Rivero RM, Ruiz JM, Garcıa PC, Lopez-Lefebre LR, Sánchez E, Romero L (2001) Resistance to cold and heat stress: accumulation of phenolic compounds in tomato and watermelon plants. Plant Sci 160:315–321.
Sadak MS, Rady M, Badr N, Gaballah M (2010) Increasing sunflower salt tolerance using nicotinamide and α--tocopherol. Int J Acad Res 2.
Sadiq M, Akram NA, Ashraf M (2017) Foliar applications of alpha-tocopherol improves the composition of fresh pods of *Vigna radiata* subjected to water deficiency. Turk J Bot 41:244–252.
Sadiq M, Akram NA, Javed MT (2016) Alpha-tocopherol alters endogenous oxidative defense system in mung bean plants under water-deficit conditions. Pak J Bot 48:2177–2182.
Sajjad H, Muhammad M, Ashfaq A, Waseem A, Hafiz MH, Mazhar A, Nasir M, Asad A, Hafiz UF, Syeda RS, Fahad S, Depeng W, Wajid N (2019) Using GIS tools to detect the land use/land cover changes during forty years in Lodhran district of Pakistan. Environ Sci Pollut Res 27:39676–39692. https://doi.org/10.1007/s11356-019-06072-3.
Saleem MH, Fahad S, Adnan M, Mohsin A, Muhammad SR, Muhammad K, Qurban A, Inas AH, Parashuram B, Mubassir A, Reem MH (2020a) Foliar application of gibberellic acid endorsed phytoextraction of copper and alleviates oxidative stress in jute (*Corchorus capsularis* L.) plant grown in highly copper-contaminated soil of China. Environ Sci Pollution Res 27:37121–37133. https://doi.org/10.1007/s11356-020-09764-3.
Saleem MH, Fahad S, Shahid UK, Mairaj D, Abid U, Ayman ELS, Akbar H, Analía L, Lijun L (2020c)

Copper-induced oxidative stress, initiation of antioxidants and phytoremediation potential of flax (Linum usitatissimum L.) seedlings grown under the mixing of two different soils of China. Environ Sci Poll Res 27:5211–5221. https://doi.org/10.1007/s11356-019-07264-7.

Saleem MH, Rehman M, Fahad S, Tung SA, Iqbal N, Hassan A, Ayub A, Wahid MA, Shaukat S, Liu L, Deng G (2020b) Leaf gas exchange, oxidative stress, and physiological attributes of rapeseed (*Brassica napus* L.) grown under different light-emitting diodes. Photosynthetica 58:836–845.

Saleh AA (2007) Amelioration of chilling injuries in mung bean (*Vigna radiata* L.) seedlings by paclobutrazol, abscisic acid and hydrogen peroxide. Am J Plant Physiol 2:318–332.

Sasse JM (2003) Physiological actions of brassinosteroids: an update. J. Plant Growth Regul. 22:276–288.

Saud S, Chen Y, Fahad S, Hussain S, Na L, Xin L, Alhussien SA (2016) Silicate application increases the photosynthesis and its associated metabolic activities in Kentucky bluegrass under drought stress and post-drought recovery. Environ Sci Pollut Res 23:17647–17655. https://doi.org/10.1007/s11356-016-6957-x.

Saud S, Chen Y, Long B, Fahad S, Sadiq A (2013) The different impact on the growth of cool season turf grass under the various conditions on salinity and drought stress. Int J Agric Sci Res 3:77–84.

Saud S, Fahad S, Cui G, Chen Y, Anwar S (2020) Determining nitrogen isotopes discrimination under drought stress on enzymatic activities, nitrogen isotope abundance and water contents of Kentucky bluegrass. Sci Rep 10:6415. https://doi.org/10.1038/s41598-020-63548-w.

Saud S, Fahad S, Yajun C, Ihsan MZ, Hammad HM, Nasim W, Amanullah Jr, Arif M, Alharby H (2017) Effects of nitrogen supply on water stress and recovery mechanisms in kentucky bluegrass plants. Front. Plant Sci 8:983. https://doi.org/10.3389/fpls.2017.00983.

Saud S, Li X, Chen Y, Zhang L, Fahad S, Hussain S, Sadiq A, Chen Y (2014) Silicon application increases drought tolerance of Kentucky bluegrass by improving plant water relations and morph physiological functions. SciWorld J 2014:1–10. https://doi.org/10.1155/2014/368694.

Scheibler C (1869) Ueber das betain, eine im safte der zuckerrüben (*Beta vulgaris*) vorkommende pflanzenbase. Berichte der deutschen chemischen Gesellschaft 2:292–295.

Schutzendubel A, Polle A (2002) Plant responses to abiotic stresses: heavy metal-induced oxidative stress and protection by mycorrhization. J Exp Bot 53:1351–1365.

Semida WM, Abd El-Mageed TA, Howladar SM, Rady MM (2016) Foliar-applied alpha-tocopherol enhances salt-tolerance in onion plants by improving antioxidant defence system. Aust J Crop Sci 10:1030.

Semida W, Taha R, Abdelhamid M, Rady M (2014a) Foliar-applied α-tocopherol enhances salt-tolerance in *Vicia faba* L. plants grown under saline conditions. S Afr J Bot 95:24–31.

Sh SM (2014) Role of ascorbic acid and α tocopherol in alleviating salinity stress on flax plant (*Linum usitatissimum* L.). J Stress Physiol Biochem 10.

Shafi MI, Adnan M, Fahad S, Fazli W, Ahsan K, Zhen Y, Subhan D, Zafar-ul-Hye M, Martin B, Rahul D (2020) Application of single superphosphate with humic acid improves the growth, yield and phosphorus uptake of wheat (*Triticum aestivum* L.) in calcareous soil. Agron 10:1224. https://doi.org/10.3390/agronomy10091224.

Shafiq S, Akram NA, Ashraf M, Arshad A (2014) Synergistic effects of drought and ascorbic acid on growth, mineral nutrients and oxidative defense system in canola (*Brassica napus* L.) plants. Acta Physiol Plant 36:1539–1553.

Shah F, Lixiao N, Kehui C, Tariq S, Wei W, Chang C, Liyang Z, Farhan A, Fahad S, Huang J (2013) Rice grain yield and component responses to near 2°C of warming. Field Crop Res 157:98–110.

Shahbaz M, Noreen N, Perveen S (2013) Triacontanol modulates photosynthesis and osmoprotectants in canola (*Brassica napus* L.) under saline stress. J Plant Interac 8:350–359.

Sharma I, Ching E, Saini S, Bhardwaj R, Pati PK (2013) Exogenous application of brassinosteroid offers tolerance to salinity by altering stress responses in rice variety Pusa Basmati-1. Plant Physiol Biochem 69:17–26.

Sharma I, Pati PK, Bhardwaj R (2010) Regulation of growth and antioxidant enzyme activities by 28-homobrassinolide in seedlings of *Raphanus sativus* L. under cadmium stress. Indian J Biochem Biophys 47:172–177.

Shi Q, Bao Z, Zhu Z, Ying Q, Qian Q (2006) Effects of different treatments of salicylic acid on heat tolerance, chlorophyll fluorescence, and antioxidant enzyme activity in seedlings of *Cucumis sativa* L. Plant Growth Regul 48:127–135.

Singer MA, Lindquist S (1998) Multiple effects of trehalose on protein folding in vitro and in vivo. Mol Cell 1:639–648.

Singh I, Shah K (2014) Exogenous application of methyl jasmonate lowers the effect of cadmium-induced oxidative injury in rice seedlings. Phytochemistry 108:57–66.

Singh B, Usha K (2003) Salicylic acid induced physiological and biochemical changes in wheat seedlings under water stress. Plant Growth Regul 39:137–141.

Sirhindi G, Mir MA, Abd-Allah EF, Ahmad P, Gucel S (2016) Jasmonic acid modulates the physio-biochemical attributes, antioxidant enzyme activity, and gene expression in Glycine max under nickel toxicity. Front Plant Sci 7:591.

Slathia S, Sharma A, Choudhary SP (2012) Influence of exogenously applied epibrassinolide and putrescine on protein content, antioxidant enzymes and lipid peroxidation in *Lycopersicon esculentum* under salinity stress. Am J Plant Sci 3:714.

Smirnoff N, Wheeler GL (2000) Ascorbic acid in plants: biosynthesis and function. Crit Rev Plant Sci 19:267–290.

Sookwong P, Nakagawa K, Murata K, Kojima Y, Miyazawa T (2007) Quantitation of tocotrienol and tocopherol in various rice brans. J Agric Food Chem 55:461–466.

Subhan D, Zafar-ul-Hye M, Fahad S, Saud S, Martin B, Tereza H, Rahul D (2020) Drought stress alleviation by ACC deaminase producing *Achromobacter xylosoxidans* and *Enterobacter cloacae*, with and without timber waste biochar in maize. Sustain 12:6286. https://doi.org/10.3390/su12156286.

Syeed S, Anjum NA, Nazar R, Iqbal N, Masood A, Khan NA (2011) Salicylic acid-mediated changes in photosynthesis, nutrients content and antioxidant metabolism in two mustard (*Brassica juncea* L.) cultivars differing in salt tolerance. Acta Physiol Plant 33:877–886. https://doi.org/10.1007/s11738-010-0614-7.

Szarka A, Tomasskovics B, Bánhegyi G (2012) The ascorbate-glutathione-α-tocopherol triad in abiotic stress response. Int J Mol Sci 13:4458–4483.

Sun T-p, Gubler F (2004) Molecular mechanism of gibberellin signaling in plants. Annu Rev Plant Biol 55:197–223.

Taha SS, Mahmoud AWM, Rad MM (2017) Effect of exogenous α-tocopherol on sweet pepper plants irrigated by diluted sea water. J Agric Stud 6:25–46.

Tanou G, Molassiotis A, Diamantidis G (2009) Hydrogen peroxide-and nitric oxide-induced systemic antioxidant prime-like activity under NaCl-stress and stress-free conditions in citrus plants. J Plant Physiol 166:1904–1913.

Tariq M, Ahmad S, Fahad S, Abbas G, Hussain S, Fatima Z, Nasim W, Mubeen M, ur Rehman MH, Khan MA, Adnan M (2018) The impact of climate warming and crop management on phenology of sunflower-based cropping systems in Punjab, Pakistan. Agri and Forest Met 256:270–282.

Thevelein JM (1984) Regulation of trehalose mobilization in fungi. Microbiol Rev 48:42.

Uchendu EE, Leonard SW, Traber MG, Reed BM (2009) Vitamins C and E improve regrowth and reduce lipid peroxidation of blackberry shoot tips following cryopreservation. Plant Cell Rep 29:25. https://doi.org/10.1007/s00299-009-0795-y.

uz Zaman Q, Zubair A, Muhammad Y, Muhammad ZI, Abdul K, Fahad S, Safder B, Ramzani PMA, Muhammad N (2017) Zinc biofortification in rice: leveraging agriculture to moderate hidden hunger in developing countries. Arch Agron Soil Sci 64:147–161. https://doi.org/10.1080/03650340.2017.1338343.

Verma A, Malik CP, Gupta VK, Bajaj BK (2011) Effects of in vitro triacontanol on growth, antioxidant enzymes, and photosynthetic characteristics in *Arachis hypogaea* L. Braz J Plant Physiol 23:271–277.

Wahid F, Fahad S, Subhan D, Adnan M,, Zhen Y, Saud S, Manzer HS, Martin B, Tereza H, Rahul D (2020) Sustainable management with mycorrhizae and phosphate solubilizing bacteria for enhanced phosphorus uptake in calcareous soils. Agri 10:334. https://doi.org/10.3390/agriculture10080334.

Wahid A, Perveen M, Gelani S, Basra SM (2007) Pretreatment of seed with H2O2 improves salt tolerance of wheat seedlings by alleviation of oxidative damage and expression of stress proteins. J Plant Physiol 164:283–294.

Wajid N, Ashfaq A, Asad A, Muhammad T, Muhammad A, Muhammad S, Khawar J, Ghulam MS, Syeda RS, Hafiz MH, Muhammad IAR, Muhammad ZH, Habib ur Rahman M, Veysel T, Fahad S, Suad S, Aziz K, Shahzad A (2017) Radiation efficiency and nitrogen fertilizer impacts on sunflower crop in contrasting environments of Punjab. Pak Environ Sci Pollut Res 25:1822–1836. https://doi.org/10.1007/s11356-017-0592-z.

Wang G, Zhang X, Li F, Luo Y, Wang W (2010) Overaccumulation of glycine betaine enhances tolerance to drought and heat stress in wheat leaves in the protection of photosynthesis. Photosynthetica 48:117–126.

Wu C, Kehui C, She T, Ganghua L, Shaohua W, Fahad S, Lixiao N,Jianliang H, Shaobing P, Yanfeng D (2020) Intensified pollination and fertilization ameliorate heat injury in rice (*Oryza sativa* L.) during the flowering stage. Field Crops Res 252:107795.

Wu C, Tang S, Li G, Wang S, Fahad S, Ding Y (2019) Roles of phytohormone changes in the grain yield of rice plants exposed to heat: a review. PeerJ 7:e7792. https://doi.org/10.7717/peerj.7792.

Wu H, Wu X, Li Z, Duan L, Zhang M (2012) Physiological evaluation of drought stress tolerance and recovery in cauliflower (*Brassica oleracea* L.) seedlings treated with methyl jasmonate and coronatine. J Plant Growth Regul 31:113–123.

Wu X, Yao X, Chen J, Zhu Z, Zhang H, Zha D (2014) Brassinosteroids protect photosynthesis and antioxidant system of eggplant seedlings from high-temperature stress. Acta Physiol Plant 36:251–261.

Yakubu H, Izge A, Hussaini M, Jibrin J, Bello O, Isyaku M (2013) Varietal response and gibberellic acid concentrations on yield and yield traits of groundnut (*Arachis hypogaea* L.) under wet and dry conditions. Acad J Agric Res 1:001–008.

Yalpani N, Shulaev V, Raskin I (1993) Endogenous salicylic acid levels correlate with accumulation of pathogenesis-related proteins and virus resistance in tobacco. Phytopathology 83:702–708.

Yamada N, Promden W, Yamane K, Tamagake H, Hibino T, Tanaka Y, Takabe T (2009) Preferential accumulation of betaine uncoupled to choline monooxygenase in young leaves of sugar beet–importance of long-distance translocation of betaine under normal and salt-stressed conditions. J Plant Physiol 166:2058–2070.

Yang Z, Zhang Z, Zhang T, Fahad S, Cui K, Nie L, Peng S, Huang J (2017) The effect of season-long temperature increases on rice cultivars grown in the central and southern regions of China. Front Plant Sci 8:1908. https://doi.org/10.3389/fpls.2017.01908.

Yildirim E, Turan M, Guvenc I (2008) Effect of foliar salicylic acid applications on growth, chlorophyll, and mineral content of cucumber grown under salt stress. J Plant Nutr 31:593–612.

Yoon JY, Hamayun M, Lee S-K, Lee I-J (2009) Methyl jasmonate alleviated salinity stress in soybean. J Crop Sci Biotechnol 12:63–68.

Yu C-W, Murphy TM, Lin C-H (2003) Hydrogen peroxide-induced chilling tolerance in mung beans mediated through ABA-independent glutathione accumulation. Funct Plant Biol 30:955–963.

Zafar-ul-Hye M, Muhammad N, Subhan D, Fahad S, Rahul D, Mazhar A, Ashfaq AR, Martin B, Jiˇrí H, Zahid HT, Muhammad N (2020a) Alleviation of cadmium adverse effects by improving nutrients uptake in bitter gourd through cadmium tolerant rhizobacteria. Environ 7:54. https://doi.org/10.3390/environments7080054.

Zafar-ul-Hye M, Tahzeeb-ul-Hassan M, Muhammad A, Fahad S, Martin B, Tereza D, Rahul D, Subhan D (2020b) Potential role of compost mixed biochar with rhizobacteria in mitigating lead toxicity in spinach. Scientific Rep 10:12159. https://doi.org/10.1038/s41598-020-69183-9.

Zahida Z, Hafiz FB, Zulfiqar AS, Ghulam MS, Fahad S, Muhammad RA, Hafiz MH, Wajid N, Muhammad S (2017) Effect of water management and silicon on germination, growth, phosphorus and arsenic uptake in rice. Ecotoxicol Environ Saf 144:11–18.

Zingg J-M (2007) Vitamin E: an overview of major research directions. Mol Asp Med 28:400–422.

Zouari M, Elloumi N, Ahmed CB, Delmail D, Rouina BB, Abdallah FB, Labrousse P (2016) Exogenous proline enhances growth, mineral uptake, antioxidant defense, and reduces cadmium-induced oxidative damage in young date palm (*Phoenix dactylifera* L.). Ecol Eng 86:202–209. https://doi.org/10.1016/j.ecoleng.2015.11.016.

5

Influence of Environmental Adversities on Physiological Changes in Plants

Usman Khalid Chaudhry, Muhammad Daniyal Junaid, and Ali Fuat Gökçe
Department of Agricultural Genetic Engineering, Ayhan Şahenk Faculty of Agricultural Sciences and Technologies, Niğde Ömer Halisdemir University, Niğde, Turkey

5.1 Overview of Environmental Adversities

Plants as being immutable have to suffer from multifarious environmental adversities in field (Fahad and Bano 2012; Zlatev and Lidon 2012; Fahad et al. 2013; Fahad et al. 2014a,b; Fahad et al. 2015a,b; Fahad et al. 2016a,b,c,d; Fahad et al. 2017; Fahad et al. 2018; Fahad et al. 2019a,b; Hesham and Fahad 2020). Various abiotic stress factors are involved in curtailing growth of plants. The most dominant around the globe are drought, soil salinity, heat, cold, and heavy metal stress which are discussed below. Intensity of these adversities has been triggered mainly due to anthropogenic activities and poor management practices. Currently, it is the main challenge to maintain plant growth and crop productivity under such stress scenario for sustainable agriculture (Mittler 2006; Fahad and Bano 2012; Fahad et al. 2013; Fahad et al. 2014a,b; Fahad et al. 2015a,b; Fahad et al. 2016a,b,c,d; Fahad et al. 2017; Fahad et al. 2018; Fahad et al. 2019a,b; Hesham and Fahad 2020). To cope with these environmental stresses, the insight into plant acclimation to such stresses has been observed for considering morphological, physiological, biochemical, and molecular approaches (Anjum et al. 2011; Saud et al. 2013; Shah et al. 2013; Saud et al. 2014; Gurel et al. 2016; Hafiz et al. 2016; Saud et al. 2016; Aziz et al. 2017a,b; Habib ur Rahman et al. 2017; Kamaran et al. 2017; Saud et al. 2017; uz Zaman et al. 2017; Wajid et al. 2017; Yang et al. 2017; Zahida et al. 2017; Adnan et al. 2018a,b; Akram et al. 2018a,b; Depeng et al. 2018; Hafiz et al. 2018; Tariq et al. 2018; Adnan et al. 2019; Ahmad et al. 2019; Baseer et al. 2019; Hafiz et al. 2019; Muhammad et al. 2019; Sajjad et al. 2019; Wu et al. 2019; Adnan et al. 2020; Farhana et al. 2020; Farhat et al. 2020; Hafiz et al. 2020a,b; Hussain et al. 2020; Ilyas et al. 2020; Jan et al. 2020; Mubeen et al. 2020; Rehman et al. 2020; Saleem et al. 2020a,b,c; Saud et al. 2020; Shafi et al. 2020; Subhan et al. 2020; Wahid et al. 2020; Wu et al. 2020; Zafar-ul-Hye et al. 2020a,b).

All the aforementioned abiotic stresses deviate normal plant physiological processes, resulting in retarded growth and declined yield. Plants have inherent buffering capacity to alleviate the harmful effects of environmental adversities which assist plants to tolerate stress environments. Physiological changes occur during abiotic stresses in plants, and general physiological response specific to certain abiotic stress is illustrated in Figure 5.1.

5.1.1 Drought Stress

Agriculture is dependent on water supply for robust plant growth and higher yield returns. Global food production is severely influenced by drought stress (Fahad et al. 2017; Saud et al. 2013; Saud et al.

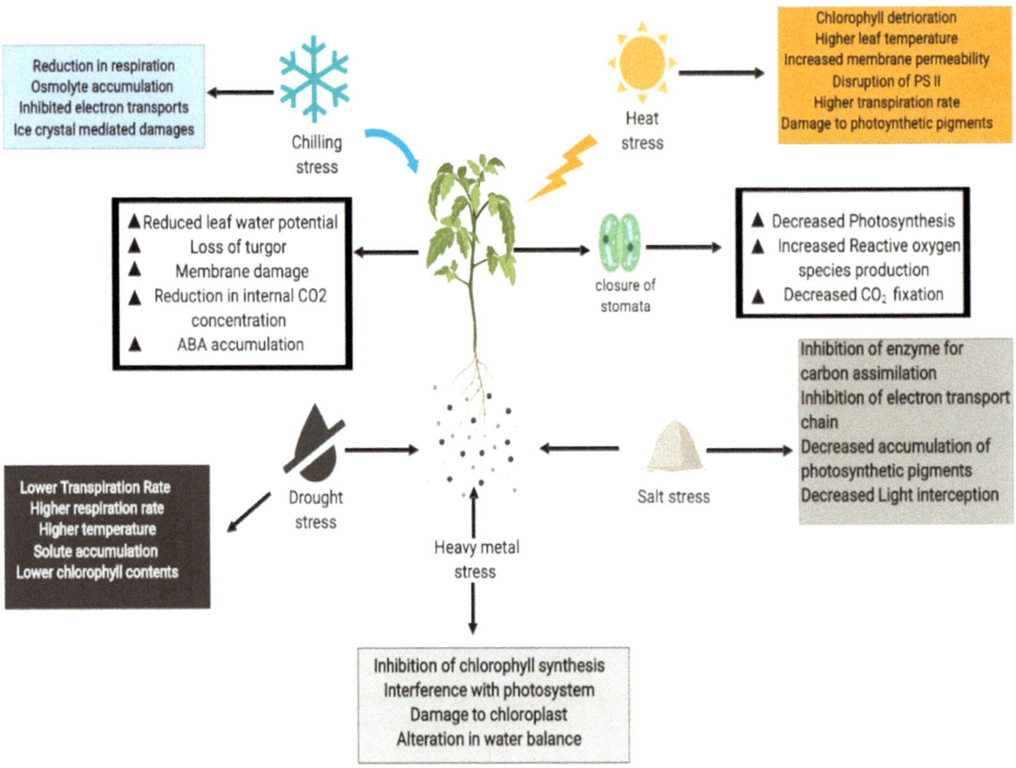

FIGURE 5.1 Major Abiotic Stresses Mediated Physiological Responses of Plant

2014; Saud et al. 2016; Saud et al. 2017). Drought can be categorized into three categories (agricultural, metrological, and hydrological drought). Agricultural drought is a condition of low soil moisture and air humidity with the rise in temperature (Saud et al. 2013; Saud et al. 2014; Saud et al. 2016; Shahzad et al. 2016; Fahad et al. 2017; Saud et al. 2017). Drought arises due to insufficient rainfall events or moisture availability in soil mostly in arid regions (Khattak et al. 2019). Drought is the most devastating stress for plants among other abiotic stresses. It limits plant growth and resultantly lower yield of crops (Aksoy et al. 2015). Plant faces drought stress when moisture availability to the root becomes restricted and transpiration rate decreases at a rapid rate. Drought is a versatile stress influencing plants at every growth stages. Drought problem is increasing at rapid rate due to several issues. Currently, it covers an area of 1–3%, and it is expected to increase up to 30% by 2090 (Kogan and Guo 2016). Drought causes several physiological changes in plants which include impaired photosynthetic machinery with decreased photosynthesis rate, decreased stomatal conductance, lower transpiration rate, and reduction in chlorophyll contents. Severe drought conditions cause drastic reduction in leaf water contents and higher leaf temperature (Farooq et al. 2009a; Blum 2017).

5.1.2 Root Signaling under Drought Stress

Root system has imperative role in coping dry soil conditions. Robust long roots seek available soil moisture shallow in soil layers, and extensive root growth has been observed under drought stress in search of water underneath soil (Schachtman and Goodger 2008). Roots play significant role by observing decrease in moisture availability, and they send signals to the upper plants in order to adapt plants' physiological processes to dry soil conditions (Jacobsen et al. 2009). Numerous hormones work as signalling molecules from roots to shoots such as abscisic acid, ethylene, and cytokinins. This complex signalling cascade results in closure of stomata which is crucial for limiting transpiration and converting plants to water-saving strategy (Liu et al. 2006). Abscisic acid promotes influx of K^+ ion from guard cells, which ultimately decreases turgor pressure and closure of stomata (Munemasa et al. 2015). Moreover, drought has direct influence on abscisic acid production as higher drought stress results in higher ABA level. Cytokinin is also an important signalling hormone that is apparently associated in response to drought stress, and its concentration is observed to be decreased in guard cells under drought stress (Sarwat and Tuteja 2017)

5.1.3 Influence of Drought Stress on Plant Photosynthesis

Drought stress influences photosynthesis of plant with a decline in gaseous exchange. It consequently results in reduction in photosynthesis machinery and oxidation of chloroplast lipids. Closure of stomata is the first response of plant against drought stress, which is source of CO_2 entry and transpires water, and both of them are essential for plant photosynthesis. Moreover, the CO_2 and photosynthetic carbon assimilation in leaves is decreased, which favours photorespiration (Zhang et al. 2016). Stoma of the leaf is the site of gaseous exchange in plants, and it is also attributed for loss of water via transpiration. The rate of transpiration is affected with stomatal conductance rate under drought stress conditions. It is well known that stomata close in response to water deficit to adapt plants to decrease water loss (Buckley 2019). The main plant hormone that transmits signals for stomatal closure is abscisic acid with the onset of mild water stress to shift plants to water-saving strategy before the significant decrease of available water contents (Huang et al. 2019). However, decrease in stomatal conductance is not a favourable trait under normal conditions as it limits plant growth. Leaf-relative water content is indicator of plant water stress, and it is reported that 70% decline in water contents irrevocably depressed photosynthesis (Liu et al. 2016). It has been reported in maize that reduction in stomatal conductance enhanced the water use efficiency of plant and productivity comparable to non-stressed plants (Kang et al. 2000).

5.1.4 Effects on Chlorophyll and Carotenoids Contents

Photosynthesis involves chlorophyll contents that are the main components of chloroplast having positive association with photosynthetic rate. Reduction in chlorophyll content is the symptom of oxidative stress causing degradation of chlorophyll and photo-oxidation of pigments. Photosynthetic pigments are essential for capturing light for photosynthesis to occur. The chlorophylls a and b both impart imperative role for plant photosynthesis to occur. Chlorophyll contents (a and b) are prone to drought conditions (Farooq et al. 2009a). Drought stress disrupts chlorophyll 'a' and 'b' contents and carotenoids. In some cases, it produces changes ratio of both chlorophyll contents as chlorophyll b content improved okra, while chlorophyll a content was unaffected, and consequently, enormous reduction of chlorophyll a:b ratio was observed against limited water regimes (Anjum et al. 2003). Majority of the studies conducted on chlorophyll loss found that its losses occur more in mesophyll cells as compared to bundle sheath cells. Carotenoids play part as antioxidant; however, they are prone to oxidative damages. Β-carotene exists in green plant part, which is solely bound to PSI and PSII. Drought stress interrupts its activity with enhanced production of ROS species in thylakoid membrane with reduction in carotenoids' concentration (Zakar et al. 2016; Trueba et al. 2019).

5.1.5 Water Relations

Plant–water relation is influenced chiefly by relative water contents (RWC) of leaves as it is observed that during drought conditions, it decreases and causes significant increase in leaf temperature. Water contents reflect status of water in plant during period of dry conditions and also provide information about dehydration tolerance. Relative water content (RWC) is associated with cell size, which depicts water balance of its supply to leaves and transpiration rate (Pour-Aboughadareh et al. 2017). It is relatively easy method to determine water contents in plants under drought stress and provides fruitful information about drought-tolerant and -susceptible genotypes. It is obvious that water contents continue to decrease under drought stress, but some plant genotypes have higher internal moisture status, thereby this characteristic makes them tolerant (Akram et al. 2013). Numerous crops have been reported to be drought tolerant that conferred higher RWC under drought stress such as wheat, rice, and soybean, and positive correlation regarding improved grain yield in wheat has been reported (Schonfeld et al. 1988; Cabuslay et al. 2002; Wang et al. 2006). RWC can be used as a selection criterion in mass selection for resistance among species with different intervals of drought stress (Ullah et al. 2017).

5.1.6 Osmotic Adjustment

In drought stress, plants maintain cell turgor by accumulating various organic solutes in cytosol which aids in lowering osmotic potential and protecting cellular machinery (Rhodes and Samaras 1994). Osmotic adjustment (OA) is achieved by accumulating organic solutes under water-deficit conditions. It is a useful trait for maintaining cellular turgor pressure when water potential of tissue declines. OA has exhibited to sustain stomatal conductance, likewise photosynthesis under drought stress conditions, ultimately delay in senescence of leaves, and better root growth to extract water from soil pockets (Jiang and Huang 2001). The aforementioned role of osmotic adjustment was observed in sorghum, wheat, and maize crops (Ludlow et al. 1990; Blum et al. 1999; Chimenti et al. 2006).

5.1.7 Cell Membrane Stability

Drought stress damages the cellular components with impaired cell membrane stability. Water-deficit conditions enhance membrane permeability and devastate cell membrane to effluence electrolytes. It can also be used as criteria for drought tolerance that how much damage has been occurred to cellular membranes by drought stress. A drought-tolerant maize cultivar exhibited higher cell membrane stability (Anjum et al. 2012; Hussain et al. 2019).

5.2 Salinity Stress

Soil salinity is major pressure for decreased productivity of crops (Haque 2006). Natural resources are becoming limited day by day due to population pressure and higher demand of food; thereby, in certain areas with least availability of good irrigation water and higher application of fertilizer for higher yield returns, farmers are unaware of long time losses of soil in the form of salinity that is increasing gradually and difficult to reclaim (Shahbaz and Ashraf 2013). Saline soil has following characteristics: electrical conductivity 4 dS m^{-1} and 15% exchangeable sodium that causes salinity stress for crop (Murtaza et al. 2006; Shahzad et al. 2017). Globally, almost all of the countries are suffering from this issue, and it covers the 3.1% of the total lands of the world (FAO 2019). Moreover, it is predicted that it will increase in number by the year 2050 (Ashraf and Akram 2009). Salt stress interrupts the physiological processes of plant (Zhang et al. 2014). Salinity stress raises the respiration rate of plant and mineral ion distribution, that is, displacement of calcium and potassium with Na^+ ion, ultimately ion toxicity (Kinraide, 1999).

5.2.1 Salinity Stress Damages to Photosynthetic Machinery

Salinity stress damages photosynthesis of plant with both short- and long-term exposure. Photosynthesis is mainly affected with a stomatal limitation due to salinity, resulting in decreased carbon assimilation (Fahad and Bano 2012). This can lead to rapid growth termination even with a couple of hours of salt stress exposure of plants (Fahad and Bano 2012; Wang et al. 2018). Continuous growth of plants under salty conditions accumulates salts in young growing leaves and reduces chlorophyll concentration (Duarte et al. 2013). Photosynthesis is essential for normal plant growth; however, due to salt stress, it is affected and decrease in rate has been observed with least cell growth of plant (Munns et al. 2006). Salt accumulation in plants limits CO_2 availability due to limited diffusion via stomata, mesophyll, and disrupting metabolism of photosynthesis (Flexas et al. 2007). Under mild salt stress, stomatal limitation occurs, which ultimately reduces photosynthesis. Severe salt stress conditions result in stomatal closure and disturbed activity of photosynthetic machinery. Salinity aggravates osmotic effect with a higher accumulation of abscisic acid (ABA) which also decreases stomatal conductance rate, concentration of CO_2, chlorophyll contents, and accumulation of sucrose with changes in electron transport. Excessive salts in photosynthetic tissues cause shrinkage of thylakoid membrane. Decreased stomatal conductance is beneficial for plants in order to restrict toxic ions delivery to the roots of plant. It has been highlighted that stomatal closure during the beginning of salt stress results in reduced flow of toxic ions to transpiration stream (Veselov et al. 2008; Vysotskaya et al. 2010). Stomatal signals are perceived by plants via roots with the accumulation/generation of ABA, which is a chief chemical compound in this regard. Reduced stomatal conductance is the adaptive strategy by plants to cope with saline conditions as well as decreasing salt load to leaves (Koyro et al. 2006). Some plants develop tolerance against salinity by closing stomata, to avoid water loss by transpiration. Reduction in stomatal conductance is directly proportional to photosynthesis. Concentration of sugar in mesophyll cells increases during plants' exposure to salt stress. Furthermore, Na^+ uptake by plant is increased under saline conditions, which limits the uptake of essential plant nutrients such as calcium, potassium, and manganese (Tuna et al. 2007). These ionic imbalances disrupt K^+ localization to chloroplast and damage photosystem II. It is relatively very sensitive component of photosynthetic machinery (Allakhverdiev et al. 2000). A remarkable decrease has been observed in effectiveness of photosystem II and CO_2 assimilation rate under the influence of salinity (Stepien and Klobus 2005).

5.2.2 Water Relations

Plant–water relations illuminate the behaviour of plants regarding cell hydration which has indispensable implications for physiological processes of plant. Water status of plant is an important physiological characteristic of plant under salinity stress (Romero-Aranda et al. 2001). Water potential, turgor, and solute potential are associated with each other in plant cell. They are all adversely affected after being exposed to salt stress conditions. Salt stress triggers osmotic stress with the reduction in uptake of water from roots. Roots are surrounded by excess Na^+ ions which reduce turgor potential and lead to stunted plant growth (Taiz and Zeiger 2002). Ultimately, it limits the supply to leaf, resulting in decreased water potential. The production of new leaves is dependent on water potential. Salts accumulate in higher concentration in older leaves as compared to younger leaves, so less damage to young growing leaves occurs during salinity stress (Munns et al. 2005).

5.2.3 Osmotic Adjustment

Osmotic stress is condition of deficit water contents in plant; however, it is also triggered with salt stress as it accumulates excessive sodium ions and causes dehydration with a loss of turgor leading to death of plant tissues (Munns 2002). The cellular and metabolic changes in plants caused due to salinity are similar as with exposure to drought stress. Under moderate salt concentration in soil, plants adjust themselves osmotically by maintaining influx of water. Compatible solutes accumulate due to salinity stress to confer osmotic tolerance. Osmotic processes are activated to assist in stomatal opening by the movement of osmolytes to enhance water movements between cells. This water movement hydrates the plant cell and absorbs more water through soil via roots.

5.3 Heat Stress

Climate is changing every year exponentially, which is causing high temperature/heat stress for plants and disrupts plants' functioning (Christensen and Christensen 2007; Fahad et al. 2016a,b,c,d; Fahad et al. 2018). Moreover, temperature is expected to increase by 2–3°C in the coming 25–45 years. Higher temperature will disturb rainfall pattern, especially in arid regions of the world. Carbon dioxide emission is increasing to trigger high temperature stress, as it is predicted to rise surface temperature from 1.4 to 5.8°C. Earth temperature is also increasing at a rapid rate, as it has been already doubled compared to 50 years ago (IPCC 2007). As the population is increasing and modernization started, industries are emitting harmful gases that also cause to raise temperature of the world. Currently, stress is the main hurdle for our planet and crop growth that needs to be addressed in order to ensure food security for burgeoning population of the world. Heat stress disrupts numerous physiological processes of plants.

5.3.1 Effects on Water Relations

Optimal leaf water content is essential to maintain cell turgidity and vital cell functions such as regulation of stomata, photosynthetic rate, and translocation of assimilates to plant parts. Rise in high temperature above than desired level affects plant water status which ultimately increases transpiration rate and loss of water from soil and plant as well. High temperature and cell water status are directly associated with each other as lower temperature favours normal growth with least transpiration rate and higher moisture contents in plant cells (Zhu et al. 2011; Machado and Paulsen 2001). Contrarily with limited water availability, water uptake by plants also becomes limited and higher temperature exacerbated the water potential of various physiological processes (Tsukaguchi et al. 2003).

5.3.2 Damages to Cell Membrane

Cell membrane is selectively permeable membrane structure surrounding the cytoplasm separating interior cell parts from outside environment. It helps in maintaining the integrity of cell for normal cell functioning. It allows the movement of selective substances in and out of the cell. High temperature causes lipid peroxidation in the membrane, which enhances the fluidity of thylakoid membrane and imparts negative influence on photosystem complex efficiency. If plants remain in a heat stress for a prolonged period, cyclic photophosphorylation rises to disperse excess energy, making photosystem II complex prone to damage (Schrader et al. 2004; Sharkey 2005; Rosli et al. 2017).

5.3.3 Effects on Photosynthesis

Photosynthesis is crucial for plant functions and sturdy plant growth. Photosynthesis utilizes light energy, water contents and carbon dioxide, and optimum temperature, and on the other hand, any deviation or excess of any of these factor results in stress to plant and impaired photosynthesis process. Heat stress exacerbates the photosynthesis process in plants (Salvucci and Crafts-Brandner 2004). In most of the crops, the optimum temperature is observed to be 28°C, while decrease in photosynthesis was reported with the increase in temperature (Snider et al. 2009). Higher temperature at vegetative growth stage damages the component of leaf photosynthesis, and reduction in CO_2 gain by leaf therefore reduced export of assimilates from leaves to developmental parts (Jiao et al. 1996). As with the exposure of every abiotic stress, the first plant response is stomatal closure, so same is under heat stress. Gaseous exchange becomes limited, and reduced CO_2 assimilation rate is associated with heat stress (Yang et al. 2006). Stomatal closure is the main reason of diminished photosynthesis (Ashraf and Hafeez 2004). Heat stress limits the stomatal conductance of leaves with a 60% decline in photosynthesis. Non-stomatal restrictions in photosynthesis have harmful effects on ribulose bisphosphate carboxylation (Greer and Weedon 2014). Leaf photosynthetic pigments and chlorophyll contents also reduce, resulting in decreased photosynthetic efficiency. Photosynthetic CO_2 fixation destructs photosystem II (PSII) and impairs mechanism of electron transport in thylakoid membrane (Ashraf and Harris 2013). It is highly prone to heat stress. Chief modifications in chloroplasts occur due to heat stress such as structural changes of thylakoids. Grana swelling and loss of grana stacking occur with wave of heat stress (Ashraf and Hafeez 2004; Yamamoto 2016). As PSII function is unstable component in electron transport (Havaux 1998), its inhibition results in enhanced chlorophyll fluorescence (Krause and Weis 1991); therefore, magnitude of heat stress-induced changes in photosynthesis mechanism is quantified on the basis of chlorophyll fluorescence (Dias et al. 2018).

5.3.4 Photorespiration

High light favours uptake of oxygen instead of CO_2, which is known as photorespiration. Increased temperature limits photosynthesis and increases photorespiration, adversely affecting plant growth and yield. Temperature is generally high during mid-day/afternoon, so plant experiences higher photorespiration. Carbon assimilation in plant species decreases resultant loss of carbohydrates (Elferjani and Soolanayakanahally 2018). During photorespiration, carbohydrates produced by photosynthesis get utilized for respiratory energy rather than to fulfil the need of plant developmental processes (Loka and Oosterhuis 2010). Furthermore, under heat stress conditions, plants are unable to accumulate and provide enough quantity of carbohydrates to match with the plants' needs. It has been documented that photorespiration significantly increased with a rise in temperature of 22–40°C. It depicts the negative role of higher temperature for reduced photosynthesis rate (Akter and Islam 2017; Dusenge et al. 2019).

5.4 Chilling Stress

Plants as being immobile have to suffer from any deviation in temperature for their normal growth and physiological changes. Fall in temperature also disrupts the plant functioning, growth, and yield. Chilling is a condition when temperature falls unconditionally, and it ranges from 0 to 15°C due to climatic changes going on around the globe. Tropical and subtropical crops are more prone to chilling stress as being unable to acclimatize for a prolonged period. However, temperate plants have the inherent ability to survive under such conditions (Riva-Roveda et al. 2016; Hussain et al. 2018a,b).

5.4.1 Influence on Plant Cell Membranes

A drop in temperature diminishes the fluidity of membranes, which is a symptom of cold stress in plants. Membrane damage occurs due to chilling stress, and changes in lipid composition of plasma membrane occur (Murata 1990). Moreover, chloroplast changes also occur in response to chilling, which ultimately destroys the thylakoid membrane, chlorophyll contents, and electron transport, and all these changes are interrelated with each other in response to stress conditions (Li et al. 2018).

5.4.2 Water Relations

Plant–water relations can be elucidated by several physiological traits which affect water contents/potential in plant cell, and the most important is change in temperature (Farooq et al. 2009b). Reduction in leaf water content was observed with a cold stress. Normally, chilling stress decreases the hydraulic water potential by roots to uptake water by plant, thereby forming lower water potential and loss of turgor pressure. As water is essential for several functions by plants, immediate decrease in water sends signals for stomatal closure with a subsequent wilting of plants (Aroca et al. 2003). Chilling stress imparts similar symptoms to water stress conditions as reduction of water contents in shoots as well due to lower water uptake by roots (Palta and Weiss 2018).

5.4.3 Photosynthesis and Stomatal Conductance

Photosynthesis is the important physiological trait that procures food for plant growth; however, under chilling stress, its rate decreases as chloroplast is the sole organelle that is severely affected during chilling stress. It is the prime site of photosynthesis and chlorophyll contents to capture light energy essential for photosynthesis. So, with a decrease in temperature, PSII imbalance also takes place with a damage to chlorophyll complexes (Fan et al. 2015). Chlorophyll (Chl) fluorescence is a tool for observing PSII efficiency (Adam and Murthy 2014). Stoma is the prime site of gaseous exchange between plant and the surrounding environment. It is essential for photosynthesis and transpiration rate in plants. Chilling stress alters the stomatal density, number of stomata, stomatal index, and epidermal cell size. Photosynthetic rate is associated with stomatal density (Hajihashemi et al. 2018), which imparts negative diminished influence on photosynthesis and photosynthetic capacity. It is attributed to poor mesophyll and stomatal conductance of carbon dioxide assimilation. Moreover, chilling causes excitation of thylakoid membrane, which triggers photo-inhibition and consequently impairs photosynthetic operations. Contrarily, chilling stress limits $NADP^+$ availability for electron transport chain (Scheibe et al. 2005).

5.5 Heavy Metal Stress

Soil parent material is the main source of heavy metals' accumulation in soil. Human activities and environmental pollution also contribute some of them, so it significantly affects the crop production grown on such types of soil (Alloway 2013). Burgeoning population is another important factor that causes contamination of urban soils by accumulating heavy metals from industrial waste, automobiles smoke, and poor dumping of municipal waste, and this type of soil is not suitable for any kind of farming (Xia et al. 2011). Industries are also growing day by day to benefit humans or speed up our work and give us good quality products within a short period of time. On the other hand, they are unaware of toxic heavy metal waste they are producing every day. Once it comes in contact with soil, it depletes soil either in the form of leaching, erosion to some other place, or plants' uptake (Dixit et al. 2015). Mostly heavy metals are toxic for human as well as agricultural lands even at minute concentrations (Blake and Goulding 2002; Shi et al. 2008; Kadi 2009), and industries normally emit Cadmium (Cd) in the form of their waste to the environment (Charlesworth et al. 2003). Heavy metals deposited to surface of soil somehow get leached and remain deposited there for thousands of years (Dijkstra et al. 2004). Heavy metal pollution problem is worse in the developing countries rather than developed countries. In these countries, they are not following proper waste management practices and generating huge volume of sewage water containing heavy metals (Qadir et al. 2010). Underdeveloped or developing countries around the globe, mostly in Africa, Latin America, and Asia, are suffering from these issues (Qadir et al. 2010), whereas some low-income countries are using treated waste water for irrigation (Al-Nakshabandi et al. 1997). Furthermore, farmers are not that much rich, and availability of clean water is also limited, so they irrigate their crops with this sewage water especially in the vicinity of urban areas.

5.5.1 Heavy Metal Toxicity Effects on Plant Photosynthesis

In plants, cell wall works as selectivity channel and metals pile up here before its dispersion to the plasma membrane (Dudev and Lim 2012). Heavy metals that enter plants in the form of ions through soil deposit in leaf cuticle, while in the form of gases, they gather in the stomata of leaves. Heavy metals that accumulate in stomata disrupt its photosynthetic machinery and photosynthetic pigments, consequently devastating effects leading to cell death (Maleva et al. 2012). Heavy metal stress effects on photosynthesis and chlorophyll contents have been thoroughly observed. Carbon dioxide fixation for photosynthesis by plants is sensitive to heavy metal stress. It affects the photosynthesis rate with a severe damage to electron transport chain and thylakoid membrane. Reduction in photosynthesis rate is due to the distortion of chloroplast structure and carotenoids. Heavy metal alters the lipid composition and swelling of thylakoid membranes. Decrease in chlorophyll content is documented with effect of cadmium and lead stress, and it affected chlorophyll b content more than chlorophyll due to inhibited chlorophyll-synthesizing enzymes (Zengin and Munzuroglu 2005). Apart from chlorophyll inhibition, cadmium restricts PSII-associated electron transports because of changes in functions of thylakoid membrane (Stiborova et al. 1987). Studies have reported a negative influence of heavy metals on reduced chlorophyll contents and photosynthesis rate. Copper exposure of 7 μM damaged the photosynthetic pigments in maize; however, similar inhibition was reported in cabbage with a exposure of $\geq$ 2 μM (Mocquot et al. 1996; Shahbaz et al. 2010).

5.5.2 Water Regime

Heavy metal accumulation in plant cells elucidated a great fall in water contents. It is because of the fact that plants' exposure to cadmium or lead retards the plant growth with a reduction is leaf area, which is the main transpiring plant organ. In certain circumstances, guard cells become smaller with the intake of heavy metal by plants to damage the stomata of leaves (Vassilev et al. 1998; Perfus-Barbeoch et al. 2002; Llamas et al. 2008).

5.6 Generation of Oxidative Stress

Multifarious abiotic stresses cause ionic imbalance that induces rapid formation of reactive oxygen species (ROS), which are produced with the onset of stress in various compartments of plant cell, but generally they are produced in chloroplast, peroxisome, and mitochondria (You and Chan 2015). Their sources include peroxides, cell wall peroxides, plasma membrane NADPH oxidase, and amine oxidase. Their production is also related to stomatal closure as it reduces carbon dioxide availability in leaves, which in turn inhibits fixation of carbon, and chloroplast dissipates unnecessary excitation energy that generates ROS (Ozgur et al. 2013). Severity of abiotic stress in the cytosol disrupts photosynthetic apparatus and decreases plants' ability to utilize light absorption by photosynthetic pigments. Moreover, abiotic stress causes water-deficit conditions or aggravates a drought stress condition that is also responsible for ROS formation in plants. The ROS that cause oxidative stress are singlet oxygen ($^1O2$), superoxide anions ($O2^{\cdot-}$), hydrogen peroxide (H_2O_2), and hydroxyl radicals ($^{\cdot}OH$) (Zheng et al. 2009). All these ROS cause devastating effects on plant functioning with damage to DNA, protein, inactivation of enzymes, and other essential constituents of plant cell (Foyer and Noctor 2005). However, oxidative stress in plants is alleviated by ROS scavenging antioxidant enzymes, superoxide dismutase (SOD) dismutases $O2^{\cdot-}$ into H_2O_2 and O_2. POD starts catalysing the H_2O_2 to H_2O and O_2 likely catalase (CAT) and ascorbate peroxidase (APX) capture H_2O_2 and convert in to H_2O (Hossain and Dietz 2016; Hussain et al. 2018). Antioxidant enzymes work as indicator of stress-tolerant and -susceptible cultivars. Increased production of ROS is obvious under any kind of stress which causes membrane damages, and cellular toxicity in numerous crops has been reported (Mittova et al. 2003; Ahmad et al. 2010a,b). Catalase was observed to render chilling tolerance after their exposure for short period of time, while there was no change in other scavenging enzyme in maize (Prasad 1997). Abiotic stress significantly increased lipid peroxidation in salt-sensitive wheat seedlings compared to salt-tolerant cultivar (Sairam et al. 2002), and further studies also confirmed raised level of lipid peroxidation and higher H_2O_2 due to abiotic stress in Brassica and wheat (Hossain et al. 2011; Alam et al. 2013; Hasanuzzaman et al. 2017a). Increase in antioxidant enzyme activities makes plant tolerant against certain abiotic stress, while susceptible shows lower activities as different enzyme activities is given in Table 5.1.

5.7 Role of Osmoprotectants

Osmoprotectants are classified into three groups, that is, compounds containing ammonium (polyamines, glycine betaine), containing amino acids (proline), and lastly containing sugars molecule (mannitol, sorbitol, fructan, and trehalose). In this chapter, we only discussed about the role of polyamines, proline, and glycine betaine. Osmoprotectants assist plants to survive under severe osmotic stress and have low-molecular-weight, neutral, and non-toxic highly soluble small molecule (Ahn et al. 2011; Lang 2007). They have the ability in stabilizing damaged membranes and protein to reduce osmotic stress and resultantly mitigate excessive dehydration of plant cells (Wani et al. 2013). Osmoprotectants accumulate in cytosol of cell for maintaining osmotic balance. Osmoprotectants confer adaptation to various harmful abiotic stress conditions with the increase in osmotic pressure in cytoplasm (Table 5.2). They render numerous positive roles in plants against abiotic stresses as their exogenous application also increases their concentration in plants which is very fruitful approach for enhancing tolerance which is given in Table 5.3.

5.7.1 Polyamines

Polyamines are positively charged low-molecular-weight compounds that regulate pH of cellular components at physiological level (Groppa and Benavides 2008). They are widely distributed in plants

TABLE 5.1
Effect of Abiotic Stresses on Antioxidant Enzymes' Activities in Various Plant Species

Plant species	Stress treatments	Enzyme activation	References
Pea	70 mM NaCl	Mn-SOD activity decreased	Hernández et al. (1993)
Pea	110 mM NaCl	APX and Mn-SOD activity increased	Gomez et al. (1999)
Tomato	100 mM	Mn-SOD and GR activity decreased	Mittova et al. (2003)
Durum wheat	50 mM	APX activity increased	Meneguzzo et al. (1998)
Bread wheat	250–500 mM	SOD activity increased	Mutlu et al. (2009)
Arabidopsis	100 mM	SOD and CAT activity increased	Leterrier et al. (2012)
Rice	100 mM	CAT and APX activity increased	Lee et al. (2013)
Bean	200 mM	APX, CAT, and GR activity increased	Taïbi et al. (2016)
Wheat	100 mM	SOD, CAT, and GR activity increased	Sairam et al. (2002)
Maize	150 mM	SOD activity increased	AbdElgawad et al. (2016)
Potato	75 mM	SOD, POD, CAT, and APX activity increased	Rahnama and Ebrahimzadeh (2005)
Potato	40% field capacity	CAT activity increased	Farhad et al. (2011)
Tomato	Water stress 35–40% field capacity	SOD, POD, and CAT activity increased	Yuan et al. (2016)
Tomato	Water stress 50% field capacity	SOD and CAT activity increased	Sánchez-Rodríguez et al. (2016)
Maize	–0.7 MPa	12 h SOD and CAT increased, 24 h APX and GR activity increased	Jiang and Zhang (2002)
Olive	30 mg $CdCl_2$ kg^{-1} soil	SOD, CAT, and GPX activity increased	Zouari et al. (2016)
Rice	500 μM Cd	SOD and CAT activity increased	Shah et al. (2001)
Rice	1000 μM Pb	CAT activity decreased, while APX and GR activity decreased	Verma and Dubey (2003)
Bean	5 μM Cd	CAT activity decreased, and APX activity increased	Chaoui et al. (1997)
Wheat	10 and 100 mM $ZnCl_2$	CAT, SOD, and GPX activity decreased	Panda et al. (2003)
Rice	40°C	SOD, POD, CAT, and APX activity increased	Zhang et al. (2017)
Cucumber	15°C	SOD and GPX activity increased	Xu et al. (2008)
Rice	8°C	SOD activity decreased	Huang and Guo (2005)

for conducting essential functions for cell growth. They confer vital role for crucial plant phases during stress conditions such as cell proliferation, development, signal regulation, and expression against certain abiotic stresses (Kusano et al. 2007). Numerous studies exhibited the positive role of polyamines' accumulation under abiotic stress, as it has potential to alleviate harmful effects of environmental adversities (Liu et al. 2007). Moreover, exogenous application of polyamines also conferred better growth and imparted protection against salinity and drought stress (Yamaguchi et al. 2007; Kuznetsov and Shevyakova 2010; Kubis et al. 2014).

5.7.2 Proline

Proline has indispensable role for plants against stress conditions as it is the imperative osmolyte that normally accumulates in cytosol to protect membrane structures, enzymes, and protein from damages from abiotic stresses. It stabilizes the protein and cell membrane and works as a scavenger against toxic derivatives of oxygen, and other important function is buffering molecule for cellular redox potential (Hoque et al. 2008). Proline plays role as signalling molecule in modifying mitochondrial functions,

TABLE 5.2

Importance of Osmoprotectants against Abiotic Stresses

Osmoprotectants	Role of osmoprotectants	References
Polyamine	It plays role in scavenging free radicals under abiotic stresses and improves plant functions	Kubis et al. (2014)
Glycine betaine	It helps in osmotic adjustment, protects thylakoid membrane, and enhances photosynthetic efficiency	Ashraf and Foolad (2007)
Proline	Its accumulation results in numerous benefits against abiotic stresses. It protects cell membranes, proteins, and enzymes, provides buffering, maintaining cellular redox potential, and modification of mitochondrial functions scavenging free radicals, and stimulates stress gene expression	Islam et al. (2009) and Anjum et al. (2011)

TABLE 5.3

Exogenous Application of Osmoprotectants Towards Abiotic Stresses

Plant species	Abiotic stresses	Effect of exogenous application	References
Tobacco	Drought	Glycine betaine alleviated water stress and improved growth	Agboma et al. (1997)
Soybean	Drought		
Wheat	Drought	Proline-treated seeds alleviated drought stress	Kamran et al. (2009)
Maize	Drought	Proline improved growth under water-deficit regime	Ali et al. (2007)
Tomato	Salt	Glycine betaine improved growth	Makela et al. (1998)
Rice	Salt	Glycine betaine assisted in reducing NaCl accumulation in plants	Lutts (2000)
Alfalfa	Salt	Proline mitigated salinity stress	Ehsanpour and Fatahian (2003)
Onion	Salt	Proline and glycine betaine alleviated NaCl effect and disruption to cell membrane	Mansour (1998)
Arabidopsis	Chilling	Glycine betaine enhanced freezing tolerance by increasing temperature from	WeiBing and Rajashekar (2001)
Potato	Chilling	Glycine betaine improved the growth	Somersalo et al. (1996)
Chickpea	Chilling	Proline enhanced level of cryoprotectants and showed better chilling tolerance	Kaur et al. (2011)
Tobacco	Heavy metal	Proline and glycine betaine enhanced antioxidant enzyme activity and render tolerance	Islam et al. (2009)
Sunflower	Heavy metal	Polyamines improved cadmium and copper tolerance	Groppa et al. (2001)
Wheat	Heavy metal	Proline and glycine betaine alleviated cadmium toxicity	Rasheed et al. (2014)
Chickpea	Heavy metal	Proline improved antioxidant enzyme activity and enhanced photosynthesis rate	Hayat et al. (2013)
Chickpea	High temperature	Proline protected essential enzymes and exhibited tolerance	Kaushal et al. (2011)
Barley	High temperature	Proline and glycine betaine mitigated adverse effects of heat stress	Oukarroum et al. (2012)

affects proliferation of cell, or sometimes leads to cell death (Szabados and Savoure 2010). Its accumulation in cytoplasm encourages water uptake and diminishes toxic ion and improvement in redox homeostasis under abiotic stresses (Ashraf and Foolad 2007; Hoque et al. 2008). Higher concentration of proline accumulation in proline is correlated with stress tolerance, while susceptible cultivars exhibit lower concentration of proline. Moreover, it causes solvation of protein and maintains integrity of membrane under abiotic stresses (Özdemir et al. 2004). Proline assists in stabilization of subcellular structures for scavenging free radicals and buffers cellular redox potential (Ashraf and Foolad 2007). Its individual role for certain abiotic stress has been observed as it decreased Na and Cl accumulation in plant cells under salinity stress conditions, and on the other hand, it plays role in hydration of plant tissues against drought stress (Ashraf and Foolad 2007; Khattab and Afifi 2009). Its exogenous application is also beneficial to increase the inherent level of plant and higher growth, and improved antioxidant defence system has been reported for several crops (Nounjana et al. 2012; Ranganayakulu et al. 2013; Shahbaz and Ashraf 2013).

5.7.3 Glycine Betaine

It is the chief osmoprotectant with a highest potential to alleviate harmful effects of various abiotic stresses. It accumulates in chloroplast of plant in response to stress conditions. Glycine betaine enhances flow of water towards cell in order to maintain osmotic equilibrium and simultaneously work as signalling molecule for stress (Ashraf and Foolad 2007; Ranganayakulu et al. 2013). Major role of glycine betaine as osmoprotectant is in maintaining turgescent cell and water potential regulation for plant–water relations like osmotic adjustment by discriminating Na^+/K^+ and triggering defence enzymes which ultimately render tolerance to plant. Numerous successful studies have been published for its positive role against major abiotic stresses; moreover, its exogenous application is also helpful to some extent in mitigating harsh conditions (Ranganayakulu et al. 2013; Sawahel 2004).

5.8 Genetic Approaches for Improving Physiological Processes against Environmental Stresses

Different types of abiotic stresses, for example, drought, salinity, extreme high and low temperatures, and oxidative stresses, directly affect crop plants and cause high loss in production, and in future, due to rapidly changing climate, these abiotic stresses might be boosted (Marco et al. 2015). Heat and drought play major role in the reduction of yield, and due to current environmental impacts, there is chance of rise in severity and existence of these factors (Lamaoui et al. 2018). As world population is increasing day by day, breeding improved quality crops having tolerance against these stresses is important. Conventional breeding shows little bit success because of the complexity of stress traits; hence, scientists have started using transgenic approaches to overcome these environmental stresses (Wang et al. 2016).

5.8.1 Transgenic Approaches against Stress Damages

It is necessary to develop plant species and varieties which show tolerance against stress, and better understanding of the mechanism of stress damages is needed. With the help of biotechnology tools, scientists can introduce desired stress tolerance traits into plant genomes. Abiotic stress-tolerant plants can be obtained by targeting phytohormones genetically. Transcription factors involved in the ABA pathway can be overexpressed to improve osmotic stress tolerance in genetically modified plants (Gao et al. 2011). In the same way, genetically modified plants having overexpressed stress inducer RD 26 show increasing sensitivity and hyperregulation of ABA and other genes that respond under stress conditions (Fujita et al. 2004). The yields of genetically modified canola are higher during period of flowering comparatively and control plants under the modest drought conditions, having overexpressed farnesyltransferase. Modifying glyoxalase pathway shows an increased tolerance in various plant species against abiotic stress. Upregulation of glyoxalase I and II and overexpression in plants discovered increased tolerance against several stresses such as extreme temperatures, drought, salinity, and metal toxicity (Hasanuzzaman et al. 2017b).

5.8.2 Engineered Enzymes for Protective Metabolites Play Role against Stress

Plants use common physiological mechanism under several stresses, and that physiological mechanism contains the gathering of organic compounds having low molecular weight, and these compounds are highly water loving and even show no toxicity even when stored at high concentrations in plants. Those compounds are usually known as osmoprotectants and play roles in plant protection against stress (Chen and Murata 2002). A large number of metabolites are discovered as osmoprotectants in which different types of sugars, sugar alcohols, and several amino acids are found. These osmoprotectants overproduced by genetically engineering used increase tolerance in plants against abiotic stresses (Bhatnagar-Mathur et al. 2008).

5.8.3 Transcription Factors

Transcription factors are necessary associates of signalling pathways that involve changes of expression of genes; these factors represent an important part of the plant genome in Arabidopsis genome 1500 transcription factors found (Ratcliffe and Riechmann 2002). These transcription factors play role in adaptation of plants in different environments (Shiu et al. 2005). So, TFs are known as one of the best targets for the generation of plants having tolerance against stress. One transcription factor controls expression of several genes for downstreaming of several genes, and one of the best options is engineered transcription factors. A transcription factor controls several genes, and a specific union of these factors formed a transcription unit known as regulon. Quite a few regulons have been defined in plants that show response towards abiotic stresses. DREB regulon is involved in cold stress response (Nakashima and Yamaguchi-Shinozaki 2006). So, these factors have been used as specific targets to produce engineered plants that show tolerance against several stresses such as cold and drought. Thus, producing transgenic lines under constitutive promoter having overexpressed DREB genes showed tolerance against drought, cold, and salt stress (Kasuga et al. 1999). This DREB gene is found in several plant species and has shown increased tolerance against abiotic stress, but their ectopic expression usually shows plants' dwarf phenotypically (Agarwal et al. 2013). Role of transcription factors against abiotic stress and regulation of gene expression is depicted in Figure 5.2.

5.8.4 Role of miRNAs

One of thc most studied classes of non-coding small RNAs is miRNAs, and these are stored and produced in plants. These regulate the expression of genes in plants by interfering in the translation of targets or by causing degradation (Sunkar 2010). In plants, miRNAs have several biological functions and play its part in development and growth along with vital towards abiotic and biotic stress responses (Khraiwesh et al. 2012). Remarkably, genes of transcription factor are the main targets for numerous stress-related miRNAs, and this makes them first choice for new transgene approaches for altering the response of stress, one study demonstrate that over expression of miRNA in tomato play role in enhancing tolerance towards drought stress (Zhang et al. 2011) in rice over expression of miRNAs shows resistance to cold stress (Yang et al. 2013).

5.9 Conclusion

Various abiotic factors are involved in the reduction of plant growth. These adversaries include drought, salinity, heat, cold, and heavy metal stress, and their effect on plant growth and physiology has been discussed above. Anthropogenic factors increase the potential of these abiotic stresses. Plant physiological mechanisms, that is, photosynthesis, metabolism activities, respiration, and cell membrane activities, are highly influenced by these abiotic stresses, ultimately resulting into reduction of economic yields. Plant species react variably against these conditions and show a range of tolerance on exposure to the aforementioned adversaries. Climate change has accelerated the incidence of hazards to plants by triggering abiotic stresses, and among all the abiotic stresses, the heat and drought stresses are of prime importance. They are responsible for huge reductions in plant production and also becoming more severe with the increase in global temperatures. To develop sustainable agricultural production with the exponential increase in population, it is the duty of geneticists and breeders to develop such cultivars that are more tolerant to the changes in climate. Conventional breeding has enabled the world to get

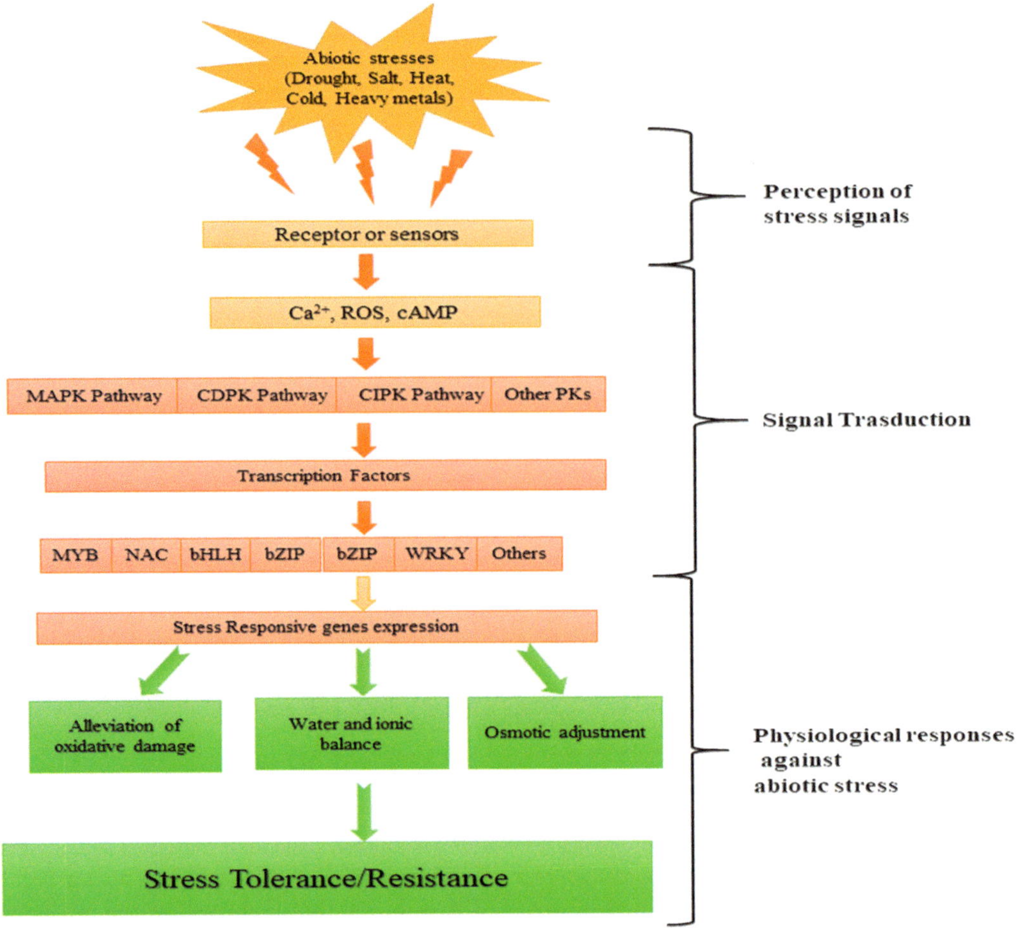

FIGURE 5.2 Transcription Factors Modulating Stress Signals and Regulating Physiological Changes to Mitigate Harmful Effects of Abiotic Stresses

tremendous yields with quality of different plant species over the years. In this era of technology, coupling conventional techniques with modern approaches is a useful tool to face climate change. Scientists have also started to use transgenic approaches in order to face climatic adversaries. Engineering of crops enzymatically and genetically for physiological improvements can enable plant species to resist climate changes.

REFERENCES

Abd Elgawad H, Zinta G, Hegab MM, Pandey R, Asard H, Abuelsoud W (2016) High salinity induces different oxidative stress and antioxidant responses in maize seedlings organs. Front Plant Sci 7:276.

Adam S, Murthy SDS (2014) Effect of cold stress on photosynthesis of plants and possible protection mechanisms. In: Approaches to plant stress and their management. Springer, New Delhi, pp 219–226.

Adnan M, Fahad S, Khan IA, Saeed M, Ihsan MZ, Saud S, Riaz M, Wang D, Wu C (2019) Integration of poultry manure and phosphate solubilizing bacteria improved availability of Ca bound P in calcareous soils. 3 Biotech 9:368.

Adnan M, Fahad S, Muhammad Z, Shahen S, Ishaq AM, Subhan D, Zafar-ul-Hye M, Martin LB, Raja MMN, Beena S, Saud S, Imran A, Zhen Y, Martin B, Jiri H, Rahul D (2020) Coupling phosphate-solubilizing bacteria with phosphorus supplements improve maize phosphorus acquisition and growth under lime induced salinity stress. Plants 9:900. https://doi.org/10.3390/plants9070900.

Adnan M, Shah Z, Sharif M, Rahman H (2018b) Liming induces carbon dioxide (CO_2) emission in PSB inoculated alkaline soil supplemented with different phosphorus sources. Environ Sci Poll Res 25:9501–9509.

Adnan M, Zahir S, Fahad S, Arif M, Mukhtar A, Imtiaz AK, Ishaq AM, Abdul B, Hidayat U, Muhammad A, Inayat-Ur R, Saud S, Muhammad ZI, Yousaf J, Amanullah, Hafiz MH, Wajid N (2018a) Phosphate-solubilizing bacteria nullify the antagonistic effect of soil calcification on bioavailability of phosphorus in alkaline soils. Sci Rep 8:4339. https://doi.org/10.1038/s41598-018-22653-7.

Agarwal P, Shukla P, Gupta K, Jha B (2013) Bioengineering for salinity tolerance in plants: state of the art. Mol Biotech 54:102–123.

Agboma PC, Peltonen-Sainio P, Hinkkanen R, Pehu E (1997) Effect of foliar application of glycinebetaine on yield components of drought-stressed tobacco plants. Exp Agric 33:345–352.

Ahmad P, Jaleel CA, Sharma S (2010a) Antioxidant defense system, lipid peroxidation, proline-metabolizing enzymes, and biochemical activities in two *Morus alba* genotypes subjected to NaCl stress. Russ J Plant Physiol 57:509–517.

Ahmad R, Kim YH, Kim MD, Kwon SY, Cho K, Lee HS, Kwak SS (2010b) Simultaneous expression of choline oxidase, superoxide dismutase and ascorbate peroxidase in potato plant chloroplasts provides synergistically enhanced protection against various abiotic stresses. Physiol Plant 138:520–533.

Ahmad S, Kamran M, Ding R, Meng X, Wang H, Ahmad I, Fahad S, Han Q (2019) Exogenous melatonin confers drought stress by promoting plant growth, photosynthetic capacity and antioxidant defense system of maize seedlings. PeerJ 7:e7793. http://doi.org/10.7717/peerj.7793.

Ahn C, Park U, Park PB (2011) Increased salt and drought tolerance by D-ononitol production in transgenic Arabidopsis thaliana. Biochem Biophys Res Commun 415:669–674.

Akram HM, Ali A, Sattar A, Rehman HSU, Bibi A (2013) Impact of water deficit stress on various physiological and agronomic traits of three basmati rice (*Oryza sativa* L.) cultivars. J Anim Plant Sci 23:1415–1423.

Akram R, Turan V, Hammad HM, Ahmad S, Hussain S, Hasnain A, Maqbool MM, Rehmani MIA, Rasool A, Masood N, Mahmood F, Mubeen M, Sultana SR, Fahad S, Amanet K, Saleem M, Abbas Y, Akhtar HM, Waseem F, Murtaza R, Amin A, Zahoor SA, ul Din MS, Nasim W (2018a) Fate of organic and inorganic pollutants in paddy soils. In: Hashmi MZ, Varma A (eds) Environmental pollution of paddy soils, soil biology. Springer International Publishing Ag, Switzerland, pp 197–214.

Akram R, Turan V, Wahid A, Ijaz M, Shahid MA, Kaleem S, Hafeez A, Maqbool MM, Chaudhary HJ, Munis, MFH, Mubeen M, Sadiq N, Murtaza R, Kazmi DH, Ali S, Khan N, Sultana SR, Fahad S, Amin A, Nasim W (2018b) Paddy land pollutants and their role in climate change. In: Hashmi MZ, Varma A (ed) Environmental pollution of paddy soils, soil biology. Springer International Publishing Ag Switzerland, pp 113–124.

Aksoy E, Demirel U, Öztürk ZN, Çalişkan S, Çalişkan ME (2015) Recent advances in potato genomics, transcriptomics, and transgenics under drought and heat stresses: a review. Turk J Bot 39:920–940.

Akter N, Islam MR (2017) Heat stress effects and management in wheat: a review. Agron Sustain Dev 37:37.

Al-Nakshabandi GA, Saqqar MM, Shatanawi MR, Fayyad M, Al-Horani H (1997) Some environmental problems associated with the use of treated wastewater for irrigation in Jordan. Agric Water Manage 34:81–94.

Alam MM, Hasanuzzaman M, Nahar K, Fujita M (2013) Exogenous salicylic acid ameliorates short-term drought stress in mustard (*Brassica juncea* L.) seedlings by upregulating the antioxidant defense and glyoxalase system. Aust J Crop Sci 7:1053.

Ali Q, Ashraf M, Athar HUR (2007) Exogenously applied proline at different growth stages enhances growth of two maize cultivars grown under water deficit conditions. Pak J Bot 39:1133–1144.

Allakhverdiev SI, Sakamoto A, Nishiyama Y, Inaba M, Murata N (2000) Ionic and osmotic effects of NaCl-induced inactivation of photosystems I and II in *Synechococcus* sp. Plant Physiol 123:1047–1056.

Alloway BJ (2013) Sources of heavy metals and metalloids in soils. In: Heavy metals in soils. Springer, Dordrecht, pp 11–50.

Anjum SA, Saleem MF, Wang LC, Bilal MF, Saeed A (2012) Protective role of glycinebetaine in maize against drought-induced lipid peroxidation by enhancing capacity of antioxidative system. Aust J Crop Sci 6:576.

Anjum SA, Xie XY, Wang LC, Saleem MF, Man C, Lei W (2011) Morphological, physiological and biochemical responses of plants to drought stress. Afr J Agric Res 6:2026–2032.

Anjum F, Yaseen M, Rasul E, Wahid A, Anjum S (2003) Water stress in barley (*Hordeum vulgare* L.). II. Effect on chemical composition and chlorophyll contents. Pak J Agric Sci 40:45–49.

Aroca R, Vernieri P, Irigoyen JJ, Sánchez-Dıaz M, Tognoni F, Pardossi A (2003) Involvement of abscisic acid in leaf and root of maize (*Zea mays* L.) in avoiding chilling-induced water stress. Plant Sci 165:671–679.

Ashraf M, Akram NA (2009) Improving salinity tolerance of plants through conventional breeding and genetic engineering: an analytical comparison. Biotechnol Adv 27:744–752.

Ashraf MFMR, Foolad M (2007) Roles of glycine betaine and proline in improving plant abiotic stress resistance. Environ Exp Bot 59:206–216.

Ashraf M, Hafeez M (2004) Thermotolerance of pearl millet and maize at early growth stages: growth and nutrient relations. Biol Plant 48:81–86.

Ashraf MHPJC , Harris PJ (2013) Photosynthesis under stressful environments: an overview. Photosynthetica 51:163–190.

Aziz K, Daniel KYT, Fazal M,Muhammad ZA, Farooq S, Fan W, Fahad S, Ruiyang Z (2017a) Nitrogen nutrition in cotton and control strategies for greenhouse gas emissions: a review. Environ Sci Pollut Res 24:23471–23487. https://doi.org/10.1007/s11356-017-0131-y.

Aziz K, Daniel KYT, Muhammad ZA, Honghai L, Shahbaz AT, Mir A, Fahad S (2017b) Nitrogen fertility and abiotic stresses management in cotton crop: a review. Environ Sci Pollut Res 24:14551–14566. https://doi.org/10.1007/s11356-017-8920-x.

Baseer M, Adnan M, Fazal M, Fahad S, Muhammad S, Fazli W, Muhammad A, Jr. Amanullah, Depeng W, Saud S, Muhammad N, Muhammad Z, Fazli S, Beena S, Mian AR, Ishaq AM (2019) Substituting urea by organic wastes for improving maize yield in alkaline soil. J Plant Nutr https://doi.org/10.1080/01904167.2019.1659344.

Bhatnagar-Mathur P, Vadez V, Sharma KK (2008) Transgenic approaches for abiotic stress tolerance in plants: retrospect and prospects. Plant Cell Rep 27:411–424.

Blake L, Goulding KWT (2002) Effects of atmospheric deposition, soil pH and acidification on heavy metal contents in soils and vegetation of semi-natural ecosystems at Rothamsted Experimental Station, UK. Plant Soil 240:235–251.

Blum A (2017) Osmotic adjustment is a prime drought stress adaptive engine in support of plant production. Plant Cell Environ 40:4–10.

Blum A, Zhang J, Nguyen HT (1999) Consistent differences among wheat cultivars in osmotic adjustment and their relationship to plant production. Field Crop Res 64:287–291.

Buckley TN (2019) How do stomata respond to water status? New Phytol 224:21–36.

Cabuslay GS, Ito O, Alejar AA (2002) Physiological evaluation of responses of rice (*Oryza sativa* L.) to water deficit. Plant Sci 163:815–827.

Chaoui A, Mazhoudi S, Ghorbal MH, El Ferjani E (1997) Cadmium and zinc induction of lipid peroxidation and effects on antioxidant enzyme activities in bean (*Phaseolus vulgaris* L.). Plant Sci 127:139–147.

Charlesworth S, Everett M, McCarthy R, Ordonez A, De Miguel E (2003) A comparative study of heavy metal concentration and distribution in deposited street dusts in a large and a small urban area: Birmingham and Coventry, West Midlands, UK. Environ Int 29:563–573.

Chen THH, Murata N (2002) Enhancement of tolerance of abiotic stress by metabolic engineering of betaines and other compatible solutes. Curr Opin Plant Biol 5:250–257.

Chimenti CA, Marcantonio M, Hall AJ (2006) Divergent selection for osmotic adjustment results in improved drought tolerance in maize (*Zea mays* L.) in both early growth and flowering phases. Field Crop Res 95:305–315.

Christensen JH, Christensen OB (2007) A summary of the PRUDENCE model projections of changes in European climate by the end of this century. Clim Change 81:7–30.

Depeng W, Fahad S, Saud S, Muhammad K, Aziz K, Mohammad NK, Hafiz MH, Wajid N (2018) Morphological acclimation to agronomic manipulation in leaf dispersion and orientation to promote “Ideotype” breeding: evidence from 3D visual modeling of “super” rice (*Oryza sativa* L.). Plant Physiol Biochem 135. https://doi.org/10.1016/j.plaphy.2018.11.010.

Dias MC, Correia S, Serôdio J, Silva AMS, Freitas H, Santos C (2018) Chlorophyll fluorescence and oxidative stress endpoints to discriminate olive cultivars tolerance to drought and heat episodes. Sci Hortic 231:31–35.

Dijkstra JJ, Meeussen JC, Comans RN (2004) Leaching of heavy metals from contaminated soils: an experimental and modeling study. Environ Sci Technol 38:4390–4395.

Dixit R, Malaviya D, Pandiyan K, Singh UB, Sahu A, Shukla R, Singh BP, Rai JP, Sharma PK, Lade H, Paul D (2015) Bioremediation of heavy metals from soil and aquatic environment: an overview of principles and criteria of fundamental processes. Sustainability 7:2189–2212.

Duarte B, Santos D, Marques JC Caçador I (2013) Ecophysiological adaptations of two halophytes to salt stress: photosynthesis, PS II photochemistry and anti-oxidant feedback–implications for resilience in climate change. Plant Physiol Biochem 67:178–188.

Dudev T, Lim C (2012) Competition among Ca2+, Mg2+, and Na+ for model ion channel selectivity filters: determinants of ion selectivity. J Physic Chem A 116:10703–10714.

Dusenge ME, Duarte AG, Way DA (2019) Plant carbon metabolism and climate change: elevated CO_2 and temperature impacts on photosynthesis, photorespiration and respiration. New Phytol 221:32–49.

Ehsanpour AA, Fatahian N (2003) Effects of salt and proline on Medicago sativa callus. Plant Cell Tiss Organ Cult 73:53–56.

Elferjani R, Soolanayakanahally R (2018) Canola responses to drought, heat, and combined stress: shared and specific effects on carbon assimilation, seed yield, and oil composition. Front Plant Sci 9:1224.

Fahad S, Adnan M, Hassan S, Saud S, Hussain S, Wu C, Wang D, Hakeem KR, Alharby HF, Turan V, Khan MA, Huang J (2019b) Rice responses and tolerance to high temperature, In: Hasanuzzaman M, Fujita M, Nahar K, Biswas JK (eds) Advances in rice research for abiotic stress tolerance. Woodhead Publ Ltd, Cambridge, England, pp 201–224.

Fahad S, Bajwa AA, Nazir U, Anjum SA, Farooq A, Zohaib A, Sadia S, NasimW, Adkins S, Saud S, Ihsan MZ, Alharby H,Wu C,Wang D, Huang J (2017) Crop production under drought and heat stress: plant responses and management options. Front Plant Sci 8:1147. https://doi.org/10.3389/fpls.2017.01147.

Fahad S, Bano A (2012) Effect of salicylic acid on physiological and biochemical characterization of maize grown in saline area. Pak J Bot 44:1433–1438.

Fahad S, Chen Y, Saud S,Wang K, Xiong D, Chen C,Wu C, Shah F, Nie L, Huang J (2013) Ultraviolet radiation effect on photosynthetic pigments, biochemical attributes, antioxidant enzyme activity and hormonal contents of wheat. J Food Agri Environ 11:1635–1641.

Fahad S, Hussain S, Bano A, Saud S, Hassan S, Shan D, Khan FA, Khan F, Chen Y, Wu C, Tabassum MA, Chun MX, Afzal M, Jan A, Jan MT, Huang J (2014a) Potential role of phytohormones and plant growth-promoting rhizobacteria in abiotic stresses: consequences for changing environment. Environ Sci Pollut Res 22:4907–4921. https://doi.org/10.1007/s11356-014-3754-2.

Fahad S, Hussain S, Matloob A, Khan FA, Khaliq A, Saud S, Hassan S, Shan D, Khan F, Ullah N, Faiq M, Khan MR, Tareen AK, Khan A, Ullah A, Ullah N, Huang J (2014b) Phytohormones and plant responses to salinity stress: A review. Plant Growth Regul 75:391–404. https://doi.org/10.1007/s10725-014-0013-y.

Fahad S, Hussain S, Saud S, Hassan S, Chauhan BS, Khan F et al (2016a) Responses of rapid viscoanalyzer profile and other rice grain qualities to exogenously applied plant growth regulators under high day and high night temperatures. PLoS One 11:e0159590. https://doi.org/10.1371/journal.pone.0159590.

Fahad S, Hussain S, Saud S, Hassan S, Ihsan Z, Shah AN,Wu C, Yousaf M, Nasim W, Alharby H, Alghabari F, Huang J (2016c) Exogenously applied plant growth regulators enhance the morphophysiological growth and yield of rice under high temperature. Front Plant Sci 7:1250. https://doi.org/10.3389/fpls.2016.01250.

Fahad S, Hussain S, Saud S, Hassan S, Tanveer M, Ihsan MZ, Shah AN, Ullah A, Nasrullah KF, Ullah S, AlharbyH NW, Wu C, Huang J (2016d) A combined application of biochar and phosphorus alleviates heat-induced adversities on physiological, agronomical and quality attributes of rice. Plant Physiol Biochem 103:191–198.

Fahad S, Hussain S, Saud S, Khan F, Hassan S, Jr A, Nasim W, Arif M, Wang F, Huang J (2016b) Exogenously applied plant growth regulators affect heat-stressed rice pollens. J Agron Crop Sci 202:139–150.

Fahad S, Hussain S, Saud S, Tanveer M, Bajwa AA, Hassan S, Shah AN, Ullah A,Wu C, Khan FA, Shah F, Ullah S, Chen Y, Huang J (2015a) A biochar application protects rice pollen from high-temperature stress. Plant Physiol Biochem 96:281–287.

Fahad S, Muhammad ZI, Abdul K, Ihsanullah D, Saud S, Saleh A, Wajid N, Muhammad A, Imtiaz AK, Chao W, Depeng W, Jianliang H (2018) Consequences of high temperature under changing climate optima for rice pollen characteristics-concepts and perspectives. Arch Agron Soil Sci 64:1479–1488. https://doi.org/10.1080/03650340.2018.1443213.

Fahad S, Nie L, Chen Y, Wu C, Xiong D, Saud S, Hongyan L, Cui K, Huang J (2015b) Crop plant hormones and environmental stress. Sustain Agric Rev 15:371–400.

Fahad S, Rehman A, Shahzad B, Tanveer M, Saud S, Kamran M, Ihtisham M, Khan SU, Turan V, Rahman MHU (2019a) Rice responses and tolerance to metal/metalloid toxicity. In: Hasanuzzaman M, Fujita M, Nahar K, Biswas JK (eds) Advances in rice research for abiotic stress tolerance. Woodhead Publ Ltd, Cambridge, England, pp 299–312.

Fan J, Hu Z, Xie Y, Chan Z, Chen K, Amombo E, Chen L, Fu J (2015) Alleviation of cold damage to photosystem II and metabolisms by melatonin in Bermudagrass. Front plant Sci 6:925.

FAO (2019) Management of salt affected soils. FAO, Italy. http://www.fao.org/soils-portal/soil-management/management-of-some-problem-soils/salt-affected-soils/more-information-on-salt-affected-soils/en/.

Farhad MS, Babak AM, Reza ZM, Hassan RSM, Afshin T (2011) Response of proline, soluble sugars, photosynthetic pigments and antioxidant enzymes in potato (*Solanum tuberosum* L.) to different irrigation regimes in greenhouse condition. Aust J Crop Sci 5:55–60.

Farhana G, Ishfaq A, Muhammad A, Dawood J, Fahad S, Xiuling L, Depeng W, Muhammad F, Muhammad F, Syed AS (2020) Use of crop growth model to simulate the impact of climate change on yield of various wheat cultivars under different agro-environmental conditions in Khyber Pakhtunkhwa, Pakistan. Arabian J Geosci 13:112. https://doi.org/10.1007/s12517-020-5118-1.

Farhat A, Hafiz MH, Wajid I, Aitazaz AF, Hafiz FB, Zahida Z, Fahad S, Wajid F, Artemi C (2020) A review of soil carbon dynamics resulting from agricultural practices. J Environ Manage 268:110319.

Farooq M, Aziz T, Wahid A, Lee DJ, Siddique KH (2009b) Chilling tolerance in maize: agronomic and physiological approaches. Crop Pasture Sci 60:501–516.

Farooq M, Wahid A, Kobayashi N, Fujita D, Basra SMA (2009a) Plant drought stress: effects, mechanisms and management. Agron Sustain Dev 29:185–212.

Flexas J, Diaz-Espejo A, Galmés J, Kaldenhoff R, Medrano H, Ribas-Carbo M (2007) Rapid variations of mesophyll conductance in response to changes in CO2 concentration around leaves. Plant Cell Environ 30:1284–1298.

Foyer CH, Noctor G (2005) Redox homeostasis and antioxidant signaling: a metabolic interface between stress perception and physiological responses. Plant Cell 17:1866–1875.

Fujita M, Fujita Y, Maruyama K, Seki M, Hiratsu K, Ohme Takagi M, Tran LSP, Yamaguchi Shinozaki K, Shinozaki K (2004) A dehydration-induced NAC protein, RD26, is involved in a novel ABA dependent stress-signaling pathway. Plant J 39:863–876.

Gao JJ, Zhang Z, Peng RH, Xiong AS, Xu J, Zhu B, Yao QH (2011) Forced expression of Mdmyb10, a myb transcription factor gene from apple, enhances tolerance to osmotic stress in transgenic Arabidopsis. Mol Biol Rep 38:205–211.

Gomez JM, Hernandez JA, Jimenez A, Del Rio LA, Sevilla F (1999) Differential response of antioxidative enzymes of chloroplasts and mitochondria to long-term NaCl stress of pea plants. Free Radic Res 31:11–18.

Greer DH, Weedon MM (2014) Does the hydrocooling of *Vitis vinifera* cv. Semillon vines protect the vegetative and reproductive growth processes and vine performance against high temperatures? Funct Plant Biol 41:620–633.

Groppa MD, Benavides MP (2008) Polyamines and abiotic stress: recent advances. Amino Acids 34:35.

Groppa MD, Tomaro ML, Benavides MP (2001) Polyamines as protectors against cadmium or copper-induced oxidative damage in sunflower leaf discs. Plant Sci 161:481–488.

Gurel F, Öztürk NZ, Yörük E, Uçarlı C, Poyraz, N (2016) Comparison of expression patterns of selected drought-responsive genes in barley (*Hordeum vulgare* L.) under shock-dehydration and slow drought treatments. Plant Growth Regul 80:183–193.

Habib ur Rahman M, Ashfaq A, Aftab W, Manzoor H, Fahd R, Wajid I, Md. Aminul I, Vakhtang S, Muhammad A, Asmat U, Abdul W, Syeda RS, Shah S, Shahbaz K, Fahad S, Manzoor H, Saddam H, Wajid N (2017) Application of CSM-CROPGRO-Cotton model for cultivars and optimum planting dates: evaluation in changing semi-arid climate. Field Crops Res 238:139–152. http://dx.doi.org/10.1016/j.fcr.2017.07.007.

Hafiz MH, Abdul K, Farhat A, Wajid F, Fahad S, Muhammad A, Ghulam MS, Wajid N, Muhammad M, Hafiz FB (2020b) Comparative effects of organic and inorganic fertilizers on soil organic carbon and wheat productivity under arid region. Commun Soil Sci Plant Anal 51:1406–1422. https://doi.org/10.1080/00103624.2020.1763385.

Hafiz MH, Farhat A, Ashfaq A, Hafiz FB, Wajid F, Carol Jo W, Fahad S, Gerrit H (2020a) Predicting kernel growth of maize under controlled water and nitrogen applications. Int J Plant Prod 14:609–620. https://doi.org/10.1007/s42106-020-00110-8.

Hafiz MH, Farhat A, Shafqat S, Fahad S, Artemi C, Wajid F, Chaves CB, Wajid N, Muhammad M, Hafiz FB (2018) Offsetting land degradation through nitrogen and water management during maize cultivation under arid conditions. Land Degrad Dev 1–10. https://doi.org/10.1002/ldr.2933.

Hafiz MH, Muhammad A, Farhat A, Hafiz FB, Saeed AQ, Muhammad M, Fahad S, Muhammad A (2019) Environmental factors affecting the frequency of road traffic accidents: a case study of sub-urban area of Pakistan. Environ Sci Pollut Res 26:11674–11685. https://doi.org/10.1007/s11356-019-04752-8.

Hafiz MH, Wajid F, Farhat A, Fahad S, Shafqat S, Wajid N, Hafiz FB (2016) Maize plant nitrogen uptake dynamics at limited irrigation water and nitrogen. Environ Sci Pollut Res 24:2549–2557. https://doi.org/10.1007/s11356-016-8031-0.

Hajihashemi S, Noedoost F, Geuns JM, Djalovic I, Siddique KH (2018) Effect of cold stress on photosynthetic traits, carbohydrates, morphology, and anatomy in nine cultivars of *Stevia rebaudiana*. Front Plant Sci 9.

Haque SA (2006) Salinity problems and crop production in coastal regions of Bangladesh. Pak J Bot 38:1359–1365.

Hasanuzzaman M, Nahar K, Gill SS, Alharb HF, Razafindrabe BH, Fujita M (2017a) Hydrogen peroxide pretreatment mitigates cadmium-induced oxidative stress in *Brassica napus* L.: an intrinsic study on antioxidant defense and glyoxalase systems. Front Plant Sci 8:115.

Hasanuzzaman M, Nahar K, Hossain M, Mahmud J, Rahman A, Inafuku M, Oku H, Fujita M (2017b) Coordinated actions of glyoxalase and antioxidant defense systems in conferring abiotic stress tolerance in plants. Int J Mol Sci 18:200.

Havaux M (1998) Carotenoids as membrane stabilizers in chloroplasts. Trends Plant Sci 3:147–151.

Hayat S, Hayat Q, Alyemeni MN, Ahmad A (2013) Proline enhances antioxidative enzyme activity, photosynthesis and yield of *Cicer arietinum* L. exposed to cadmium stress. Acta Bot Croatica 72:323–335.

Hernández JA, Corpas FJ, Gómez M, del Rio LA, Sevilla F (1993) Salt induced oxidative stress mediated by activated oxygen species in pea leaf mitochondria. Physiol Plant 89:103–110.

Hesham FA, Fahad S (2020) Melatonin application enhances biochar efficiency for drought tolerance in maize varieties: modifications in physio-biochemical machinery. Agron J 1–22.

Hoque MA, Banu MNA, Nakamura Y, Shimoishi Y, Murata Y (2008) Proline and glycine betaine enhance antioxidant defense and methyl glyoxal detoxification systems and reduce NaCl-induced damage in cultured tobacco cells. J Plant Physiol 165:813–824.

Hossain MS, Dietz KJ (2016) Tuning of redox regulatory mechanisms, reactive oxygen species and redox homeostasis under salinity stress. Front Plant Sci 7:548.

Hossain MA, Hasanuzzaman M, Fujita M (2011) Coordinate induction of antioxidant defense and glyoxalase system by exogenous proline and glycinebetaine is correlated with salt tolerance in mung bean. Front Agric China 5:1–14.

Huang M, Guo Z (2005) Responses of antioxidative system to chilling stress in two rice cultivars differing in sensitivity. Biol Plant 49:81–84.

Huang S, Waadt R, Nuhkat M, Kollist H, Hedrich R, Roelfsema MRG (2019) Ca2+ signals in guard cells enhance the efficiency by which ABA triggers stomatal closure. New Phytol 224:177–187.

Hussain MA, Fahad S, Rahat S, Muhammad FJ, Muhammad M, Qasid A, Ali A, Husain A, Nooral A, Babatope SA, Changbao S, Liya G, Ibrar A, Zhanmei J, Juncai H (2020) Multifunctional role of brassinosteroid and its analogues in plants. Plant Growth Regul 92:141–156. https://doi.org/10.1007/s10725-020-00647-8.

Hussain HA, Hussain S, Khaliq A, Ashraf U, Anjum SA, Men S, Wang L (2018a) Chilling and drought stresses in crop plants: implications, cross talk, and potential management opportunities. Front Plant Sci 9:393.

Hussain S, Khalid MF, Saqib M, Ahmad S, Zafar W, Rao MJ, Morillon R, Anjum MA (2018b) Drought tolerance in citrus rootstocks is associated with better antioxidant defense mechanism. Acta Physiol plant 40:135.

Hussain S, Rao MJ, Anjum MA, Ejaz S, Zakir I, Ali MA, Ahmad N, Ahmad S (2019) Oxidative stress and antioxidant defense in plants under drought conditions. In: Plant Abiotic Stress Tolerance. Springer, Cham, pp 207–219.

Ilyas M, Mohammad N, Nadeem K, Ali H, Aamir HK, Kashif H, Fahad S, Aziz K, Abid U (2020) Drought tolerance strategies in plants: a mechanistic approach. J Plant Growth Regul. https://doi.org/10.1007/s00344-020-10174-5.

IPCC (2007) Intergovernmental panel on climate change. Fourth Assessment Report: Climate Change 2007: Synthesis Rep 76.

Islam MM, Hoque MA, Okuma E, Banu MNA, Shimoishi, Y, Nakamura Y, Murata Y (2009) Exogenous proline and glycinebetaine increase antioxidant enzyme activities and confer tolerance to cadmium stress in cultured tobacco cells. J Plant Physiol 166:1587–1597.

Jacobsen SE, Liu F, Jensen CR (2009) Does root-sourced ABA play a role for regulation of stomata under drought in quinoa (*Chenopodium quinoa* Willd.). Sci Hort 122:281–287.

Jan M, Anwar-ul-Haq M, Adnan NS, Muhammad Y, Javaid I, Xiuling L, Depeng W, Fahad S (2019) Modulation in growth, gas exchange, and antioxidant activities of salt-stressed rice (*Oryza sativa* L.) genotypes by zinc fertilization. Arabian J Geosci 12:775. https://doi.org/10.1007/s12517-019-4939-2.

Jiang Y, Huang B (2001) Osmotic adjustment and root growth associated with drought preconditioning-enhanced heat tolerance in Kentucky bluegrass. Crop Sci 41:1168–1173.

Jiang M, Zhang J (2002) Water stress induced abscisic acid accumulation triggers the increased generation of reactive oxygen species and up regulates the activities of antioxidant enzymes in maize leaves. J Exp Bot 53:2401–2410.

Jiao D, Benhua J, Zuo B (1996) Changes of the photosynthetic electron transport and photosyntheic enzyme activities of two rice varieties under photooxidation condition. Acta Agron Sinica 22:43–49.

Kadi MW (2009) Soil pollution hazardous to environment: a case study on the chemical composition and correlation to automobile traffic of the roadside soil of Jeddah city, Saudi Arabia. J Haz Mater 168:1280–1283.

Kamaran M, Wenwen C, Irshad A, Xiangping M, Xudong Z, Wennan S, Junzhi C, Shakeel A, Fahad S, Qingfang H, Tiening L (2017) Effect of paclobutrazol, a potential growth regulator on stalk mechanical strength, lignin accumulation and its relation with lodging resistance of maize. Plant Growth Regul 84:317–332. https://doi.org/10.1007/s10725-017-0342-8.

Kamran M, Shahbaz M, Ashraf M, Akram NA (2009) Alleviation of drought-induced adverse effects in spring wheat (*Triticum aestivum* L.) using proline as a pre-sowing seed treatment. Pak J Bot 41:621–632.

Kang S, Shi W, Zhang J (2000) An improved water-use efficiency for maize grown under regulated deficit irrigation. Field Crop Res 67:207–214.

Kasuga M, Liu Q, Miura S, Yamaguchi-Shinozaki K, Shinozaki K (1999) Improving plant drought, salt, and freezing tolerance by gene transfer of a single stress inducible transcription factor. Nat Biotechnol 17:287–291.

Kaur G, Kumar S, Thakur P, Malik JA, Bhandhari K, Sharma KD, Nayyar H (2011) Involvement of proline in response of chickpea (*Cicer arietinum* L.) to chilling stress at reproductive stage. Sci Hort 128:174–181.

Kaushal N, Gupta K, Bhandhari K, Kumar S, Thakur P, Nayyar H (2011) Proline induces heat tolerance in chickpea (*Cicer arietinum* L.) plants by protecting vital enzymes of carbon and antioxidative metabolism. Physiol Mol Biol Plant 17:203.

Khattab EA, Afifi MH (2009) Effect of proline and glycinebetain on canola plants grown under salinity stress condition. Mod J App Biol Sci Crop Sci 3:42–51.

Khattak MS, Khan A, Khan MA, Ahmad W, Sharif M, Ahmad S (2019) Investigation of characteristics of hydrological droughts in Indus basin. Sarhad J Agric 35:48–56.

Khraiwesh B, Zhu JK, Zhu J (2012) Role of miRNAs and siRNAs in biotic and abiotic stress responses of plants – gene regulatory mechanisms. Biochim Biophysic Acta 1819:137–148.

Kinraide TB (1999) Interactions among Ca2+, Na+ and K+ in salinity toxicity: quantitative resolution of multiple toxic and ameliorative effects. J Exp Bot 50:1495–1505.

Kogan F, Guo W (2016) Early twenty-first-century droughts during the warmest climate. Geomat Nat Haz Risk 7:127–137.

Koyro HW (2006) Effect of salinity on growth, photosynthesis, water relations and solute composition of the potential cash crop halophyte *Plantago coronopus* (L.). Environ Exp Bot 56:136–146.

Krause GH, Weis E (1991) Chlorophyll fluorescence and photosynthesis: the basics. Annu Rev Plant Biol 42:313–349.

Kubis J, Floryszak-Wieczorek J, Arasimowicz-Jelonek M (2014) Polyamines induce adaptive responses in water deficit stressed cucumber roots. J Pant Res 127:151–158.

Kusano T, Yamaguchi K, Berberich T, Takahashi Y (2007) Advances in polyamine research in 2007. J Plant Res 120:345–350.

Kuznetsov V, Shevyakova NI (2010) Polyamines and plant adaptation to saline environments. In: Remawat KG (ed) Desert plants. Springer, Berlin, pp 261–298.

Lamaoui M, Jemo M, Datla R, Bekkaoui F (2018) Heat and drought stresses in crops and approaches for their mitigation. Front Chem 6:26.

Lang F (2007) Mechanisms and significance of cell volume regulation. J Am Coll Nutr 26:613–623.

Lee MH, Cho EJ, Wi SG, Bae H, Kim JE, Cho JY, Lee S, Kim JH, Chung BY (2013) Divergences in morphological changes and antioxidant responses in salt-tolerant and salt-sensitive rice seedlings after salt stress. Plant Physiol Biochem 70:325–335.

Leterrier M, Barroso JB, Valderrama R, Palma JM, Corpas FJ (2012) NADP-dependent isocitrate dehydrogenase from Arabidopsis roots contributes in the mechanism of defence against the nitro-oxidative stress induced by salinity. Sci World J 2012.

Li SL, Li ZG, Yang LT, Li YR, He ZL (2018) Differential effects of cold stress on chloroplasts structures and photosynthetic characteristics in cold-sensitive and cold-tolerant cultivars of sugarcane. Sugar Tech 20:11–20.

Liu JH, Kitashiba H, Wang J, Ban Y, Moriguchi T (2007) Polyamines and their ability to provide environmental stress tolerance to plants. Plant Biotechnol 24:117–126.

Liu EK, Mei XR, Yan CR, Gong DZ, Zhang YQ (2016) Effects of water stress on photosynthetic characteristics, dry matter translocation and WUE in two winter wheat genotypes. Agric Water Manage 167:75–85.

Liu F, Shahnazari A, Andersen MN, Jacobsen SE, Jensen CR (2006) Physiological responses of potato (*Solanum tuberosum* L.) to partial root-zone drying: ABA signalling, leaf gas exchange, and water use efficiency. J Exp Bot 57:3727–3735.

Llamas A, Ullrich CI Sanz A (2008) Ni2+ toxicity in rice: effect on membrane functionality and plant water content. Plant Physiol Biochem 46:905–910.

Loka DA, Oosterhuis DM (2010) Effect of high night temperatures on cotton respiration, ATP levels and carbohydrate content. Environ Exp Bot 68:258–263.

Ludlow MM, Santamaria JM, Fukai S (1990) Contribution of osmotic adjustment to grain yield in *Sorghum bicolor* (L.) Moench under water-limited conditions. II. Water stress after anthesis. Aust J Agric Res 41:67–78.

Lutts S (2000) Exogenous glycine betaine reduces sodium accumulation in salt-stressed rice plants. Int Rice Res Notes 25:39–40.

Machado S, Paulsen GM (2001) Combined effects of drought and high temperature on water relations of wheat and sorghum. Plant Soil 233:179–187.

Makela P, Peltonen-Sainio P, Jokinen K, Pehu E, Setala H, Hinkkanen R, Somersalo S (1998) Effect of foliar application of glycine betaine on stomatal conductance, abscisic acid and solute concentrations in leaves of salt- and drought-stressed tomato. Aust J Plant Physiol 25:655–663.

Maleva MG, Nekrasova GF, Borisova GG, Chukina NV, Ushakova OS (2012) Effect of heavy metals on photosynthetic apparatus and antioxidant status of Elodea. Russ J Plant Physiol 59:190–197.

Mansour MMF (1998) Protection of plasma membrane of onion epidermal cells by glycine betaine and proline against NaCl stress. Plant Physiol Biochem 36:767–772.

Marco F, Bitrián M, Carrasco P, Rajam MV, Alcázar R, Tiburcio AF (2015) Genetic engineering strategies for abiotic stress tolerance in plants. Plant Biol Biotechnol 579–609.

Meneguzzo S, Sgherri CL, Navari Izzo F, Izzo R (1998) Stromal and thylakoid bound ascorbate peroxidases in NaCl treated wheat. Physiol Plant 104:735–740.

Mittler R (2006) Abiotic stress, the field environment and stress combination. Trends Plant Sci 11:15–19.

Mittova V, Tal M, Volokita M, Guy M (2003) Up regulation of the leaf mitochondrial and peroxisomal antioxidative systems in response to salt induced oxidative stress in the wild salt tolerant tomato species *Lycopersicon pennellii*. Plant Cell Environ 26:845–856.

Mocquot B, Vangronsveld J, Clijsters H, Mench M (1996) Copper toxicity in young maize (*Zea mays* L.) plants: effects on growth, mineral and chlorophyll contents, and enzyme activities. Plant Soil 182:287–300.

Mubeen M, Ashfaq A, Hafiz MH, Muhammad A, Hafiz UF, Mazhar S, Muhammad Sami ul Din, Asad A, Amjed A, Fahad S, Wajid N (2020) Evaluating the climate change impact on water use efficiency of cotton-wheat in semi-arid conditions using DSSAT model. J Water Climate Chang 11:1661–1675. https://doi.org/10.2166/wcc.2019.179/622035/jwc2019179.pdf

Muhammad Z, Abdul MK, Abdul MS, Kenneth BM, Muhammad S, Shahen S, Ibadullah J, Fahad S (2019) Performance of *Aeluropus lagopoides* (mangrove grass) ecotypes, a potential turfgrass, under high

saline conditions. Environ Sci Pollut Res 26:13410–13421. https://doi.org/10.1007/s11356-019-0483 8-3.
Munemasa S, Hauser F, Park J, Waadt R, Brandt B, Schroeder JI (2015) Mechanisms of abscisic acid-mediated control of stomatal aperture. Cur Opin Plant Biol 28:154–162.
Munns R (2002) Comparative physiology of salt and water stress. Plant Cell Environ 25:239–250.
Munns R, Goyal S, Passioura J (2005) Salinity stress and its mitigation. Plant Stress Website. In: Blum A (ed). http://www.plantstress.com/Articles/index.asp.
Munns R, James RA, Läuchli A (2006) Approaches to increasing the salt tolerance of wheat and other cereals. J Exp Bot 57:1025–1043.
Murata T (1990) Relation of chilling stress to membrane permeability. Chilling Injury Hortic Crops 201–209.
Murtaza G, Ghafoor A, Qadir M (2006) Irrigation and soil management strategies for using saline-sodic water in a cotton–wheat rotation. Agric Water Manage 81:98–114.
Mutlu S, Atici Ö, Nalbantoglu B (2009) Effects of salicylic acid and salinity on apoplastic antioxidant enzymes in two wheat cultivars differing in salt tolerance. Biol Plant 53:334–338.
Nakashima K, Yamaguchi-Shinozaki K (2006) Regulons involved in osmotic stress-responsive and cold stress responsive gene expression in plants. Physiol Plant 126:62–71.
Nounjana N, Nghiab PT, Theerakulpisuta P (2012) Exogenous proline and trehalose promote recovery of rice seedlings from salt-stress and differentially modulate antioxidant enzymes and expression of related genes. J Plant Physiol 169:596–604.
Oukarroum A, El Madidi S, Strasser RJ (2012) Exogenous glycine betaine and proline play a protective role in heat-stressed barley leaves (*Hordeum vulgare* L.): a chlorophyll a fluorescence study. Plant Biosyst Int J Dealing Asp Plant Biol 146:1037–1043.
Özdemir F, Bor M, Demiral T, Türkan İ (2004) Effects of 24-epibrassinolide on seed germination, seedling growth, lipid peroxidation, proline content and antioxidative system of rice (*Oryza sativa* L.) under salinity stress. Plant Growth Regul 42:203–211.
Ozgur R, Uzilday B, Sekmen AH, Turkan I (2013) Reactive oxygen species regulation and antioxidant defence in halophytes. Funct Plant Biol 40:832–847.
Palta JP, Weiss LS (2018) Ice formation and freezing injury: an overview on the survival mechanisms and molecular aspects of injury and cold acclimation in herbaceous plants. In: Advances in plant cold hardiness. CRC Press, Florida, pp 143–176.
Panda SK, Chaudhury I, Khan MH (2003) Heavy metals induce lipid peroxidation and affect antioxidants in wheat leaves. Biol Plant 46:289–294.
Perfus-Barbeoch L, Leonhardt N, Vavasseur A, Forestier C (2002) Heavy metal toxicity: cadmium permeates through calcium channels and disturbs the plant water status. Plant J 32:539–548.
Pour-Aboughadareh A, Ahmadi J, Mehrabi AA, Etminan A, Moghaddam M, Siddique KH (2017) Physiological responses to drought stress in wild relatives of wheat: implications for wheat improvement. Acta Physiol Plant 39:106.
Prasad TK (1997) Role of catalase in inducing chilling tolerance in preemergence maize seedlings. Plant Physiol 114:1369–1376.
Qadir M, Wichelns D, Raschid-Sally L, McCornick PG, Drechsel P, Bahri A, Minhas PS (2010) The challenges of wastewater irrigation in developing countries. Agric Water Manage 97:561–568.
Rahnama H, Ebrahimzadeh H (2005) The effect of NaCl on antioxidant enzyme activities in potato seedlings. Biol Plant 49:93–97.
Ranganayakulu GS, Veeranagamallaiah G, Sudhakar C (2013) Effect of salt stress on osmolyte accumulation in two groundnut cultivars (*Arachis hypogaea* L.) with contrasting salt tolerance. Afr J Plant Sci 12:586–592.
Rasheed R, Ashraf MA, Hussain I, Haider MZ, Kanwal U, Iqbal M (2014) Exogenous proline and glycinebetaine mitigate cadmium stress in two genetically different spring wheat (*Triticum aestivum* L.) cultivars. Braz J Bot 37:399–406.
Ratcliffe OJ, Riechmann JL (2002) Arabidopsis transcription factors and the regulation of flowering time: a genomic perspective. Curr Issues Mol Biol 4:77–91.
Rehman M, Fahad S, Saleem MH, Hafeez M, Habib ur Rahman M, Liu F, Deng G (2020) Red light optimized physiological traits and enhanced the growth of ramie (*Boehmeria nivea* L.). Photosynthetica 58:922–931.
Rhodes D, Samaras Y (1994) Genetic control of osmoregulation in plants. Cellu Mol Physiol Cell Vol Reg 416.

Riva-Roveda L, Escale B, Giauffret C, Périlleux C (2016) Maize plants can enter a standby mode to cope with chilling stress. BMC Plant Biol 16:212.

Romero-Aranda R, Soria T, Cuartero J (2001) Tomato plant-water uptake and plant-water relationships under saline growth conditions. Plant Sci 160:265–272.

Rosli RE, Sulong AB, Daud WRW, Zulkifley MA, Husaini T, Rosli MI, Majlan, EH, Haque MA (2017) A review of high-temperature proton exchange membrane fuel cell (HT-PEMFC) system. Int J Hydrogen Energy 42:9293–9314.

Sairam RK, Rao KV, Srivastava GC (2002) Differential response of wheat genotypes to long term salinity stress in relation to oxidative stress, antioxidant activity and osmolyte concentration. Plant Sci 163:1037–1046.

Sajjad H, Muhammad M, Ashfaq A, Waseem A, Hafiz MH, Mazhar A, Nasir M, Asad A, Hafiz UF, Syeda RS, Fahad S, Depeng W, Wajid N (2019) Using GIS tools to detect the land use/land cover changes during forty years in Lodhran district of Pakistan. Environ Sci Pollut Res 27:39676–39692. https://doi.org/10.1007/s11356-019-06072-3.

Saleem MH, Fahad S, Adnan M, Mohsin A, Muhammad SR, Muhammad K, Qurban A, Inas AH, Parashuram B, Mubassir A, Reem MH (2020a) Foliar application of gibberellic acid endorsed phytoextraction of copper and alleviates oxidative stress in jute (*Corchorus capsularis* L.) plant grown in highly copper-contaminated soil of China. Environ Sci Pollution Res 27:37121–37133. https://doi.org/10.1007/s11356-020-09764-3.

Saleem MH, Fahad S, Shahid UK, Mairaj D, Abid U, Ayman ELS, Akbar H, Analía L, Lijun L (2020c) Copper-induced oxidative stress, initiation of antioxidants and phytoremediation potential of flax (Linum usitatissimum L.) seedlings grown under the mixing of two different soils of China. Environ Sci Poll Res 27:5211–5221. https://doi.org/10.1007/s11356-019-07264-7.

Saleem MH, Rehman M, Fahad S, Tung SA, Iqbal N, Hassan A, Ayub A, Wahid MA, Shaukat S, Liu L, Deng G (2020b) Leaf gas exchange, oxidative stress, and physiological attributes of rapeseed (*Brassica napus* L.) grown under different light-emitting diodes. Photosynthetica 58:836–845.

Salvucci ME, Crafts-Brandner SJ (2004) Inhibition of photosynthesis by heat stress: the activation state of rubisco as a limiting factor in photosynthesis. Physiol Plant 120:179–186.

Sarwat M, Tuteja N (2017) Hormonal signaling to control stomatal movement during drought stress. Plant Gene 11:143–153.

Saud S, Chen Y, Fahad S, Hussain S, Na L, Xin L, Alhussien SA (2016) Silicate application increases the photosynthesis and its associated metabolic activities in Kentucky bluegrass under drought stress and post-drought recovery. Environ Sci Pollut Res 23:17647–17655. https://doi.org/10.1007/s11356-016-6957-x.

Saud S, Chen Y, Long B, Fahad S, Sadiq A (2013) The different impact on the growth of cool season turf grass under the various conditions on salinity and drought stress. Int J Agric Sci Res 3:77–84.

Saud S, Fahad S, Cui G, Chen Y, Anwar S (2020) Determining nitrogen isotopes discrimination under drought stress on enzymatic activities, nitrogen isotope abundance and water contents of Kentucky bluegrass. Sci Rep 10:6415. https://doi.org/10.1038/s41598-020-63548-w.

Saud S, Fahad S, Yajun C, Ihsan MZ, Hammad HM, Nasim W, Amanullah Jr, Arif M, Alharby H (2017) Effects of nitrogen supply on water stress and recovery mechanisms in kentucky bluegrass plants. Front. Plant Sci 8:983. https://doi.org/10.3389/fpls.2017.00983.

Saud S, Li X, Chen Y, Zhang L, Fahad S, Hussain S, Sadiq A, Chen Y (2014) Silicon application increases drought tolerance of Kentucky bluegrass by improving plant water relations and morph physiological functions. SciWorld J 2014:1–10. https://doi.org/10.1155/2014/368694.

Sawahel W (2004) Improved performance of transgenic glycine betaine-accumulating rice plants under drought stress. Biol Plant 47:39–44.

Schachtman DP, Goodger JQ (2008) Chemical root to shoot signaling under drought. Trends Plant Sci 13:281–287.

Scheibe R, Backhausen JE, Emmerlich V, Holtgrefe S (2005) Strategies to maintain redox homeostasis during photosynthesis under changing conditions. J Exp Bot 56:1481–1489.

Schonfeld MA, Johnson RC, Carver BF, Mornhinweg DW (1988) Water relations in winter wheat as drought resistance indicators. Crop Sci 28:526–531.

Schrader SM, RR Wise, WF Wacholtz, DR Ort, TD Sharkey (2004) Thylakoid membrane responses to moderately high leaf temperature in pima cotton. Plant Cell Environ 27:725–735.

Shafi MI, Adnan M, Fahad S, Fazli W, Ahsan K, Zhen Y, Subhan D, Zafar-ul-Hye M, Martin B, Rahul D (2020) Application of single superphosphate with humic acid improves the growth, yield and phosphorus uptake of wheat (*Triticum aestivum* L.) in calcareous soil. Agron 10:1224. https://doi.org/10.3390/agronomy10091224.
Shah K, Kumar RG, Verma S, Dubey RS (2001) Effect of cadmium on lipid peroxidation, superoxide anion generation and activities of antioxidant enzymes in growing rice seedlings. Plant Sci 161:1135–1144.
Shah F, Lixiao N, Kehui C, Tariq S, Wei W, Chang C, Liyang Z, Farhan A, Fahad S, Huang J (2013) Rice grain yield and component responses to near 2°C of warming. Field Crop Res 157:98–110.
Shahbaz M, Ashraf M (2013) Improving salinity tolerance in cereals. Critic Rev Plant Sci 32:237–249.
Shahbaz M, Tseng MH, Stuiver CEE, Koralewska A, Posthumus FS, Venema JH, Parmar S, Schat H, Hawkesford MJ, De Kok LJ (2010) Copper exposure interferes with the regulation of the uptake, distribution and metabolism of sulfate in Chinese cabbage. J Plant Physiol 167:438–446.
Shahzad S, Chaudhry UK, Anwar B, Saboor A, Yousaf MF, Saeed F, Yaqoob S (2016) Drought stress effect on morphological and physiological characteristics of different varieties of annual verbena (Verbena hybrid). J Biod Environ Sci 9:32–46.
Shahzad S, Khan MY, Zahir ZA, Asghar HN, Chaudhry UK (2017) Comparative effectiveness of different carriers to improve the efficacy of bacterial consortium for enhancing wheat production under salt affected field conditions. Pak J Bot 49:1523–1530.
Sharkey TD (2005) Effects of moderate heat stress on photosynthesis: importance of thylakoid reactions, rubisco deactivation, reactive oxygen species, and thermotolerance provided by isoprene. Plant Cell Environ 28:269–277.
Shi G, Chen Z, Xu S, Zhang J, Wang L, Bi C, Teng J (2008) Potentially toxic metal contamination of urban soils and roadside dust in Shanghai, China. Environ Pollut 156:251–260.
Shiu SH, Shih MC, Li WH (2005) Transcription factor families have much higher expansion rates in plants than in animals. Plant Physiol 139:18–26.
Snider JL, Oosterhuis DM, Skulman BW, Kawakami EM (2009) Heat stress-induced limitations to reproductive success in *Gossypium hirsutum*. Physiol Plant 137:125–138.
Somersalo S, Kyei-Boahen S, Pehu E (1996) Exogenous glycine betaine application as a possibility to increase low temperature tolerance of crop plants. Nordisk Jordbruksforskning 78:10.
Stepien P, Klobus G (2005) Antioxidant defense in the leaves of C3 and C4 plants under salinity stress. Physiol Plant 125: 31–40.
Stiborova M, Ditrichová M, BŘEzinová A (1987) Effect of heavy metal ions on growth and biochemical characteristics of photosynthesis of barley and maize seedlings. Biol Plant 29:453.
Subhan D, Zafar-ul-Hye M, Fahad S, Saud S, Martin B, Tereza H, Rahul D (2020) Drought stress alleviation by ACC deaminase producing *Achromobacter xylosoxidans* and *Enterobacter cloacae*, with and without timber waste biochar in maize. Sustain 12:6286. https://doi.org/10.3390/su12156286.
Sunkar R (2010) MicroRNAs with macro-effects on plant stress responses. Sem Cell Dev Biol 21:805–811.
Szabados L, Savoure A (2010) Proline: a multifunctional amino acid. Trends Plant Sci 15:89–97.
Sánchez-Rodríguez E, Romero L, Ruiz JM (2016) Accumulation of free polyamines enhances the antioxidant response in fruits of grafted tomato plants under water stress. J Plant Physiol 190:72–78.
Taiz L, Zeiger E (2002) Plant physiology, 3rd edn. Sinauer Associates, England
Tariq M, Ahmad S, Fahad S, Abbas G, Hussain S, Fatima Z, Nasim W, Mubeen M, Hamid ur Rehman M, Khan MA, Adnan M (2018) The impact of climate warming and crop management on phenology of sunflower-based cropping systems in Punjab, Pakistan. Agri For Met 15:270–282.
Taïbi K, Taïbi F, Abderrahim LA, Ennajah A, Belkhodja M, Mulet JM (2016) Effect of salt stress on growth, chlorophyll content, lipid peroxidation and antioxidant defence systems in *Phaseolus vulgaris* L. S Afr J Bot 105:306–312.
Trueba S, Pan R, Scoffoni C, John GP, Davis SD, Sack L (2019) Thresholds for leaf damage due to dehydration: declines of hydraulic function, stomatal conductance and cellular integrity precede those for photochemistry. New Phytol 223:134–149.
Tsukaguchi T, Kawamitsu Y, Takeda H, Suzuki K, Egawa Y (2003) Water status of flower buds and leaves as affected by high temperature in heat-tolerant and heat-sensitive cultivars of snap bean (*Phaseolus vulgaris* L.). Plant Prod Sci 6:24–27.
Tuna AL, Kaya C, Ashraf M, Altunlu H, Yokas I, Yagmur B (2007) The effects of calcium sulphate on growth, membrane stability and nutrient uptake of tomato plants grown under salt stress. Environ Exp Bot 59:173–178.

Ullah A, Sun H, Yang X, Zhang X (2017) Drought coping strategies in cotton: increased crop per drop. Plant Biotechnol J 15:271–284.
uz Zaman Q, Zubair A, Muhammad Y, Muhammad ZI, Abdul K, Fahad S, Safder B, Ramzani PMA, Muhammad N (2017) Zinc biofortification in rice: leveraging agriculture to moderate hidden hunger in developing countries. Arch Agron Soil Sci 64:147–161. https://doi.org/10.1080/03650340.2017.1338343.
Vassilev A, Yordanov I, Tsonev T (1998) Effects of Cd2+ on the physiological state and photosynthetic activity of young barley plants. Photosynthetica 34:293–302.
Verma S, Dubey RS (2003) Lead toxicity induces lipid peroxidation and alters the activities of antioxidant enzymes in growing rice plants. Plant Sci 164:645–655.
Veselov DS, Sharipova GV, Veselov SU, Kudoyarova GR (2008) The effects of NaCl treatment on water relations, growth, and ABA content in barley cultivars differing in drought tolerance. J Plant Growth Regul 27:380.
Vysotskaya L, Hedley PE, Sharipova G, Veselov D, Kudoyarova G, Morris J, Jones HG (2010) Effect of salinity on water relations of wild barley plants differing in salt tolerance. AoB Plants 2010
Wahid F, Fahad S, Subhan D, Adnan M,, Zhen Y, Saud S, Manzer HS, Martin B, Tereza H, Rahul D (2020) Sustainable management with mycorrhizae and phosphate solubilizing bacteria for enhanced phosphorus uptake in calcareous soils. Agri 10:334. https://doi.org/10.3390/agriculture10080334.
Wajid N, Ashfaq A, Asad A, Muhammad T, Muhammad A,Muhammad S, Khawar J, Ghulam MS, Syeda RS, Hafiz MH, Muhammad IAR, Muhammad ZH, Muhammad Habib ur R, Veysel T, Fahad S, Suad S, Aziz K, Shahzad A (2017) Radiation efficiency and nitrogen fertilizer impacts on sunflower crop in contrasting environments of Punjab. Pak Environ Sci Pollut Res 25:1822–1836. https://doi.org/10.1007/s11356-017-0592-z.
Wang X, Wang W, Huang J, Peng S, Xiong D (2018) Diffusional conductance to CO_2 is the key limitation to photosynthesis in salt stressed leaves of rice (*Oryza sativa*). Physiol Plant 163:45–58.
Wang H, Wang H, Shao H, Tang X (2016) Recent advances in utilizing transcription factors to improve plant abiotic stress tolerance by transgenic technology. Front Plant Sci 7:67.
Wang L, Zhang T, Ding S (2006) Effect of drought and rewatering on photosynthetic physioecological characteristics of soybean. Acta Ecol Sinica 26:2073–2078.
Wani HS, Singh NB, Haribhushan A, Iqbal Mir J (2013) Compatible solute engineering in plants for abiotic stress tolerance-role of glycine betaine. Curr Genom 14:157–165.
WeiBing X, Rajashekar CB (2001) Glycine betaine involvement in freezing tolerance and water stress in Arabidopsis thaliana. Environ Exp Bot 46:21–28.
Wu C, Kehui C, She T, Ganghua L, Shaohua W, Fahad S, Lixiao N,Jianliang H, Shaobing P, Yanfeng D (2020) Intensified pollination and fertilization ameliorate heat injury in rice (*Oryza sativa* L.) during the flowering stage. Field Crops Res 252:107795.
Wu C, Tang S, Li G, Wang S, Fahad S, Ding Y (2019) Roles of phytohormone changes in the grain yield of rice plants exposed to heat: a review. PeerJ 7:e7792. https://doi.org/10.7717/peerj.7792.
Xia X, Chen X, Liu R, Liu H (2011) Heavy metals in urban soils with various types of land use in Beijing, China. J Haz Mat 186:2043–2050.
Xu PL, Guo YK, Bai JG, Shang L, Wang XJ (2008) Effects of long-term chilling on ultrastructure and antioxidant activity in leaves of two cucumber cultivars under low light. Physiol Plant 132:467–478.
Yamaguchi K, Takahashi Y, Berberich T, Imai A, Takahashi T, Michael AJ, Kusano TA (2007) Protective role for the polyamine spermine against drought stress in Arabidopsis. Biochem Biophysic Res Commun 352:486–490.
Yamamoto Y (2016) Quality control of photosystem II: the mechanisms for avoidance and tolerance of light and heat stresses are closely linked to membrane fluidity of the thylakoids. Front Plant Sci 7:1136.
Yang X, Chen X, Ge Q, Li B, Tong Y, Zhang A, Li Z, Kuang T, Lu C (2006) Tolerance of photosynthesis to photoinhibition, high temperature and drought stress in flag leaves of wheat: a comparison between a hybridization line and its parents grown under field conditions. Plant Sci 171:389–397.
Yang C, Li D, Mao D, Liu XUE, Ji C, Li X, Zhao X, Cheng Z, Chen C, Zhu L (2013) Overexpression of microRNA319 impacts leaf morphogenesis and leads to enhanced cold tolerance in rice (*Oryza sativa*). Plant Cell Environ 36:2207–2218.
Yang Z, Zhang Z, Zhang T, Fahad S, Cui K, Nie L, Peng S, Huang J (2017) The effect of season-long temperature increases on rice cultivars grown in the central and southern regions of China. Front Plant Sci 8:1908. https://doi.org/10.3389/fpls.2017.01908.

You J, Chan Z (2015) ROS regulation during abiotic stress responses in crop plants. Front Plant Sci 6:1092.

Yuan XK, Yang ZQ, Li YX, Liu Q, Han W (2016) Effects of different levels of water stress on leaf photosynthetic characteristics and antioxidant enzyme activities of greenhouse tomato. Photosynthetica 54:28–39.

Zafar-ul-Hye M, Muhammad N, Subhan D, Fahad S, Rahul D, Mazhar A, Ashfaq AR, Martin B, Jiˇrí H, Zahid HT, Muhammad N (2020a) Alleviation of cadmium adverse effects by improving nutrients uptake in bitter gourd through cadmium tolerant rhizobacteria. Environ 7:54. https://doi.org/10.3390/environments7080054.

Zafar-ul-Hye M, T ahzeeb-ul-Hassan M, Muhammad A, Fahad S,, Martin B, Tereza D, Rahul D, Subhan D (2020b) Potential role of compost mixed biochar with rhizobacteria in mitigating lead toxicity in spinach. Scientific Rep 10:12159. https://doi.org/10.1038/s41598-020-69183-9.

Zahida Z, Hafiz FB, Zulfiqar AS, Ghulam MS, Fahad S, Muhammad RA, Hafiz MH, Wajid N, Muhammad S (2017) Effect of water management and silicon on germination, growth, phosphorus and arsenic uptake in rice. Ecotoxicol Environ Saf 144:11–18.

Zakar T, Laczko-Dobos H, Toth TN, Gombos Z (2016) Carotenoids assist in cyanobacterial photosystem II assembly and function. Front Plant Sci 7:295.

Zengin FK, Munzuroglu O (2005) Effects of some heavy metals on content of chlorophyll, proline and some antioxidant chemicals in bean (*Phaseolus vulgaris* L.) seedlings. Acta Biol Cracoviensia Series Bot 47:157–164.

Zhang CX, Feng BH, Chen TT, Zhang XF, Tao LX, Fu GF (2017) Sugars, antioxidant enzymes and IAA mediate salicylic acid to prevent rice spikelet degeneration caused by heat stress. Plant Growth Regul 83:313–323.

Zhang L, Ma H, Chen T, Pen J, Yu S, Zhao X (2014) Morphological and physiological responses of cotton (*Gossypium hirsutum* L.) plants to salinity. PLoS One 9:112807.

Zhang C, Zhan DX, Luo HH, Zhang YL, Zhang WF (2016) Photorespiration and photoinhibition in the bracts of cotton under water stress. Photosynthetica 54:12–18.

Zhang X, Zou Z, Gong P, Zhang J, Ziaf K, Li H, Xiao F, Ye Z (2011) Over-expression of microRNA169 confers enhanced drought tolerance to tomato. Biotechnol Lett 33:403–409.

Zheng C, Jiang D, Liu F, Dai T, Jing Q, Weixing C (2009) Effects of salt and water logging stresses and their combination on leaf photosynthesis, chloroplast ATP synthesis, and antioxidant capacity in wheat. Plant Sci 176:575–582.

Zhu XC, Song FB, Liu SQ, Liu TD (2011) Effects of arbuscular mycorrhizal fungus on photosynthesis and water status of maize under high temperature stress. Plant Soil 346:189–199.

Zlatev Z, Lidon FC (2012) An overview on drought induced changes in plant growth, water relations and photosynthesis. Emir J Food Agric 57–72.

Zouari M, Ahmed CB, Elloumi N, Bellassoued K, Delmail D, Labrousse P, Abdallah FB, Rouina, BB (2016) Impact of proline application on cadmium accumulation, mineral nutrition and enzymatic antioxidant defense system of *Olea europaea* L. cv Chemlali exposed to cadmium stress. Ecotox Environ Safe 128:195–205.

6

Climate Change and Role of Genetics and Genomics in Climate-Resilient Sorghum

Shakra Jamil[1], Rahil Shahzad[1], Shakeel Ahmad[2], Zulfiqar Ali[3], Sana Shaheen[4], Hamna Shahzadee[4,5], Noreen Fatima[4], Iram Sharif[5], Muqadas Aleem[6], Razia Sultana[1], Rida Fatima[4], Sajid Ur Rahman[1], and Muhammad Zaffar Iqbal[1]

[1]Agricultural Biotechnology Research Institute, Ayub Agricultural Research Institute, Faisalabad, Pakistan
[2]State Key Laboratory of Rice Biology, China National Rice Research Institute, Hangzhou, China
[3]Muhammad Nawaz Sharif, University of Agriculture, Multan, Pakistan
[4]University of Agriculture, Faisalabad, Pakistan
[5]Cotton Research Station, Ayub agricultural Research Institute, Faisalabad, Pakistan
[6]Nanjing Agricultural University, Nanjing, China

6.1 Introduction

Sorghum (*Sorghum bicolor*), also known as great millet, kafir corn, and guinea corn, is the fifth major crop after wheat (*Triticum aestivum* L.), rice (*Oryza sativa* L.), maize (*Zea mays* L.), and barley (*Hordeum vulgare* L.) growing worldwide (Fernández-Aparicio et al. 2013). Globally, sorghum is a staple food of more than 500 million individuals and is an important crop in the semiarid regions of Africa along with other crops like pearl millet and cassava. It is classified under the family *Poaceae*, tribe *Andopogoneae*, subtribe *Sorghinae*, and genus sorghum (Prasad and Staggenborg 2009). The genus *sorghum* is further classified into five subgenera: Eusorghum, Chaetosorghum, Heterosorghum, Para-sorghum, and Stiposorghum. Para-sorghum comprised of seven species that are from Africa, Asia, Australia, and Central America. Stiposorghum includes ten northern Australian species (Berenji et al. 2011). The Eusorghum includes cultivated species *S. bicolor* and is originated from Africa or Asia (Duvall and Doebley 1990). *S. bicolor* consists of three subspecies: *S. bicolor* subspecie *bicolor*, *S. bicolor* subspecie drummondii, and *S. bicolor* subspecie verticilliflorum. *S. bicolor* subspecie *bicolor* is considered as cultivated sorghum. This cultivated type is represented by different agronomic names such as grain sorghum, sweet sorghum, Sudan grass, and broomcorn (Berenji and Dahlberg 2004). Subgenera Chaetosorghum and Heterosorghum comprise of *Sorghum macrospermum* and *Sorghum laxiflorum*. Moreover, both of these species are annuals and have polyploidy genomes (Lazarides et al. 1991).

6.2 Origin and Domestication

Sorghum was originated near Lake chad in Africa and domesticated about 700 BP in West Africa, 1500 BC in India, 900 AD in China, and was introduced in America a century ago (Prasad and Staggenborg 2009).

It is being consumed as a staple food in more than 30 countries and also an important crop in semiarid regions of Africa. The Eusorghum contains cultivated species *S. bicolor,* which was originated either from Africa or Asia (Duvall and Doebley 1990). Largest genetic diversity of cultivated and wild sorghum was found in Africa as a result of disruptive selection and separation and recombination in extremely different habitats of northern Africa and movement of species throughout the continent (Miller 1982). All races of sorghum are from same biological species, and when hybridized, they are fully fertile. Sorghum has four wild and five cultivated races. Arundinaceum, virgatum, aethiopicum, and verticilliflorum are four wild races and are placed under *S. bicolor* subspecies verticilliflorum as interpreted by De Wet and Prasada. The cultivated races of sorghum are bicolor, guinea, kafir, caudatum, and durra. These cultivated races are placed under subspecies bicolor. Cultivated races durra and bicolor originated from wild race aethiopicum, guinea from arundinaceum, and kafir from verticilliflorum. De Wet and Huckabay agreed with these centres of origin of Snowdon with the exception of durra about which they believe that kafir gave rise to durra (De Wet and Huckabay 1967).

Today, bicolor is not only distributed in Africa but also in Asia, India, Indonesia, and China. The guinea race originated from West Africa but also found in the high rainfall receiving mountains of Eastern Africa. Guinea is considered one of the oldest races because of its wide distribution and diversity throughout the world. It is distributed in east and Southeastern Africa and then moved to India through east African ports. Caudatum is believed to be originated from bicolor ancestral lines and domesticated after bicolor and guinea. Stemler and her colleagues described that caudatums of the Ethiopian highlands are acquired by trade. It is well adapted to extreme conditions and found mostly in areas which receive annual rainfall of 250–1300 mm (Stemler et al. 1975). Kafir is thought to be derived from an ancient bicolor race that was moved to east and south from the Savanna belt. Durra race is considered most important grain in Ethiopia and India and thought to be domesticated there frequently. It has compact panicle and predominantly white seeds which show their adaptation to low rainfall environment with minimum risk of exposure to grain mould and is found in north of equator in Africa. Durra was mostly cultivated by Muslim Oromo (Gallo) in Ethiopia about 500 years ago, and this race was used as a foundation of their agricultural system. Today, durras are present in mid-altitude highlands of Ethiopia, the Nile Valley of Sudan and Egypt, and 10–15 ° north latitude from Ethiopia to Mauritania (Harlan et al. 1973).

6.3 Morphology

Sorghum is a C4 plant like other grasses, that is, maize, sugarcane, millet, and switch grass, and gives high forage biomass yields per unit of land (Rooney et al. 2007). Sorghum is a perennial crop by nature and hence is very useful as multicut forage crop and also grown as annual rain-fed crop (Prasad and Staggenborg 2009). Sorghum is an erect plant, with an average height of 0.8–5 m (measured as a sum of panicle and stalk), depending upon its sensitivity to photoperiod. Stalk is made up of nodes, internodes, and peduncle (upper most internode). Based on plant height, sorghum has been divided into two major categories: Standard (tall) and dwarf. Leaf consists of leaf sheath (attached to each node and spans around internode above that node) and leaf blade. Internode is usually larger than the leaf sheath. Typically, sorghum leaf blades are on average 8–12 cm wide and 50–90 cm long, with a prominent midrib. Leaf sheath and stem are usually surrounded by a waxy bloom. Panicles grow out from the top of main stem and sometime in tillers also. Panicle comprises of peduncle located at the top node of the stalk, rachis, and fibres (cluster of fibrous braches). Seeds are usually red to brown to white and enclosed in the glumes of colours tan, red, or mahogany and contain awns and pubescence (Berenji et al. 2011). Sorghum has a deep and extensive fibrous root system that facilitates plants in nutrient and water uptake and plays important role in drought tolerance. Roots sometime emerge from the aerial parts of the last node without significant agronomic contribution (Prasad and Staggenborg 2009).

6.4 How Important Is It?

6.4.1 Feed for Humans

Grain sorghum is one of the Africa's most important indigenous cereal crops that has been used as major food security crop from centuries (Dicko et al. 2006). It contains approximately 80–90% dry matter (DM), 8.9–15% crude protein (CP), 2.8% ether extract (EE), 1.5–1.7% ash, 2.1–2.3% crude fibre (CF), and 71.7–72.3% nitrogen-free extract (NFE). The sorghum crude protein content is equal to that of wheat but more than maize. Depending on the livestock specie, sorghum is preferred over maize in terms of energy value. A metabolizable energy of 13.96, 14.04, and 13.70 MJ/kg was reported for all grains, kaffir, and milo types, respectively. *S. guineense* contains high level of phosphorus and manganese, *S. margaritiferum* contains iron and zinc, *S. durra* contains potassium, and *S. caudatum* contains moderate level of all minerals (Palavecino et al. 2016). However, sorghum is not recommended as a feed due to various nutritional shortcomings. Firstly, the dominant storage proteins (kafirins) present in sorghum are difficult to digest and also lack essential amino acids, that is, lysine, methionine, and tryptophan (Wong et al. 2009). Secondly, some grains of sorghum contain antinutrients such as tannins and phytate (Gilani et al. 2012), which results in reduced bioavailability and nutrient uptake by binding with proteins and essential minerals. However, efforts are underway for enhancement of nutritional value of sorghum.

One of the approaches was RNA interference technology (RNAi) that was being used to knock out kafirin protein, which resulted in increased protein digestibility by 53%. However, these transgenic lines have not yet been commercialized and cultivated because of strict legislative criteria involved in the approval of genetically modified (GM) crops before commercial release. In the past, sorghum was used in production of unfermented and fermented breads, porridges, couscous, rice-like products, snacks, and malted alcoholic and non-alcoholic drinks in many African and Asian countries. Several attempts were made to introduce food products of sorghum to developed markets, but only a few products got access to Europe and the USA (Carson et al. 2000; Suhendro et al. 2000).

6.4.2 Forage for Animal and Poultry Feed

Like human feed, the use of sorghum as forage crop is also gaining importance in many parts of the world (Zerbini and Thomas 2003). Approximately 51% of sorghum crop is used to feed livestock, while remaining 49% is used as human food and other purposes. Moreover, 48% of sorghum grain is fed to animals, which is comparable to maize and is considered its best alternative (Reddy et al. 2016). Sorghum competes well with other feed grains in terms of total carbohydrates and has better protein profile than maize but lower in vitamin A. Stalks of sweet sorghum contain sugars, mainly sucrose and glucose (55%) and dry matter (3.2%). They also contain fibre content like cellulose (12.4%) and hemicelluloses (10.2%) (Ramya et al. 2017). Taking into consideration the nutrition value, cost, and availability of sorghum grain, it is considered next alternative to maize in poultry feed also. However, carcasses of birds were lighter in weight which fed sorghum than maize. Nevertheless, it was reported that sorghum can replace 50% of maize without any deleterious effects on animal growth and development, but there was 10% reduction in weight gain of animals.

6.4.3 Medicinal Usages

The World health organization (WHO) states that 80% of the population of underdeveloped countries used traditional medicines as their primary health care source, mostly obtained from plant source (Schmincke 2003). Plant-based medicines are attracting masses due to lack of accessibility to expensive pharmacy medicines by poor people and bioprospecting of new plant-derived drugs. *S. bicolor* is being used as traditional medicine throughout the drier areas of Africa and India (Ogwumike 2002). It is a source of primary cure of anaemia (Oladiji et al. 2007), cancer, and viral diseases in developing countries. Sorghum leaf sheath extract has hepatoprotective and hematopoietic effects that include generation of haemoglobin content of erythrocytes (Akande et al. 2010). It also helps alleviate chronic pain, and its leaf base is thought to have central antinociceptive properties. Efforts are underway to use sorghum leaf sheath for treatment of anaemia and to obtain its boosting effect on blood haemoglobin and red cells' concentration (Sènou et al. 2017). Sorghum leaves were thought to have haematinic property in ethno-veterinary treatment when used with *Telfaria occdentials* (Adedapo et al. 2002). Sorghum is a safe food product for celiac patients because it did not show toxic reaction as with other gluten-containing diets, that is, wheat, rye, and barely (Ciacci et al. 2007).

Different research groups such as Lloyd Rooney at Texas A&M University, Scott Bean, United States Department of Agriculture, Agriculture Research Service (ASDA-ARS), Manhattan, Kansas, and Ron Prior, ASDA-ARS, Little Rock, Arkansas, are working on exploring the antioxidant activities and other nutritional aspects of sorghum. Tannin sorghums have significantly higher antioxidants, anti-inflammatory, and anticolon cancer characteristics (Awika and Rooney 2004). Sorghum leaf sheath contains high level of 3-deoxy anthocyanins that have colour stability and numerous health applications (Geera et al. 2012). It acts as a strong chemoprotective and anticancer agent by inhibiting proliferating effects on gastrointestinal cancer cell lines (Liu 2007). In addition to high concentration of anti-inflammatory compounds, sorghum contains several other bioactive compounds that can induce pro-inflammatory immune responses. One of the examples is water-soluble beta-glucans that initiate macrophage activation. Moreover, sorghum seeds also contain an antiviral peptide, capable of inhibiting infection, replication, virus spread, including Herpes simplex, and to some extent the non-enveloped polio virus (Camargo et al. 2008). Sorghum is also a rich source of flavonoids (Borneo and León 2012) which acts as antioxidants and signalling molecules, and it lowers incidence of cardiovascular disease, cancer, type II diabetes, and neurodegenerative disease (Rio et al. 2013).

6.5 Cultivated Area and Production of Sorghum

In the past half century, sorghum-cultivated area has increased 60% worldwide, and yield increased to 233%. Between the years 1992 and 1994, 27% of the world's total sorghum was produced in Africa. Some important sorghum-producing countries are the United States, China, India, Ethiopia, Nigeria, Sudan, Mexico, Brazil, and Burkina. The United States ranks first in sorghum production worldwide. However, it is also the world's largest exporter of sorghum grains and constitutes 70% in the world trade as well (Prasad and Staggenborg 2009). According to FAO (2018), Africa and Asia are the major sorghum-producing regions covering maximum of the harvested area (Figure 6.1). In August 2019, the United States Department of Agriculture (USDA) estimates that the World Sorghum Production 2019–2020 will be 59.48 million metric tons, around 1.15 million tons more than previous month's projection. Last year the sorghum production was 59.35 million tons. This year's estimated 59.48

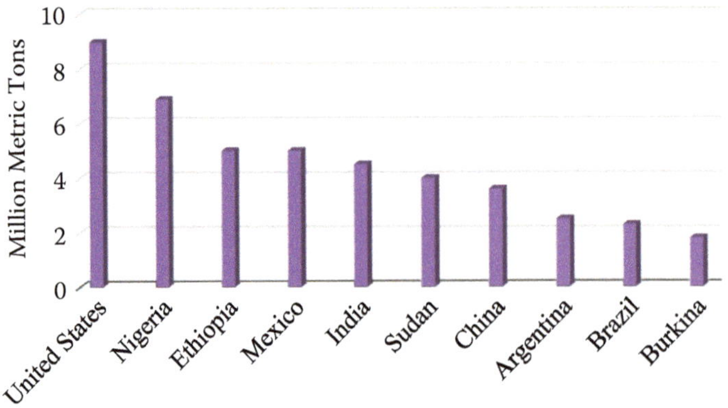

FIGURE 6.1 Top Sorghum-Producing Countries in 2018–2019.

millions of tons could represent an increase of 0.13 million tons or 0.21% in sorghum production around the globe (http://www.worldagriculturalproduction.com/crops/sorghum.aspx).

6.6 Effect of Climate Change on Sorghum Production

Stresses are unfavourable environmental conditions that physiologically damage plants and lead to significant yield losses (Shao et al. 2008; Saud et al. 2013; Shah et al. 2013; Saud et al. 2014; Hafiz et al. 2016; Saud et al. 2016; Aziz et al. 2017a,b; Kamaran et al. 2017; Saud et al. 2017; Uz Zaman et al. 2017; Wajid et al. 2017; Yang et al. 2017; Zahida et al. 2017; Adnan et al. 2018a,b; Akram et al. 2018a,b; Depeng et al. 2018; Hafiz et al. 2018; Tariq et al. 2018; Adnan et al. 2019; Ahmad et al. 2019; Baseer et al. 2019; Hafiz et al. 2019; Muhammad et al. 2019; Sajjad et al. 2019; Wu et al. 2019; Adnan et al. 2020; Farhana et al. 2020; Farhat et al. 2020; Hafiz et al. 2020a,b; Hussain et al. 2020; Ilyas et al. 2020; Mubeen et al. 2020; Rehman et al. 2020; Saleem et al. 2020a,b,c; Shafi et al. 2020; Subhan et al. 2020; Wahid et al. 2020; Wu et al. (2020); Zafar-ul-Hye et al. 2020a,b). Stress is impact of both biotic and abiotic factors that adversely affect growth of individual plant, and population growth rate declined due to reduced survival and repair process of plants (Vinebrooke et al. 2004; Fahad and Bano 2012; Fahad et al. 2013; Fahad et al. 2014a,b; Fahad et al. 2015a,b; Fahad et al. 2016a,b,c,d; Fahad et al. 2017; Fahad et al. 2018; Fahad et al. 2019a,b; Hesham and Fahad 2020). All the biotic and abiotic factors that limit sorghum productivity are discussed here in detail.

6.6.1 Abiotic Stresses

Abiotic stresses are major obstacles of sugar production in sweet sorghum which negatively affect its biomass production (Zegada-Lizarazu and Monti 2013). Drought, salinity, heat, cold, waterlogging, and heavy metal stress limit plant yield to a considerable level depending on the duration and intensity of stress. Plant yield decreased up to 75% due to adverse effects of climatic stresses. Biotic stresses are negative impacts of living components of environment such as insect, bacteria, fungi, viruses, or weeds on the plants (Guo et al. 2011). Effect of different abiotic stresses on sorghum productivity is discussed in detail below and in Figure 6.2.

6.6.2 Drought Stress

Drought stress is an alarming threat than rest of climatic changes. It is a condition when there is not enough water available for the plant to carry out essential chemical processes including photosynthesis. In this case, rate of water loss from the plant surpasses the rate of water acquisition by roots. Drought depends on many factors including duration and frequency of rainfall and the capacity of soil to hold moisture, so severity of drought stress is always unpredictable (Tari et al. 2013). It can induce nutrient deficiency by reducing the diffusion rate of nutrients from soil to plant roots, and as a result, plant may appear weaker. Many sorghum diseases and pests such as stalk rot and sorghum ergot (Bandyopadhyay et al. 1996) provoke drought stress, resulting in weak plant defence system. The average yield of sorghum grown in drylands is half than that of irrigated sorghum. Decrease of soil water potential up to −13 bars from vegetative to booting stage may reduce plant yield up to 17% (Xie et al. 2010). In Africa, during La Nina period, drought drastically reduced sorghum production in semiarid regions. Similarly, during 2011 in Somalia, the total sorghum production was 25 kilotons with 80% reduction in yield due to onset of drought (Anyamba et al. 2014).

Drought tolerance is the ability of plant to grow and reproduce under drought stress adopting specific mechanisms (Beck et al. 2007). Plant responds to drought stress by different morphological, physiological, and biochemical mechanisms. Dense root system helps plant for survival under drought conditions. The roots can extract water from deep soils and made it available to plant. Sorghum has more secondary roots as an adaptation to survive under harsh climate (Wasaya et al. 2018). The ability of plant to adjust osmotic potential, stomatal opening, and photosynthesis at low water potential also aids in bringing drought tolerance in sorghum (Bibi et al. 2012). Some late-flowering sorghum cultivars can

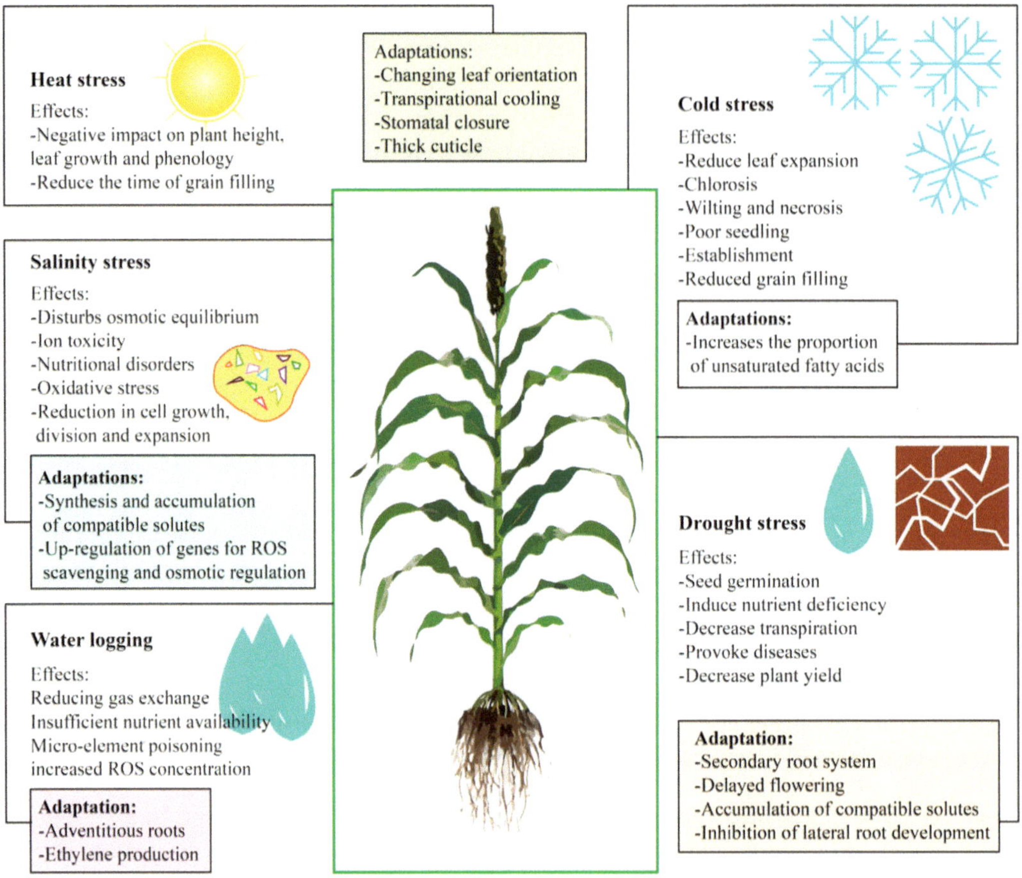

FIGURE 6.2 Effect of Abiotic Stresses on Sorghum Plant and Different Adoptive Mechanisms.

effectively tolerate water deficiency. High water extraction efficiency from soil, less nodal roots, and fewer metaxylem vessels in nodal roots were also reported and found associated with drought tolerance. Transpiration rate was found very low due to small surface area of leaf and well-developed sclerenchyma in plant leaf tissues (Salih et al. 1999). Another distinctive feature of drought-tolerant cultivar of sorghum is the coating of epicuticular wax on leaf surface that results in decreased transpiration rate (Surwenshi et al. 2010).

Response of Sorghum plant to drought stress is also dependent on plant developmental stage. Vegetative stage and early reproductive stage were more prone to drought as compared to reproductive stage because of low transpiration rate (Tingting et al. 2010). Under drought stress, there was accumulation of soluble carbohydrates such as glucose, sucrose, and fructose in leaf tissues during vegetative and reproductive stage and a substantial decrease in soluble sugar level during maturation stage. It is an indication that plants were able to maintain their osmotic balance by degradation of starch and accumulation of soluble sugars during vegetative and early reproductive stages (Oliveira et al. 2009). At anthesis, starch and sucrose were accumulated in stem in drought-stressed and -unstressed plants. However, during panicle maturation, there were more sucrose and starch contents in drought-stressed plants as compared to unstressed plants (Massacci et al. 1996). Further application of silicon (Si)-based fertilizers helps alleviate drought stress by uptake of more water from soil through accumulation of the soluble sugars and amino acids (alanine and glutamic acid). Silica also becomes accumulated in endodermis tissues of roots to prevent them from being collapsed under severe drought conditions (Sonobe et al. 2010).

Different quantitative trait loci linked to stay-green trait, root system, plant height, and plant yield under drought conditions were identified and mapped for ten sorghum linkage groups (Sanchez et al. 2002).

Drought-related genes were classified into two groups depending upon the timing of induction and expression of genes. The first group of genes produces proteins that are directly involved in bringing drought tolerance in sorghum, and the second group includes the genes that produce proteins involved in signal transduction and expression of drought-tolerant genes (Shinozaki and Yamaguchi-Shinozaki 2000). One of the mechanisms of drought tolerance in plants is inhibition of lateral root development through secretion of abscisic acid (Xiong et al. 2006). Plants that are sensitive to abscisic acid are more tolerant to drought stress than those that are insensitive. The plants that are insensitive to ABA have high transpiration rates and lose more water than ABA-sensitive plants (Shinozaki and Yamaguchi-Shinozaki 2000).

6.6.3 Salinity Stress

Salinity stress is characterized by accumulation of various ions, that is, Ca^{2+}, Na^{+}, Cl^{-}, and Mg^{2+}, to toxic levels. Sodium chloride is relatively more harmful as sodium interferes with potassium uptake and disturbs stomatal regulation, ultimately causing water loss and necrosis (Zuo and Zhang 2011). On the other hand, chloride ion causes chlorotic toxicity due to impaired production of chlorophyll (Tavakkoli et al. 2010). Irrigated, arid, and semiarid regions are equally vulnerable to this problem. In irrigated areas, poor quality of water, improper drainage system, and mixing of sea water cause salinity, while in arid and semiarid regions, soil becomes saline due to excessive evaporation of water or insufficient leaching of ions due to low rainfall and accumulation of salts in root zone. Currently, more than 20% of the world's agricultural irrigated land is affected by excess salt concentrations, and this problem continues to worsen worldwide because of improper application of fertilizers, industrial pollution, and poor irrigation practices (Zhu et al. 2019). Salinity becomes more severe in areas affected by drought, and high temperature stress due to an increase in upward movement of water- and water-soluble salts becomes concentrated in root zone of plants. Salt stress causes ion toxicity, nutritional disorders, oxidative stress, and reduction in cell growth, division, and expansion (Munns 2002).

Sorghum is generally grown in areas affected by salinity, low rainfall, and high temperature. It can grow well in moderately salt-affected soils (Almodares and Sharif 2007). However, differences may occur among various cultivars depending upon their genetic makeup (Zuo and Zhang 2011). Synthesis and accumulation of compatible solutes such as soluble sugars, amino acids, and proline is a common response to salt stress. In the cell, toxic ions accumulate in vacuoles, and osmotic equilibrium between vacuole and cytoplasm may be established by an increase in the synthesis of osmolytes. The osmolytes in return stabilize protein molecules and membranes (Hamilton and Heckathorn 2001) or may act as reserve in plant metabolism. Under saline conditions, most of the plant species show a significant increase in proline synthesis. It decreases the water potential and helps maintaining water content in leaves (Sabir et al. 2011). Availability of various nitrogen forms especially NH_4^+ and NO_3^- can also interfere with plant sensitivity and salt stress tolerance (Kant et al. 2007). In sorghum, fertilization with NH_4^+ limits sodium concentration and greatly enhances K^+/Na^+ balance in response to salt stress. Moreover, the increased activity of transporter proteins such as Na^+/H^+ greatly reduces Na^+ accumulation (Olías et al. 2009; Mansour 2014).

Gene expression under salt stress involves upregulation or downregulation of specific genes (Gupta and Huang 2014). These salt stress-responsive genes are classified in different categories: ion transporter or homeostasis (salt overly sensitive, SOS, gene family), molecular chaperones (heat shock proteins, HSP, genes), and dehydration-related transcription factors (dehydration responsive element-binding, DREB, gene family). Among all, SOS genes play an important role in ion homeostasis (Liu et al. 2000; Shi et al. 2000). Genes for ROS scavenging and osmotic regulation are also upregulated. Transcription factors such as OsNAC5 and ZFP179 are upregulated in response to salinity which regulate synthesis and accumulation of proline, sugar, and LEA proteins that in turn confer stress tolerance (Song et al. 2011). There are a number of functional groups of proteins affected by salt stress, that is, ion transports, homeostasis proteins, ROS scavenging enzymes, and osmotic stress-related proteins (Kosová et al. 2013). Glutamate ammonia ligase accumulation also plays an important role in nitrogen assimilation and amino acid synthesis, which in turn is responsible for bringing salt tolerance in sorghum (Swami et al. 2011).

6.6.4 Heat Stress

Heat stress is defined as an increase in temperature beyond an ambient temperature for a period of time sufficient to cause severe damage (Fahad et al. 2016a,b,c,d; Fahad et al. 2017). The effects caused by heat stress depend on its intensity, duration, and rate of the increase in temperature (Wahid et al. 2007; Fahad et al. 2016a,b,c,d; Fahad et al. 2017). Typically, 10–15 °C temperature rise above the prevailing temperature is considered as heat stress (Fahad et al. 2016a,b,c,d; Fahad et al. 2017). In the 20th century, the temperature rose approximately 0.74 °C as compared to 19th century as assessed by the authors of the Intergovernmental Panel on Climatic Change (Tadross et al. 2007). It is expected that temperature would be 2.5–5.4 °C higher in 2100 coupled with 15% decline in rainfall (Ciscar 2012). Sorghum is an important crop in the semiarid regions. Optimum vegetative temperature for its production is 26–34 °C, whereas optimum reproductive growth takes place at 25–28 °C. Sorghum is very sensitive to high temperature especially around reproductive stage (Singh et al. 2015b). Whereas Hatfield et al. reported an annual decrease of 7.8% in global sorghum production by 1 °C rise in temperature (Adhikari et al. 2015).

Temperature above 28–33 °C (night time and day time) during panicle development induces embryo abortion. High temperature also reduces grain filling period resulting in small seeds. Various long-term phenological and morphological mechanism and short-term acclimatization or avoidance mechanisms are adopted by plants for survival. Plants exposed to heat stress usually close stomata to prevent transpiration (Srivastava et al. 2012). Small hairs as well as thick cuticle or waxy layers present on leaf surface help plant in avoiding exposure to solar radiations. Rolling of leaf blades and their orientation parallel to solar radiations to avoid transpiration is also an adoptive mechanism. In well-hydrated plant, increased transpiration keeps leaf temperature 10–15 °C lower than the seasonal temperature. Some major heat stress tolerance mechanisms include ion transport, production of late embryogenic abundant (LEA) proteins, osmoprotectants, and antioxidant (Rodríguez et al. 2005).

Heat stress induces oxidative stress which is mitigated by enzymatic and non-enzymatic ROS scavengers (Apel and Hirt 2004). Antioxidant activity is maximum at 35–40 °C in tolerant varieties and at 30 °C in susceptible ones (Chakraborty and Pradhan 2011). Exogenous application of osmoprotectants and phytohormones has beneficial effects. High temperature activates heat shock genes that encode heat shock proteins (HSPs, master players of heat stress tolerance). These proteins express constitutively under heat stress and protect plants from heat stroke. HSPs act as chaperones protected by maintaining integrity and functional stability of intracellular proteins through protein folding. Broadly, HSPs are classified in five different families, that is, HSP100, HSP90, HSP70, HSP60, and HSP20 (Swindell et al. 2007). The expression of HSP is restricted to important plant developmental stages, that is, seed germination, embryogenesis, microsporogenesis, and fruit development. These are also involved in rehabilitation of proteins' damage due to heat stress (Török et al. 2001).

6.6.5 Cold Stress

Plants that are adapted to arid and semiarid regions exhibit cold stress injury when exposed to temperature below 10–15 °C. Reduced leaf expansion, chlorosis (leaf yellowing), wilting, and necrosis (plant tissue death) are early responses of plants to cold stress. The onset of cold stress at reproductive stage results in pollen sterility and reduction in grain yield. Another adverse effect of cold stress on plants is solidification of unsaturated fatty acids from plasma membrane. Sorghum is generally grown in semiarid tropical regions and sensitive to low (20 °C) temperature (Wen et al. 2002). However, temperature below 10–15 °C adversely affects sorghum productivity (Yu and Tuinstra 2001). Low temperature exposure during first week results in poor germination and poor seedling stand establishment. When grown at low temperature, sorghum undergoes more vegetative growth, delayed flowering, male sterility, reduced stomatal opening, and reduced photosynthesis (Ercoli et al. 2004). Cold stress induces ice formation in the apoplastic space because of low concentration of solutes. As vapour pressure of ice is much lower than that of water thus, vapour pressure gradient is established between cell cytosol and apoplast. As a result, water moves out from cell cytosol to the apoplastic space and cell undergoes mechanical strain and rupturing (Uemura and Steponkus 1997).

Response to cold stress in sorghum is stage dependent, that is, at early stage, germination and emergence reduce, at maturity stage, grain filling rate reduces which may lead to significant grain yield loss. By manipulating planting dates and cultivation of short duration cultivars, cold stress may be avoided. The environmental factors may also activate cold tolerance mechanisms (cold acclimatization) (Mahajan and Tuteja 2005). The main function of cold acclimation is to maintain the integrity of plasma membrane against chilling injury. It induces physical and biochemical changes in plasma membrane and lipid composition, that is, proportion of unsaturated fatty acids increases (Williams et al. 1988). Non-enzymatic proteins are synthesized; solutes are added to plant tissues, thus preventing formation of ice in plant tissues (Mahajan and Tuteja 2005). The structure, catalytic activity, and function of enzymes change (Kubien et al. 2003). Plant tends to normalize their metabolism in two ways, that is, by restoring metabolism to avoid cold stress or synthesis of osmolytes (organic acids, amino acids, sugars, etc.) (Fernie et al. 2005). Osmolytes function to maintain cellular water relations and reduce cellular dehydration. They also behave as compatible solutes to maintain the stability of enzymes, membranes, and cellular structures (Nayyar et al. 2005; Farooq et al. 2009).

Cold stress induces synthesis of proline which regulates expression of many stress-responsive genes that have proline-responsive elements ACTCAT in their promoters (Oono et al. 2003). It further involves regulation of different transcriptional factors (TFs), that is, cis-elements and abscisic acid responsive elements and cold-regulated genes (Viswanathan and Zhu 2002). Moreover, cold stress regulates expression of various kinases of mitogen-activated proteins kinase (MAPK) pathways. Reactive oxygen species (ROS) activate MAPK cascades that are involved in signal transduction pathways and help plant in cold acclimation (Teige et al. 2004a,b). Burow et al. (2011) identified simple sequence repeat (SSR) molecular markers associated with various traits for early-season cold tolerance in sorghum.

6.6.6 Waterlogging/Submergence Stress

Waterlogging drastically changes soil properties, and it no longer supports plant growth. It affects plants by reducing gas exchange, insufficient nutrient availability, and microelement poisoning (Setter and Waters 2003). When plants are grown in waterlogged soils, several anatomical and morphological alterations occur in root system, that is, rate of root respirations is highly reduced. The crops sometimes encounter short or prolonged period of submergence due to heavy rainfall, flooding, and high-water table (Waheed 2014). The gaseous exchange in case of oxygen is 10,000 times slower in water than air. So dry land crops growing in waterlogged conditions suffer anaerobic conditions (Armstrong et al. 1994).

Sorghum is susceptible to waterlogging especially at seedlings stage due to lack of aerenchyma tissues (aerenchyma are the soft tissues with large intercellular spaces that ensure gaseous exchange between aerobic shoots and anaerobic roots) and does not get enough oxygen (Zhang et al. 2016a,b). Short or prolonged period of submergence stress decreases stomatal conductance, transpiration, and photosynthesis rate, which reduces leaf elongation rate, leaf area, leaf number plant and stalk height, and grain yield ultimately (Henshaw et al. 2007). Due to oxygen limitation, respiration and electron transport chain are also inhibited resulting in low ATP production. Leaf lengthening rate and its total size are also reduced (Malik et al. 2002). Further stomatal resistance to water increases, resulting in water deficiency in plant tissues. The absence of oxygen reduced chlorophyll content, leaf area, and leaf senescence, leading to reduction in net photosynthesis rate (Ashraf et al. 2011).

Oxygen is necessary for survival of living organisms on the earth; its reduction under waterlogging stress generates reactive oxygen species including hydrogen peroxide, oxygen-free radical, singlet oxygen, and hydroxyl ion. These lethal ROS are highly reactive in nature and damage several cellular biomolecules such as proteins, lipids, and DNA (Ashraf et al. 2010). Although ROS are produced in plants under normal conditions, their concentration is very low. Almost all plant species have ability to overcome harmful effects of ROS by producing enzymatic and non-enzymatic antioxidants. Superoxide dismutase (SOD), peroxidase (POD), glutathione reductase (GR), and catalase (CAT) are categorized as enzymatic antioxidants, while non-enzymatic antioxidants include tocopherols, glutathione, carotenoids, and ascorbic acid (Gupta et al. 2005). Ethylene production stimulates formation of aerenchyma which is the major adaptive feature in response to waterlogging. The presence of aerenchyma in plant roots avoids anaerobic conditions by increasing oxygen availability. Aerenchyma provides a pathway through which oxygen moves between roots and surrounding environment. Thus, a small oxygenated environmental region makes the oxygen available to the plant roots and prevents the formation of toxic products of iron, copper, and manganese (Jackson and Armstrong 1999). Portion of plant stem develops adventitious roots that replace the old root system, as new roots are closer to water surface and aerenchyma tissues and have large air spaces; hence, oxygen is easily accessible to plants (Mergemann and Sauter 2000).

Aquaporins (AQPs) are a group of integral proteins and are important player under water stress. These form channels for the movement of water from one plant compartment to another. They have crucial role in different developmental and physiological processes, that is, photosynthesis, stomatal conductance, germination, reproduction, and responses to abiotic stresses (Ariani and Gepts 2015). Aquaporins are highly expressed in plant roots playing role in root water transport. Among all AQPs, Plasma membrane Intrinsic Protein (PIPs) and Tonoplast Intrinsic Protein (TIPs) are extensively studied with reference to their function in plant water relations (Kadam et al. 2017). TIPs are localized in vacuolar membranes, function to maintain osmotic potential, and are responsible for movement of water across plant subcellular compartments. In addition to water, they also transport a variety of small solutes such as hydrogen peroxide and urea (Bienert et al. 2007). PIPs are localized in plasma membranes. Analysis of RNA-seq data revealed transcript abundance of PIP1–6, PIP2–5, PIP2–6, PIP2–7, TIP2-1, TIP2-2, TIP4-4, TIP5-1, and NIP4-1 in sorghum roots under waterlogging conditions (Kadam et al. 2017).

6.6.7 Heavy Metal Stress (HMS)

During last few decades, our environment is getting contaminated with heavy metals due to increased anthropogenic activities, rapid industrialization, and modern agricultural techniques. The elements of block d have been categorized as heavy metals depending upon the density, that is, >5 g/cm^3 (Jarup 2003). The accumulation of heavy metals in the environment causes toxicity in living organisms (Miransari 2011). Many biological processes such as photosynthesis, chlorophyll biosynthesis, redox reactions, DNA synthesis, post-translational mechanism, and nitrogen fixation are all aided by macro- and micronutrients (Ricachenevsky et al. 2013). However, at elevated concentrations, these heavy metals are very toxic to plants (Fidalgo et al. 2013). Some heavy metals such as lead, mercury, cadmium, and chromium are non-essential and very toxic even at low concentrations and do not take part in any physiological process (Wang et al. 2013). Reduced biomass accumulation, chlorosis, altered nutrient assimilation, senescence, reduction in plant growth, and photosynthesis are main symptoms of HMS (Singh et al. 2015b). Plant roots are the first organ exposed to the heavy metals. Soils rich in heavy metals do not support plant growth, and as a result, plants suffer from decreased growth and lower yield (Keunen et al. 2011).

Plants undergo several changes in protein and peptide structures such as the synthesis of metallothioneins after coming in contact with heavy metals. The binding of metal ions to sulfhydryl group of proteins brought toxicity leading to disruption of structure and activity of enzymes (Morelli and Scarano 2004). Cu and Cd accumulation produces low-molecular-weight peptides in many plants (Mittra et al. 2008). Although Copper is an essential element for all plants and participates in many physiological processes, that is, oxidative phosphorylation, ATP synthesis, and prosthetic group of different enzymes (Yruela 2005), its excessive accumulation causes toxicity and results in reduced shoot and root growth through inhibition of cell division and cell elongation (Ouzounidou 1993).

Some metabolites were found effective against HMS and instigate tolerance mechanisms. Amino acids, that is, proline and histidine, were reported to chelate heavy metals in plant cells and xylem sap (Sharma and Dietz 2006). Organic compounds, that is, proline, citrate, malate, and oxalate, are also involved in sequestration of heavy metal ions in vacuoles and transport them through xylem (Siripornadulsil et al. 2002). Alpha tocopherol (an active form of vitamin E) has a supportive role in overcoming HMS as it scavenges reactive oxygen species and lipid peroxides by quenching oxygen in chloroplast. Phenolic compounds also chelate metal ions due to presence of carboxyl and hydroxyl groups (Pawlak-Sprada et al. 2011). Different types of proteins, that is, glutathione-S-transferase (GST), flavonoid3′, 5′-hydroxylase, and cytochrome P450, were produced in cadmium-treated sorghum seedlings, indicating their putative role in overcoming HMS (Bogs et al. 2006). Ferritin (a multimeric protein) is synthesized in plants under iron toxicity and serves as first defence line against iron-induced oxidative damage. Ferritin is not only a store house of iron but is also involved in scavenging of reactive forms of iron (Ravet et al. 2009). Furthermore, there are different gene families that are involved in HMS tolerance, that is, ATP-binding cassette (ABC), metal tolerance protein (MTP), zinc transporters (ZIP) (Koźmińska et al. 2018), aluminium tolerance locus (Alt_{SB}), and Al-activated root citrate transporter (SbMATE) are involved in regulating HMS in sorghum. Alt_{SB} in particular increases aluminium tolerance by ten folds by exclusion of aluminium from root apics (Magalhaes et al. 2007).

6.6.8 Biotic Stresses

Any negative effect on plants caused by bacteria, viruses, fungi, insects, weeds, or native plants and influence of geography and climate may be defined as biotic stress. There are about 10,000 species of insects, 30,000 species of weeds, 100,000 diseases (caused by bacteria, viruses, fungi, and other micro-organisms), and 1000 species of nematodes that are damaging plants worldwide. However, less than 10% pest species are causing serious damage, so they are considered as major crop pests (Dhaliwal et al. 2010). An approximate 30% yield losses were observed in Sorghum due to pests and pathogens (Reddy and Zehr 2004). Major biotic stresses of sorghum are leaf anthracnose (caused by a fungus, *Colletotrichum graminicola*), sooty stripe (caused by a fungus, *Ramulispora sorghi*), grain mould (caused by fungal species belonging to Fusarium, Curvularia, Alternaria, Aspergillus, and Phoma genera), and striga disease (caused by parasitic plant *Striga hermonthica*). Different insect species, that is, *Contarinia sorghicola, Sitodiplosis mosellana,* and *Stenodiplosis sorghicola*, also attack developing sorghum grains (Sharma et al. 2003).

6.6.9 Insect Pests

Sweet sorghum grows tall and fast, is important for sugar production, and is more vulnerable to insects' pests. About 150 insect pests have been identified that attack sorghum worldwide (Kruger et al. 2008). Approximate yield losses of up to 86% have been recorded in sorghum in different regions. Major disease-causing insect pests of sorghum are green bug (*Schizaphis graminum*), sorghum midge (*Stenodiplosis sorghicola*), fall armyworm (*Spodoptera frugiperda*), and corn borers (*Ostrinia furnacalis*) (Damte et al. 2009). Panicle insect pests are of particular importance as they damage sorghum grains. Panicle insects decrease seed germination and grain yield, resulting in discolouration and punctures. The quality of infected grains deteriorates and becomes unfit for human consumption (Malgwi and Dunuwel 2011).

The early larvae of Asian corn borer (*Ostrinia furnacalis*) severely damage young leaves, tassels, bracts, and filaments of sorghum. After fourth instar, larvae penetrate into the stalk and pupate there. The larvae are mainly attracted by light, moisture, and sugar. Asian corn borers (*Ostrinia furnacalis*) cause an annual yield loss of 10–30% in sorghum (Dua et al. 2008). These losses are mainly due to the reduction in photosynthesis rates of affected plants because this pest mainly damages leaves and stem. Reduced photosynthesis leads to poor grain filling and lodging (Guo et al. 2011). Sorghum midge is also an economically important insect pest of sorghum. Female midge oviposits eggs into the flowering sorghum spikelets where larvae feed and abort kernel development, resulting in 100% destruction of the crop. Sorghum midge was originated in Africa and is unique among all pests because it is specific to sorghum (Peterson et al. 1989). Sorghum strip borer (*Chilo sacchariphagus*) also known as sugarcane stem borer is another damaging insect pest of sorghum. The larvae feed on young leaves, and after 4th instar, they penetrate into stalk and feed there until pupation and cause approximate 10–40% yield losses due to reduced photosynthesis and stalk lodging (Guo et al. 2011).

Aphids are soft-bodied insects that have piercing mouth parts for sucking juice from soft and succulent plant tissues. Yellow sugarcane aphids are lemon yellow in colour and are main damaging pests throughout China and many other countries (Singh et al. 2004). Both nymph and adult stages cause damage throughout the life cycle of sorghum. Larvae first occur at bottom leaves and later on move towards upper canopy. With the help of piercing mouth parts, aphid sucks juice from plant tissues and slows down plant growth (Guo et al. 2011). Honeydew moth (*Cryptoblabes gnidiella*), mainly found in China, occurs in drought conditions. The female lays eggs on ears and glumes of sorghum. Larvae feed on ear till pupation and cause up to 40% yield losses. Corn earworm (*Helicoverpa zea*) and fall armyworm (*Spodoptera frugiperda*) are also major insect pests of sorghum that feed on developing sorghum seeds to cause significant yield losses. Mixed species of *S. frugiperda* and *H. zea* feed on sorghum panicle in US sorghums (Pendleton et al. 2000). Both species are polyphagous, and female lays eggs in whorl stage or silking corn. Plant blossoms, buds, and fruits are particularly affected by larvae (Sloderbeck et al. 2008).

Different agronomic and entomological practices are used to control insect pests of sorghum, that is, an experiment was carried out on rabbi sorghum to determine the effectiveness of seed-treated chemicals against shoot fly, aphid, and shoot bug (*Peregrinus maidis*). Sorghum seeds were treated with thiamethoxam 70 WS at the rate of 3 g/kg, and significant control was observed (Balikai and Bhagwat 2010). Similarly, tillage, crop rotation, intercropping, and fertilizer application were also found effective for controlling insect pests (Spurthi et al. 2009). Different botanicals may also be used for controlling insect's pests. Lesser grain borers were controlled by help of calamus rhizomes (1%) up to 180 days after treatment (Kudachi and Balikai 2009). The spraying of leaves extract of Dursban or Neem (*Azadirachta indica*) in quantity of 5.3 mg/ml on sorghum proved very effective in controlling aphids. Biorational pesticides, that is, endosulfan, carbofuran, Neem extract, and *Bt* (*Bacillus thuringiensis*), were applied for the control of stem borers. Bt gave best leaf protection, while stem was best protected by Neem extract (Jose and Patil 2010).

An effective strategy for insect control is development and cultivation of resistant varieties of sorghum. Resistance against stem borer *C. partellus* was observed in SSV-7073 variety (Marulasiddesha et al. 2007). Similarly, three genotypes ASSVT-8, ASSVT-9, and ASSVT-10 were found resistant against various borers, and two genotypes NSSH-104 and RSSV-9 were resistant against sorghum top borer (Gupta et al. 2007). Resistance is mainly coded by different genes. An aphid resistance locus was located on linkage group 9; similarly, genes linked to different insect resistance may be present on other linkage groups (Chang et al. 2006). During last few years, focus of researchers is on developing transgenic insect-resistant plants. Sorghum plant has been transformed to develop resistance against stem borers using Cry genes from *Bacillus thuringiensis* and Protease inhibitor gene that take part in plant's natural defence system (O'Kennedy et al. 2006). Protease inhibitor gene produces cumulative protection in plants against fungal, viral, bacterial, and insect diseases. The strategy of using more than one inhibitor genes seems to be more effective as it affects different digestive inhibitors in the insects (Babu et al. 2003).

6.6.10 Sorghum Diseases

6.6.10.1 Fungal Diseases

There are about 40 fungal genera that affect sorghum yield including *Alternaria spp., Cladosporium spp., Fusarium spp., Rhizopus spp., Penicillium spp., Aspergillus spp.*, and *Helminthosporium* (da Silva et al. 2000). Most commonly isolated endophytic fungi from sorghum are *Fusarium spp.,* which produces fumonisis (a special group of mycotoxins). The consumption of such contaminated feed causes diseases in animals with high risk of oesophageal cancer in humans (Cavaglieri et al. 2004). Grain moulds of sorghum are caused by complex pathogenic fungi. It disturbs caryopsis and results in deterioration of stored grains due to presence of mycotoxins. Favourable conditions for spread of grain moulds are 85–90% humidity, rainy weather, and 25–35°C temperature (Sharma et al. 2011). The estimated annual losses due to grain moulds are $130 million (Bandyopadhyay et al. 2002). Chitinases are of particular importance as they provide resistance against mould infection. These destroy fungal cell wall by breaking β-1, 4-N acetylglucosamine linkage. Resistant cultivars have mechanisms for synthesis and accumulation of higher level of Chitinases as compared to susceptible one.

Alternaria spp. are other economically important fungi that appear as dark hyphal growth on surface of sorghum kernels. *S. reilianum*, a causative agent of head smut, attacks on sorghum plant at seedling stage, but visual symptoms do not appear till flowering. Sorghum resistance to head smut is outcome of two or more different host mechanisms. Response of sorghum to *S. reilianum* was classified into four categories, that is, R1, R2, R3, and S1 based on exposure of plant to head smut. *Claviceps Africana*, one of the ergot pathogens, causes sorghum ergot or sugary disease. It targets unfertilized ovary by penetrating through stigma. Hyphae reaches ovary just like germinating pollen and produces a structure called sphacelia that secretes opaque substance (Honeydew). Sphacelia contains sugar and large number of conidia that become source of secondary infection in field. Atmospheric temperature in the range of 10 to 22.5 °C and relative humidity around 30% promote ergot rot of sorghum (Montes-García et al. 2009). Plants respond either through pollen-based resistance (through self-pollination by resistance plants) or non-pollen-based resistance (a combination of short duration exposure of stigma to inoculum before pollination followed by rapid drying of stigma after pollination) (Dahlberg et al. 2001).

Another fungus, *Colletotrichum sublineolum,* causes anthracnose of sorghum and is widely distributed in humid regions of Asia, Africa, and America (Figueiredo et al. 2010). The symptoms include decreased photosynthate accumulation and stalk lodging, resulting in decreased grain quantity and quality. It causes approximately 50% yield losses under high temperature and dry environments (Ngugi et al. 2002). Various physical (surface and structural barrier) and chemical (phytoalexins) strategies are used to protect anthracnose. Phytoalexins have fungicidal activity in response to invasion (Singh and Boora 2008). Four resistant lines CMSXS657, ATF14, ATF08, and CMSXS210 have been identified against 20 virulence isolates of ergot rot collected from different sorghum-producing regions of Brazil (Pereira et al. 2011). Downy mildew caused by oomycete *Peronosclerospora sorghi* also damages sorghum crop. The pathogen infects roots, and plants become chlorotic, show stunted growth, and are partially or fully sterile with approximately 78% yield losses (Thakur et al. 2007). Different agronomic (i.e. crop rotation, proper planting time, deep ploughing, trap crops) and entomological (use of chemical compositions, i.e., metalaxyl or fosetyle-Al) measures were found effective in controlling downy mildew of sorghum. Long-term effective solution is development of downy mildew-resistant genotypes as two genotypes (SC170-6-17 and 9910296) showed immune response against downy mildew (Barbosa et al. 2005).

6.6.10.2 Weeds

Weeds compete with agricultural crops for essential factors like light, water, and nutrients, or by providing a reservoir for insects and diseases contaminate the seed lots. Weeds are controlled by use of herbicides. For instance, herbicide-tolerant transgenic resistant crops having 5-enolpyruvylshikimate-3-phosphate synthase (cp4) gene resulted in the increased use of glyphosate which has now become most widely used herbicide in agriculture because of its effectiveness in controlling recalcitrant weeds such as Johnsongrass (Duke and Powles 2008). Nevertheless, the continuous use of these herbicides for a long period of time results in the development of resistance in several weed species. Because of development of resistance in weeds and tougher registration criteria and environmental regulations, there is lack of novel herbicides particularly in Europe, which is a threat to crop production worldwide (Busi et al. 2013).

Witchweed (*Striga spp.*) is one of the limiting factors of sorghum grain which has a direct effect on its production and causes 20–100% yield loss in case of severe infestation in many areas of Africa and Asia (Parker 2009). It competes with host for nutrients and water for survival and produces toxic materials called phytotoxins which are harmful to host. It ultimately results in 'witch appearance of host plant' characterized by stunting and withering growth. Striga seeds have the ability to remain dormant and viable in the soil for 20 years resulting in increased infestation. Low germination stimulant production, low production of the haustorial initiation factor, avoidance mechanisms, physical barriers, hypersensitive response, and antibiosis are different control measures to control infestation of witchweed. Low germination stimulant controlled by a single recessive gene (Igs) is most effective. Markers-assisted selection of Igs gene in the background may enhance the development of Striga-resistant cultivars. Igs gene was mapped on SBI-05 between two tightly linked microsatellite markers SB3344 and SB3352 at a distance of 0.5 and 1.5 cm, respectively (Satish et al. 2012).

Beyond low germination stimulant production mechanism, there are many other resistant phenotypes that have been identified and to some extent being used in breeding programmes. Wild and cultivated sorghums were screened by using a laboratory method for their ability to cause haustorial initiation of germinated *S. asiatica* (L.) Kuntze, and wild accessions showed less haustorial formation in comparison to cultivated sorghum (Gamar et al. 2018; Mohemed et al. 2018). Moreover, pyramiding genes for various mechanisms of striga resistance have been exploited. For striga resistance, different traits of LGS, LHF, HR, and IR genes were stacked in improved varieties and African landraces. These genes have successfully been transferred to several sorghum cultivars and exploited in different countries (Mbuvi et al. 2017).

6.6.10.3 Viruses

Maize mosaic virus (MMV), also known as sorghum chlorosis virus, is an RNA virus that was first rhabdovirus identified in plants. There are many strains of this virus that cause diseases in the sorghum (Lim et al. 2018). Infection starts when the vector *Peregrines maidis* (Ashmead) enters the virus and virus then attacks plant after getting into insect nerves system. Initial attack results in appearance of chlorotic streaks on veins which later on become necrotic on sheath and stalk and leads to stunted growth. Maize streak virus is another major constraint of sorghum especially in Sub-Saharan Africa where it was identified in 1900 and named as 'mealie variegation' and later renamed as maize streak. Maize stripe virus attacks sorghum in tropical and subtropical regions where it causes significant yield loss. Virus affects the mesophyll, phloem, and cytoplasmic fluid where its particle is commonly observed. Later on, chlorotic stripes and bands appear on leaves, which results in stunted panicles, and in severe cases, plant death may occur. Sorghum mosaic virus, which was thought to be a strain of sugarcane mosaic virus, was later on identified as new specie. It causes mosaic and red leaf and sometime yellow green streak, which turns to reddish brown streak, strips, and finally reddish-brown necrosis. Virus perpetuates through aphids and spreads through ratoon crop of sugarcane in the adjacent field of maize and sorghum (Zhang et al. 2016a,b).

6.11 Strategies to Cope with Changing Climate in Sorghum

Conventional and modern plant breeding techniques have helped breeders in crop improvement through development of climate-resilient crops, particularly crops tolerant to abiotic stress, that is, drought, heat, waterlogging, cold, and heavy metal stress. Similarly, crops resistant to different biotic stresses, that is, diseases, insects, viruses, and weeds, were developed by combined use of conventional as well as modern plant breeding tools as described below.

6.11.1 Cultural Practices for Stress Tolerance

Farmers are trying with different cultural practices to tackle the climatic variation for plant adaptation since the birth of this issue. There are many useful approaches adopted by farmers, including abiotic factors such as altering planting and harvesting time, a collection of crops with short life cycles, crop rotation, irrigation techniques, and variation in cropping schemes. Under climatic stress conditions, all of these approaches have been proved beneficial for crop adaptability (Marcinkowski and Piniewski 2018; Teixeira et al. 2018; Deligios et al. 2019). However, the alteration in sowing time, cultivation of drought resistant cultivars, and the domestication of new crops are, thus, some important strategies to lessen the climatic variability danger and provide better adaptability to crop plants for assuring food safety and security.

6.11.2 Conventional Breeding Approaches for Stress Tolerance

Conventional breeding approaches include different breeding procedures, that is, selection (pure line selection, mass selection), introduction, hybridization, wild hybridization, backcrossing, composite crossing, multiline breeding, polyploidy, and heterosis breeding. Hybridization is followed by different breeding methods for the achievement of homozygosity in the filial generation such as pedigree method, bulk, single seed descent, ear to row, and many other methods which take eight to ten years for development and release of commercial cultivars for field testing as described briefly by Breseghello and Coelho (2013). All these techniques are used to develop ideotype plants. In ideotype breeding, all simpler traits are targeted that are directly or indirectly related to our trait of interest, and best combinations are selected for testing in the advanced generations. Detail of different varieties developed through breeding approaches against different stresses is given in Table 6.1. However, conventional breeding approaches are too much time and resources consuming.

6.11.3 Molecular Breeding Approaches for Stress Tolerance

6.11.3.1 Tissue Culture and Mutation Breeding

Modern breeding techniques include use of tissue culture, mutation breeding, genomics, marker-assisted breeding, QTL mapping, bioinformatics tools, and omics approaches. Tissue culture can be opted to create genetic variation with respect to different biotic and abiotic stresses. It can be through somaclonal variation or through use of mutagenic agent, that is, ethyl methane sulphonate and other DNA chelating chemical mutagen followed by screening for desired mutations through selection media (Khalil et al. 2018). Mutation breeding is used to speed up the process of normal breeding as it creates large amount of genetic variation through use of physical and chemical mutagens and selection of desired mutants in the M_2 generation followed by different breeding procedures for development of cultivars and achievement of homozygosity (Bado et al. 2015).

6.11.3.2 Genomics Approaches

Genomics approaches for the improvement of crop plant include the use of different marker systems for selection of trait of interest (gene of interest) through linked DNA markers. There is a range of markers system being used in the improvement and markers-assisted selection of different traits. Markers were broadly classified into PCR based and non-PCR-based systems. Non-PCR-based markers, RFPLs, are now considered very old and obsolete. Now days, PCR-based markers are used for gene tagging and marker-assisted selection. Among these, the robust and highly reproducible and reliable markers are SSR markers. However, now a days, focus is shifting towards single-nucleotide polymorphisms' (SNPs) marker system as these are reliable markers for marker-assisted breeding (MAB). However, genomics has progressed one step ahead, and genotyping by sequencing (GBS) is fulfilling the objective of marker-assisted breeding and is relatively more efficient.

6.11.3.3 Bioinformatics and Omics Approaches

Sequencing of sorghum genome has laid the foundation of different bioinformatics studies which are helping in identification of gene families involved in different biotic and abiotic stress responses. This is being accomplished by comparing the genome of sorghum with model organism. Use of *in silico*

TABLE 6.1

Resistance Cultivars Against Different Abiotic Stresses

Resistant cultivars	Abiotic resistance	Reference
PGRI-191, PGRI-35, PARC-SS-1, and H-18	Drought tolerance	Jagtap et al. (1998)
IS 22380, PA 59, and M 35-1	Drought tolerance	Jagtap et al. (1998)
E 36-1	Heat, drought, and light stress tolerance	Jagtap et al. (1998)
'Shan Qui Red' SQR	Cold tolerance	Knoll and Ejeta (2008)
JS-2002 and Sandalbar	Salt tolerance	Abbass (2012)
CSF20	Salt tolerant	Alves dos Reis et al. (2005)
ATF10B	Aluminium tolerance	Magalhaes et al. (2007)

expression studies and RNA-seq data has played significant role in identification of different gene families and their active members that are helpful in overcoming different biotic and abiotic stresses (Baek et al. 2019). Omics has emerged as a new era of system biology in the recent past. It refers to comprehensive analysis of biological systems and has transformed cell biology study from one gene or protein analysis to understanding of whole organelle and pathway simultaneously. Omics involves high-throughput measurements of collection of protein in a cell (the proteome), the collection of RNA transcribed from a gene (the transcriptome), or collection of metabolites (the metabolome). These approaches have greatly enhanced the understanding of plant breeders about the different metabolic pathways involved in response to certain stress and have made significant contribution towards variety development (Varshney et al. 2009). These techniques must be used in harmony for development of stress-tolerant and climate-resilient sorghum plants.

6.11.3.4 Genetic Engineering for Stress Tolerance

Biotechnology is an influential approach for genetic manipulation of the genome for the betterment of human beings. The genetic modification through biotechnology is a powerful strategy. Encouraging data is collected from genetics which can be exploited significantly to various biotic and abiotic stresses such as salinity, drought, heat, and cold. Identification of stress-responsive genes/transcription factors (TFs) is powerful finding to develop stress-resistant crop cultivars. These genes/TFs can control the phenotypes of genes in genetic-engineered crops associated with various stresses (Reynolds et al. 2015). There are numerous transgenic plants which have been established by genetic engineering to tackle the biotic and abiotic stresses. These genetically engineered plants demonstrate significant resistance against climatic variations compared to normal plants (Shah et al. 2016; Nejat and Mantri 2017; Ali et al. 2018).

6.11.3.5 Genome Editing Strategies for Stress Tolerance

In recent years, genome editing (GE) has emerged as one of the most powerful strategies to combat against the factors affecting crops' productivity. It has revolutionized the plant sciences and played a significant role in the improvement and development of new varieties. GE techniques manipulate the plant genome by means of sequence-specific nucleases. GE for crop improvement has the remarkable ability to tackle food insecurity and develop a climate-smart agriculture system globally. In traditional plant breeding approaches, genes are discovered to be associated with various important traits by means of mutation and conventional breeding strategies, which has been recognized as a significant technique for the development of elite and high yielding germplasm. Genetic diversity of various elite varieties has been substantially decreased due to the exploitation of important crops extensively that has, in various circumstances, been associated with the enhancement of the susceptibility to several abiotic and biotic stresses.

Plant breeding approaches have been greatly influenced by the GE tools in exploring new strategies for fast and accurate manipulations in crop genomes to protect them against different stresses and improve crop quality and yield (Fiaz et al. 2019). Thus, developing the novel modifications in the gene pool of various plant germplasm is required under abiotic and biotic stresses for the improvement of elite crop varieties with great ability to produce high yielding crops. In genome editing technology, site-specific endonucleases are used comprising of zinc-finger nucleases (ZFNs), transcription activator like effector nucleases (TALENs), and CRISPR-Cas9. In contrast to the ZENs and TALENs genome editing tools, the CRISPR/Cas9 system is emerging as the most powerful GE strategy because it is economical, rapid, accurate, and enables multiple site-specific editing within the genome. Many success stories have been published that witness the efficiency and the reproducibility of the results through these technologies. In addition, base editing technology (Yan et al. 2018), in which base A converts into base G and base C converts into T, and speeding breeding (Watson et al., 2018) is an emerging technology which can support CRISPR/Cas9 in coming days to facilitate the development of climate-resilient crops. Certainly, we shall witness many reports on GE in near future.

6.12 Conclusion

Changes in climate are seriously threatening not only to the agriculture and its products but also to the people on earth. Global warming is increasing day by day due to intensive industrialization and heavy production of poisonous gases, which ultimately disturbs the world's environment. Different changes in climate are directly affecting sorghum growth and yield. Being sessile in nature, plants particularly sorghum suffer a lot due to different stresses, but among those, the abiotic stresses (e.g. heat, cold, drought, salinity, etc.) are the major ones. So, there is the need of time to understand the genetic basis and the mechanisms involved in different responses against different stresses. Changing climate has both positive and negative outcomes, but the negative effects are more thought-provoking. Rapidly changing climate is yielding unstoppable losses. So, to tackle this issue, some sound strategies must be opted. Besides conventional breeding, new breeding technologies (NBTs) such as genome editing and speed breeding are urgently needed. Though different approaches of biotechnology have been significantly applied to develop genetically modified (GM) plants with enhanced resistance against different a/biotic stress responses. But due to the label of 'GM organisms,' their future is ambiguous. Briefly, in future, we have to make eco-friendly modifications and edits in the genome of plants (e.g. sorghum), for example, genome editing via CRISPR/Cas systems can produce transgene-free plants which can be publicly accepted. So, NBTs can help the scientists and farmers to battle against climate change and develop climate-resilient sorghum verities.

REFERENCES

Abbass Q (2012) Evaluation of sorghum varieties/lines for salt tolerance using physiological indices as screening tool. Pak J Bot 44:47–52.

Adedapo, A, Dina O, Saba A, Oladipo O (2002) Evaluation of *Telfaria occidentalis* and *Sorghum bicolor* extracts as potent haematinics in domestic rabbits. Nigerian J Anim Prod 29:88–93.

Adhikari U, Pouyan A, Nejadhashemi, Sean AW (2015) Climate change and eastern Africa: a review of impact on major crops. Food Energy Sec 4:110–132.

Adnan M, Fahad S, Khan IA, Saeed M, Ihsan MZ, Saud S, Riaz M, Wang D, Wu C (2019) Integration of poultry manure and phosphate solubilizing bacteria improved availability of Ca bound P in calcareous soils. 3 Biotech 9:368.

Adnan M, Fahad S, Muhammad Z, Shahen S, Ishaq AM, Subhan D, Zafar-ul-Hye M, Martin LB, Raja MMN, Beena S, Saud S, Imran A, Zhen Y, Martin B, Jiri H, Rahul D (2020) Coupling phosphate-solubilizing bacteria with phosphorus supplements improve maize phosphorus acquisition and growth under lime induced salinity stress. Plants 9:900. https://doi.org/10.3390/plants9070900.

Adnan M, Shah Z, Sharif M, Rahman H (2018b) Liming induces carbon dioxide (CO_2) emission in PSB inoculated alkaline soil supplemented with different phosphorus sources. Environ Sci Poll Res 25:9501–9509.

Adnan M, Zahir S, Fahad S, Arif M, Mukhtar A, Imtiaz AK, Ishaq AM, Abdul B, Hidayat U, Muhammad A, Inayat-Ur R, Saud S, Muhammad ZI, Yousaf J, Amanullah, Hafiz MH, Wajid N (2018a) Phosphate-solubilizing bacteria nullify the antagonistic effect of soil calcification on bioavailability of phosphorus in alkaline soils. Sci Rep 8:4339. https://doi.org/10.1038/s41598-018-22653-7.

Ahmad S, Kamran M, Ding R, Meng X, Wang H, Ahmad I, Fahad S, Han Q (2019) Exogenous melatonin confers drought stress by promoting plant growth, photosynthetic capacity and antioxidant defense system of maize seedlings. PeerJ 7:e7793. http://doi.org/10.7717/peerj.7793.

Akande I, Oseni A, Biobaku O (2010) Effects of aqueous extract of *Sorghum bicolor* on hepatic, histological and haematological indices in rats. J Cell Anim Biol 4:137.

Akram R, Turan V, Hammad HM, Ahmad S, Hussain S, Hasnain A, Maqbool MM, Rehmani MIA, Rasool A, Masood N, Mahmood F, Mubeen M, Sultana SR, Fahad S, Amanet K, Saleem M, Abbas Y, Akhtar HM, Waseem F, Murtaza R, Amin A, Zahoor SA, ul Din MS, Nasim W (2018a) Fate of organic and inorganic pollutants in paddy soils. In: Hashmi MZ, Varma A (eds) Environmental pollution of paddy soils, soil biology. Springer International Publishing Ag, Switzerland, pp 197–214.

Akram R, Turan V, Wahid A, Ijaz M, Shahid MA, Kaleem S, Hafeez A, Maqbool MM, Chaudhary HJ, Munis, MFH, Mubeen M, Sadiq N, Murtaza R, Kazmi DH, Ali S, Khan N, Sultana SR, Fahad S, Amin A,

Nasim W (2018b) Paddy land pollutants and their role in climate change. In: Hashmi MZ, Varma A (ed) Environmental pollution of paddy soils, soil biology. Springer International Publishing Ag Switzerland, pp 113–124.

Ali S, Xu Y, Jia Q, Ma X, Ahmad I, Adnan M, Gerard R, Ren X, Zhang P, Cai T, Zhang J (2018) Interactive effects of plastic film mulching with supplemental irrigation on winter wheat photosynthesis, chlorophyll fluorescence and yield under simulated precipitation conditions. Agric Water Manage 30:1–4.

Almodares A, Sharif M (2007) Effects of irrigation water qualities on biomass and sugar contents of sugar beet and sweet sorghum cultivars. J Env Bio 28:213–218.

Alves dos Reis T, Zorzete P, Pozzi CR, Nascimento da Silva V, Ortega E, Corrêa B (2010) Mycoflora and fumonisin contamination in Brazilian sorghum from sowing to harvest. J Sci Food Agri 90:1445–1451.

Anyamba A, Small JL, Britch SC, Tucker CJ, Pak EW, Reynolds CA, Crutchfield J, Linthicum KJ (2014) Recent weather extremes and impacts on agricultural production and vector-borne disease outbreak patterns. PLoS One 9:e92538.

Apel K, Hirt H (2004) Reactive oxygen species: metabolism, oxidative stress, and signal transduction. Annu Rev Plant Biol 55:373–399.

Ariani A, Gepts P (2015) Genome-wide identification and characterization of aquaporin gene family in common bean (*Phaseolus vulgaris* L.). Mol Genet Genom 290:1771.

Armstrong W, Brändle R, Jackson MB (1994) Mechanisms of flood tolerance in plants. Acta Bot Neerlandica 43:307–358.

Ashraf MA, Ahmad MS, Ashraf M, Al-Qurainy F, Ashraf MY (2011) Alleviation of waterlogging stress in upland cotton (*Gossypium hirsutum* L.) by exogenous application of potassium in soil and as a foliar spray. Crop Pasture Sci 62:25–38.

Ashraf MA, Ashraf M, Ali Q (2010) Response of two genetically diverse wheat cultivars to salt stress at different growth stages: leaf lipid peroxidation and phenolic contents. Pak J Bot 42:559–565.

Awika JM, Rooney LW (2004) Sorghum phytochemicals and their potential impact on human health. Phytochem 65:1199–1221.

Aziz K, Daniel KYT, Fazal M, Muhammad ZA, Farooq S, Fan W, Fahad S, Ruiyang Z (2017a) Nitrogen nutrition in cotton and control strategies for greenhouse gas emissions: a review. Environ Sci Pollut Res 24:23471–23487. https://doi.org/10.1007/s11356-017-0131-y.

Aziz K, Daniel KYT, Muhammad ZA, Honghai L, Shahbaz AT, Mir A, Fahad S (2017b) Nitrogen fertility and abiotic stresses management in cotton crop: a review. Environ Sci Pollut Res 24:14551–14566. https://doi.org/10.1007/s11356-017-8920-x.

Babu RM, Sajeena A, Seetharaman K, Reddy MS (2003) Advances in genetically engineered (transgenic) plants in pest management—an over view. Crop Prot 22:1071–1086.

Bado S, Forster BP, Nielen S, Ghanim A, Lagoda PJ, Till BJ, Laimer M (2015) Plant mutation breeding: current progress and future assessment. Plant Breed Rev 39:23–88.

Baek YS, Goodrich LV, Brown PJ, James BT, Moose SP, Lambert KN, Riechers DE (2019) Transcriptome profiling and genome-wide association studies reveal GSTs and other defense genes involved in multiple signaling pathways induced by Herbicide Safener in grain sorghum. Front Plant Sci 10:192.

Balikai R, Bhagwat V (2010) Evaluation of integrated pest managment components for the management of shoot fly, shoot bug and aphid in rabi sorghum. Karna J Agri Sci 22:532–534.

Bandyopadhyay R, Frederickson DE, McLaren NW, Odvody GN (1996) Ergot-A global threat to sorghum. Int Sorghum Millet Newsl 37:1–32.

Bandyopadhyay R, Little CR, Waniska RD, Butler DR (2002) Sorghum grain mold: through the 1990s into the new millennium. Sorghum Millet Dis 173–183.

Barbosa FC, Casela CR, Pfenning LH, Santos FG (2005) Identification of sources of resistance in sorghum to *Peronosclerospora sorghi*. Fitopatol Bras 30:522–524.

Baseer M, Adnan M, Fazal M, Fahad S, Muhammad S, Fazli W, Muhammad A, Jr. Amanullah, Depeng W, Saud S, Muhammad N, Muhammad Z, Fazli S, Beena S, Mian AR, Ishaq AM (2019) Substituting urea by organic wastes for improving maize yield in alkaline soil. J Plant Nutr https://doi.org/10.1080/01904167.2019.1659344.

Beck P, Hutchison S, Gunter S, Losi T, Stewart C, Capps P, Phillips J (2007) Chemical composition and in situ dry matter and fiber disappearance of sorghum× Sudangrass hybrids. J Anim Sci 85:545–555.

Berenji J, Dahlberg J (2004) Perspectives of sorghum in Europe. J Agr Crop Sci 190:332–338.

Berenji J, Dahlberg J, Sikora V, Latkovi D (2011) Origin, history, morphology, production, improvement, and utilization of broomcorn [*Sorghum bicolor* (L.) Moench] in Serbia. Econ Bot 65:190–208.

Bibi A, Sadaqat HA, Tahir MHN, Akram HM (2012) Screening of sorghum (*Sorghum bicolor* var Moench) for drought tolerance at seedling stage in polyethylene glycol. J Anim Plant Sci 22:671–678.

Bienert GP, Møller AL, Kristiansen KA, Schulz A, Møller IM, Schjoerring JK, Jahn TP (2007) Specific aquaporins facilitate the diffusion of hydrogen peroxide across membranes. J Biol Chem 282:1183–1192.

Bogs J, Ebadi A, McDavid D, Robinson SP (2006) Identification of the flavonoid hydroxylases from grapevine and their regulation during fruit development. Plant Physiol 140:279–291.

Borneo R, León AE (2012) Whole grain cereals: functional components and health benefits. Food and Fun 3:110–119.

Breseghello F, Coelho AS (2013) Traditional and modern plant breeding methods with examples in rice (*Oryza sativa* L.). J Agri Food Chem 61:8277–8286.

Burow G, Burke JJ, Xin Z, Franks CD (2011) Genetic dissection of early-season cold tolerance in sorghum (*Sorghum bicolor* (L.) Moench). Mol Breed 28:391–402.

Busi R, Vila-Aiub MM, Beckie HJ, Gaines TA, Goggin DE, Kaundun SS, Lacoste M, Neve P, Nissen SJ, Norsworthy JK, Renton M (2013) Herbicide-resistant weeds: from research and knowledge to future needs. Evo App 6:1218–1221.

Camargo FI, Cortez D, Ueda-Nakamura T, Nakamura C, Dias FB (2008) Antiviral activity and mode of action of a peptide isolated from *Sorghum bicolor*. Phytomed 15:202–208.

Carson L, Setser C, Sun XS (2000) Sensory characteristics of sorghum composite bread. Int J Food Sci Tech 35:465–471.

Cavaglieri L, Passone A, Etcheverry M (2004) Screening procedures for selecting rhizobacteria with biocontrol effects upon *Fusarium verticillioides* growth and fumonisin B1 production. Res Microbiol 155:747–754.

Chakraborty U, Pradhan D (2011) High temperature-induced oxidative stress in Lens culinaris, role of antioxidants and amelioration of stress by chemical pre-treatments. J Plant Int 6:43–52.

Chang J, Xia X, Zhang L (2006) Analysis of the resistance gene to the sorghum aphid, *Melanaphis sacchari*, with SSR marker in *Sorghum bicolor*. Acta Pratacu Sin 15:113.

Ciacci C, Maiuri L, Caporaso N, Bucci C, Del GL, Massardo DR, Loerger B (2007) Celiac disease: in vitro and in vivo safety and palatability of wheat-free sorghum food products. Clin Nut 26:799–805.

Ciscar JC (2012) The impacts of climate change in Europe (the PESETA research project). Climat Chang 112:1–6.

da Silva JB, Pozzi CR, Mallozzi MA, Ortega EM, Corrêa B (2000) Mycoflora and occurrence of aflatoxin B1 and fumonisin B1 during storage of Brazilian sorghum. J Agri Food Chem 48:4352–4356.

Dahlberg JA, Bandyopadhyay R, Rooney WL, Odvody GN, Madera-Torres P (2001) Evaluation of sorghum germplasm used in US breeding programmes for sources of sugary disease resistance. Plant Pathol 50:681–689.

Damte T, Pendleton BB, Almas LK (2009) Cost-benefit analysis of sorghum midge, *Stenodiplosis sorghicola* 1 (Coquillett)-resistant sorghum hybrid research and development in Texas. Southwest Entomol 34:395–405.

De Wet JD, Huckabay J (1967) The origin of *Sorghum bicolor*. II. Distribution and domestication. Evo 21:787–802.

Deligios PA, Chergia AP, Sanna G, Solinas S, Todde G, Narvarte L, Ledda L (2019) Climate change adaptation and water saving by innovative irrigation management applied on open field globe artichoke. Sci Total Environ 649:461–472.

Depeng W, Fahad S, Saud S, Muhammad K, Aziz K, Mohammad NK, Hafiz MH, Wajid N (2018) Morphological acclimation to agronomic manipulation in leaf dispersion and orientation to promote "Ideotype" breeding: Evidence from 3D visual modeling of "super" rice (*Oryza sativa* L.). Plant Physiol Biochem 135. https://doi.org/10.1016/j.plaphy.2018.11.010.

Dhaliwal GS, Jindal V, Dhawan AK (2010) Insect pest problems and crop losses: changing trends. Ind J Ecol 37:1–7.

Dicko MH, Gruppen H, Traoré AS, Voragen AG, Van BWJ (2006) Sorghum grain as human food in Africa: relevance of content of starch and amylase activities. Afr J Biotech 5:384–395.

Dua YH, Sun GZ, Zou JQ, Lu F (2008) Distribution of holes harmed by Asian corn borer in sorghum and its relation with yield damage. Liaoning Agri Sci 4:004.

Duke SO, Powles SB (2008) Glyphosate: a once-in-a-century herbicide. Pest Manag Sci 64:319–325.

Duvall MR, Doebley JF (1990) Restriction site variation in the chloroplast genome of Sorghum (Poaceae). Syst Bot 472–480.

Ercoli L, Mariotti M, Masoni A, Arduini I (2004) Growth responses of sorghum plants to chilling temperature and duration of exposure. Eur J Agron 2:93–103.

Fahad S, Adnan M, Hassan S, Saud S, Hussain S, Wu C, Wang D, Hakeem KR, Alharby HF, Turan V, Khan MA, Huang J (2019b) Rice responses and tolerance to high temperature, In: Hasanuzzaman M, Fujita M, Nahar K, Biswas JK (eds) Advances in rice research for abiotic stress tolerance. Woodhead Publ Ltd, Cambridge, England, pp 201–224.

Fahad S, Bajwa AA, Nazir U, Anjum SA, Farooq A, Zohaib A, Sadia S, NasimW, Adkins S, Saud S, Ihsan MZ, Alharby H, Wu C, Wang D, Huang J (2017) Crop production under drought and heat stress: plant responses and management options. Front Plant Sci 8:1147. https://doi.org/10.3389/fpls.2017.01147.

Fahad S, Bano A (2012) Effect of salicylic acid on physiological and biochemical characterization of maize grown in saline area. Pak J Bot 44:1433–1438.

Fahad S, Chen Y, Saud S, Wang K, Xiong D, Chen C, Wu C, Shah F, Nie L, Huang J (2013) Ultraviolet radiation effect on photosynthetic pigments, biochemical attributes, antioxidant enzyme activity and hormonal contents of wheat. J Food Agri Environ 11:1635–1641.

Fahad S, Hussain S, Bano A, Saud S, Hassan S, Shan D, Khan FA, Khan F, Chen Y, Wu C, Tabassum MA, Chun MX, Afzal M, Jan A, Jan MT, Huang J (2014a) Potential role of phytohormones and plant growth-promoting rhizobacteria in abiotic stresses: consequences for changing environment. Environ Sci Pollut Res 22:4907–4921. https://doi.org/10.1007/s11356-014-3754-2.

Fahad S, Hussain S, Matloob A, Khan FA, Khaliq A, Saud S, Hassan S, Shan D, Khan F, Ullah N, Faiq M, Khan MR, Tareen AK, Khan A, Ullah A, Ullah N, Huang J (2014b) Phytohormones and plant responses to salinity stress: A review. Plant Growth Regul 75:391–404. https://doi.org/10.1007/s10725-014-0013-y.

Fahad S, Hussain S, Saud S, Hassan S, Chauhan BS, Khan F et al (2016a) Responses of rapid viscoanalyzer profile and other rice grain qualities to exogenously applied plant growth regulators under high day and high night temperatures. PLoS One 11:e0159590. https://doi.org/10.1371/journal.pone.0159590.

Fahad S, Hussain S, Saud S, Hassan S, Ihsan Z, Shah AN, Wu C, Yousaf M, Nasim W, Alharby H, Alghabari F, Huang J (2016c) Exogenously applied plant growth regulators enhance the morphophysiological growth and yield of rice under high temperature. Front Plant Sci 7:1250. https://doi.org/10.3389/fpls.2016.01250.

Fahad S, Hussain S, Saud S, Hassan S, Tanveer M, Ihsan MZ, Shah AN, Ullah A, Nasrullah KF, Ullah S, AlharbyH NW, Wu C, Huang J (2016d) A combined application of biochar and phosphorus alleviates heat-induced adversities on physiological, agronomical and quality attributes of rice. Plant Physiol Biochem 103:191–198.

Fahad S, Hussain S, Saud S, Khan F, Hassan S, Jr A, Nasim W, Arif M, Wang F, Huang J (2016b) Exogenously applied plant growth regulators affect heat-stressed rice pollens. J Agron Crop Sci 202:139–150.

Fahad S, Hussain S, Saud S, Tanveer M, Bajwa AA, Hassan S, Shah AN, Ullah A, Wu C, Khan FA, Shah F, Ullah S, Chen Y, Huang J (2015a) A biochar application protects rice pollen from high-temperature stress. Plant Physiol Biochem 96:281–287.

Fahad S, Muhammad ZI, Abdul K, Ihsanullah D, Saud S, Saleh A, Wajid N, Muhammad A, Imtiaz AK, Chao W, Depeng W, Jianliang H (2018) Consequences of high temperature under changing climate optima for rice pollen characteristics-concepts and perspectives. Arch Agron Soil Sci 64:1479–1488. https://doi.org/10.1080/03650340.2018.1443213.

Fahad S, Nie L, Chen Y, Wu C, Xiong D, Saud S, Hongyan L, Cui K, Huang J (2015b) Crop plant hormones and environmental stress. Sustain Agric Rev 15:371–400.

Fahad S, Rehman A, Shahzad B, Tanveer M, Saud S, Kamran M, Ihtisham M, Khan SU, Turan V, Rahman MHU (2019a) Rice responses and tolerance to metal/metalloid toxicity. In: Hasanuzzaman M, Fujita M, Nahar K, Biswas JK (eds) Advances in rice research for abiotic stress tolerance. Woodhead Publ Ltd, Cambridge, England, pp 299–312.

FAO (2018) FAO statistical pocketbook: World Food and Agriculture 2018; Food and Agriculture Organization of the United Nations: Rome, Italy. Available online: http://www.fao.org/3/CA1796EN/ca1796en.pdf. Accessed 16 March 2019.

Farhana G, Ishfaq A, Muhammad A, Dawood J, Fahad S, Xiuling L, Depeng W, Muhammad F, Muhammad F, Syed AS (2020) Use of crop growth model to simulate the impact of climate change on yield of various wheat cultivars under different agro-environmental conditions in Khyber Pakhtunkhwa, Pakistan. Arabian J Geosci 13:112. https://doi.org/10.1007/s12517-020-5118-1.

Farhat A, Hafiz MH, Wajid I, Aitazaz AF, Hafiz FB, Zahida Z, Fahad S, Wajid F, Artemi C (2020) A review of soil carbon dynamics resulting from agricultural practices. J Environ Manage 268:110319.

Farooq M, Wahid A, Kobayashi N, Fujita DB, Basra SM (2009) Plant drought stress: effects, mechanisms and management. In: Sustainable agriculture. Springer, Dordrecht, pp 153–188.

Fernie AR, Geigenberger P, Stitt M (2005) Flux an important, but neglected, component of functional genomics. Cur Opin Plant Biom 8:174–182.

Fernández-Aparicio M, Huang K, Wafula K, Honaas E, Wickett L, Timko M, Westwood M, (2013) Application of qRT-PCR and RNA-Seq analysis for the identification of housekeeping genes useful for normalization of gene expression values during *Striga hermonthica* development. Mol Bio Rep 40:3395–3407.

Fiaz S, Ahmad S, Noor MA, Wang X, Younas A, Riaz A, Riaz A, Ali F (2019) Applications of the CRISPR/Cas9 System for rice grain quality improvement: perspectives and opportunities. Int J Mol Sci 20:888.

Fidalgo F, Azenha M, Silva AF, de Sousa A, Santiago A, Ferraz P, Teixeira J (2013) Copper-induced stress in *Solanum nigrum* L. and antioxidant defense system responses. Food Energy Sec 2:70–80.

Figueiredo JE, Depaoli HC, Coelho VD, Casela CR, Ferreira AD, Guimaraes CT, Gomes EA, Bressan W (2010) Genetic diversity among Colletotrichum sublineolum pathotypes isolated from sorghum (*Sorghum bicolor*). Revista Brasileira de Milho e Sorgo 5.

Gamar YA, Omer AB, Hatim GM, Elfadil MA (2018) Assessment of Striga resistance in wild relatives of sorghum under field conditions. J Agri Sci 63:367–377.

Geera B, Ojwang LO, Awika JM (2012) New highly stable dimeric 3-deoxyanthocyanidin pigments from *Sorghum bicolor* leaf sheath. J Food Sci 77:566–572.

Gilani GS, Xiao CW, Cockell KA (2012) Impact of antinutritional factors in food proteins on the digestibility of protein and the bioavailability of amino acids and on protein quality. British J Nutr 108:S315–S332.

Guo C, Cui W, Feng X, Zhao J, Lu G (2011) Sorghum insect problems and management. J Integ Plant Biol 53:178–192.

Gupta B, Huang B (2014) Mechanism of salinity tolerance in plants: physiological, biochemical, and molecular characterization. Int J Genom 2014. http://dx.doi.org/10.1155/2014/701596.

Gupta SK, Singh L, Nazir SB, Krishna R (2007) Performance of sweet sorghum genotypes in North India. Cooperative Sugar 39:59.

Gupta KJ, Stoimenova M, Kaiser WM (2005) In higher plants, only root mitochondria, but not leaf mitochondria reduce nitrite to NO, in vitro and in situ. J Exp Bot 56:2601–2609.

Habib ur Rahman M, Ashfaq A, Aftab W, Manzoor H, Fahd R, Wajid I, Md. Aminul I, Vakhtang S, Muhammad A, Asmat U, Abdul W, Syeda RS, Shah S, Shahbaz K, Fahad S, Manzoor H, Saddam H, Wajid N (2017) Application of CSM-CROPGRO-Cotton model for cultivars and optimum planting dates: Evaluation in changing semi-arid climate. Field Crops Res 238:139–152. http://dx.doi.org/10.1016/j.fcr.2017.07.007.

Hafiz MH, Farhat A, Ashfaq A, Hafiz FB, Wajid F, Carol Jo W, Fahad S, Gerrit H (2020a) Predicting kernel growth of maize under controlled water and nitrogen applications. Int J Plant Prod 14:609–620. https://doi.org/10.1007/s42106-020-00110-8.

Hafiz MH, Farhat A, Shafqat S, Fahad S, Artemi C, Wajid F, Chaves CB, Wajid N, Muhammad M, Hafiz FB (2018) Offsetting land degradation through nitrogen and water management during maize cultivation under arid conditions. Land Degrad Dev 1–10. https://doi.org/10.1002/ldr.2933.

Hafiz MH, Muhammad A, Farhat A, Hafiz FB, Saeed AQ, Muhammad M, Fahad S, Muhammad A (2019) Environmental factors affecting the frequency of road traffic accidents: a case study of sub-urban area of Pakistan. Environ Sci Pollut Res 26:11674–11685. https://doi.org/10.1007/s11356-019-04752-8.

Hafiz MH, Wajid F, Farhat A, Fahad S, Shafqat S, Wajid N, Hafiz FB (2016) Maize plant nitrogen uptake dynamics at limited irrigation water and nitrogen. Environ Sci Pollut Res 24:2549–2557. https://doi.org/10.1007/s11356-016-8031-0.

Hamilton EW, Heckathorn SA (2001) Mitochondrial adaptations to NaCl. Complex I is protected by antioxidants and small heat shock proteins, whereas complex II is protected by proline and betaine. Plant Physiol 126:1266–1274.

Harlan JR, Wet JD, Price EG (1973) Comparative evolution of cereals. Evo 27:311–325.

Henshaw TL, Gilbert RA, Scholberg JM, Sinclair TR (2007) Soya bean (*Glycine max* L. Merr.) genotype response to early-season flooding: I. Root and nodule development. J Agron Crop Sci 193:177–188.
Hesham FA, Fahad S (2020) Melatonin application enhances biochar efficiency for drought tolerance in maize varieties: modifications in physio-biochemical machinery. Agron J 1–22.
Hussain MA, Fahad S, Rahat S, Muhammad FJ, Muhammad M, Qasid A, Ali A, Husain A, Nooral A, Babatope SA, Changbao S, Liya G, Ibrar A, Zhanmei J, Juncai H (2020) Multifunctional role of brassinosteroid and its analogues in plants. Plant Growth Regul 92:141–156. https://doi.org/10.1007/s10725-020-00647-8.
Ilyas M, Mohammad N, Nadeem K, Ali H, Aamir HK, Kashif H, Fahad S, Aziz K, Abid U (2020) Drought tolerance strategies in plants: a mechanistic approach. J Plant Growth Regul. https://doi.org/10.1007/s00344-020-10174-5.
Jackson M, Armstrong W (1999) Formation of aerenchyma and the processes of plant ventilation in relation to soil flooding and submergence. Plant Biol 1:274–287.
Jagtap V, Bhargava S, Streb P, Feierabend J (1998) Comparative effect of water, heat and light stresses on photosynthetic reactions in *Sorghum bicolor* (L.) Moench. J Exp Bot 49:1715–1721.
Jan M, Muhammad Anwar-ul-Haq, Adnan NS, Muhammad Y, Javaid I, Xiuling L, Depeng W, Fahad S (2019) Modulation in growth, gas exchange, and antioxidant activities of salt-stressed rice (*Oryza sativa* L.) genotypes by zinc fertilization. Arabian J Geosci 12:775. https://doi.org/10.1007/s12517-019-4939-2.
Jarup L (2003) Hazards of heavy metal contamination. Br Med Bull 68:167–182.
Jose D, Patil R (2010) Evaluation of biorational pesticides for the management of stem borer, Chilo partellus Swinhoe in sweet sorghum. Karnat J Agri Sci 21.
Kadam S, Abril A, Dhanapal AP, Koester RP, Vermerris W, Jose S, Fritschi FB (2017) Characterization and regulation of aquaporin genes of sorghum [*Sorghum bicolor* (L.) Moench] in response to waterlogging stress. Front Plant Sci 30:862.
Kamaran M, Wenwen C, Irshad A, Xiangping M, Xudong Z, Wennan S, Junzhi C, Shakeel A, Fahad S, Qingfang H, Tiening L (2017) Effect of paclobutrazol, a potential growth regulator on stalk mechanical strength, lignin accumulation and its relation with lodging resistance of maize. Plant Growth Regul 84:317–332. https://doi.org/10.1007/ s10725-017-0342-8.
Kant S, Kant P, Lips H, Barak S (2007) Partial substitution of NO^{3-} by NH^{4+} fertilization increases ammonium assimilating enzyme activities and reduces the deleterious effects of salinity on the growth of barley. J Plant Physiol 164:303–311.
Keunen E, Remans T, Bohler S, Vangronsveld J, Cuypers A (2011) Metal-induced oxidative stress and plant mitochondria. Int J Mol Sci 12:6894–6918.
Khalil F, Naiyan X, Tayyab M, Pinghua C (2018) Screening of EMS-induced drought-tolerant sugarcane mutants employing physiological, molecular and enzymatic approaches. Agron 8:226.
Knoll J, Ejeta G (2008) Marker-assisted selection for early-season cold tolerance in sorghum: QTL validation across populations and environments. Theor App Genet 116:541–553.
Kosová K, Prášil I, Vítámvás P (2013) Protein contribution to plant salinity response and tolerance acquisition. Int J Mol Sci 14:6757–6789.
Koźmińska A, Wiszniewska A, Hanus-Fajerska E, Muszyńska E (2018) Recent strategies of increasing metal tolerance and phytoremediation potential using genetic transformation of plants. Plant Biotech Rep 12:1–14.
Kruger M, Van Den Berg J, Du Plessis H (2008) Diversity and seasonal abundance of sorghum panicle-feeding Hemiptera in South Africa. Crop Prot 27:444–451.
Kubien DS, Von Caemmerer S, Furbank RT, Sage RF (2003) C4 photosynthesis at low temperature. A study using transgenic plants with reduced amounts of Rubisco. Plant Physiol 132:1577–1585.
Kudachi D, Balikai R (2009) Efficacy of botanicals for the management of lesser grain borer, *Rhizopertha dominica* Fab. in sorghum during storage. Karnat J Agri Sci 22:487–490.
Lazarides M, Hacker J, Andrew M (1991) Taxonomy, cytology and ecology of indigenous Australian sorghums (*Sorghum Moench*: Andropogoneae: Poaceae). Aus Syst Bot 4:591–635.
Lim S, Yoon Y, Jang YW, Bae DH, Kim BS, Maharjan R, Yi H, Bae S, Lee YH, Lee BC, Park CY (2018) First report of maize yellow mosaic virus infecting *Panicum miliaceum* and *Sorghum bicolor* in South Korea. Plant Dis 102:689.
Liu RH (2007) Whole grain phytochemicals and health. J Cereal Sci 46:207–219.

Liu J, Ishitani M, Halfter U, Kim CS, Zhu JK (2000) The Arabidopsis thaliana SOS_2 gene encodes a protein kinase that is required for salt tolerance. Proc Natl Acad Sci 97:3730–3734.

Magalhaes JV, Liu J, Guimaraes CT, Lana UG, Alves VM, Wang YH, Schaffert RE, Hoekenga OA, Pineros MA, Shaff JE, Klein PE (2007) A gene in the multidrug and toxic compound extrusion (MATE) family confers aluminum tolerance in sorghum. Nat Genet 39:1156–1161.

Mahajan S, Tuteja N (2005) Cold, salinity and drought stresses: an overview. Arch Biochem Biophysics 444:139–158.

Malgwi A, Dunuwel D (2011) Damage and control of panicle insect pests of sorghum (*Sorghum bicolor* (L.) Moench) in the North Eastern Nigeria. ARPN J Sci Technol 3:195–202.

Malik AI, Colmer TD, Lambers H, Setter TL, Schortemeyer M (2002) Short-term waterlogging has long-term effects on the growth and physiology of wheat. New Phytol 153:225–236.

Mansour MMF (2014) The plasma membrane transport systems and adaptation to salinity. J Plant Physiol 171:1787–1800.

Marcinkowski P, Piniewski M (2018) Effect of climate change on sowing and harvest dates of spring barley and maize in Poland. Int Agrophys 32:265–271.

Marulasiddesha KN, Sankar M, Gouda GR (2007) Screening of sorghum genotypes for resistance to damage caused by the stem borer *Chilo partellus* (Swinhoe). Span J Agri Res 5:79–81.

Massacci A, Battistelli A, Loreto F (1996) Effect of drought stress on photosynthetic characteristics, growth and sugar accumulation of field-grown sweet sorghum. Funct Plant Bio 23:331–340.

Mbuvi DA, Masiga CW, Kuria EK, Masanga J, Wamalwa M, Mohamed A, Odeny D, Hamza N, Timko MP, Runo SM (2017) Novel sources of witchweed (Striga) resistance from wild sorghum accessions. Front Plant Sci 8:116.

Mergemann H, Sauter M (2000) Ethylene induces epidermal cell death at the site of adventitious root emergence in rice. Plant Physiol 124:609–614.

Miller F (1982) Genetic and environmental response characteristics of sorghum. In: House LR, Munghogho LK, Peacock JM (eds) Sorghum in the eighties. ICRISAT, Patancheru, India, pp 393–402.

Miransari M (2011) Hyperaccumulators, arbuscular mycorrhizal fungi and stress of heavy metals. Biotech Adv 29:645–653.

Mittra B, Sharma S, Das AB, Henry SL, Das TK, Ghosh P, Ghosh S, Mohanty PA (2008) A novel cadmium induced protein in wheat: characterization and localization in root tissue. Biol Plant 52:343–346.

Mohemed N, Charnikhova T, Fradin EF, Rienstra J, Babiker AG, Bouwmeester HJ (2018) Genetic variation in *Sorghum bicolor* strigolactones and their role in resistance against *Striga hermonthica*. J Exp Bot 69:2415–2430.

Montes-García N, Prom LK, Williams-Alanis H, Isakeit T (2009) Effect of temperature and relative humidity on sorghum ergot development in northern Mexico. Aust Plant Pathol 38:632–637.

Morelli E, Scarano G (2004) Copper-induced changes of non-protein thiols and antioxidant enzymes in the marine microalga *Phaeodactylum tricornutum*. Plant Sci 167:289–296.

Mubeen M, Ashfaq A, Hafiz MH, Muhammad A, Hafiz UF, Mazhar S, Sami ul Din M, Asad A, Amjed A, Fahad S, Wajid N (2020) Evaluating the climate change impact on water use efficiency of cotton-wheat in semi-arid conditions using DSSAT model. J Water Climate Chang 11:1661–1675. https://doi.org/10.2166/wcc.2019.179/622035/jwc2019179.pdf.

Muhammad Z, Abdul MK, Abdul MS, Kenneth BM, Muhammad S, Shahen S, Ibadullah J, Fahad S (2019) Performance of *Aeluropus lagopoides* (mangrove grass) ecotypes, a potential turfgrass, under high saline conditions. Environ Sci Pollut Res 26:13410–13421. https://doi.org/10.1007/s11356-019-04838-3.

Munns R (2002) Comparative physiology of salt and water stress. Plant Cell Env 25:239–250.

Nayyar H, Chander K, Kumar S, Bains T (2005) Glycine betaine mitigates cold stress damage in chickpea. Agron Sustainable Dev 25:381–388.

Nejat N, Mantri N (2017) Plant immune system: crosstalk between responses to biotic and abiotic stresses the missing link in understanding plant defence. Curr Issues Mol Biol 23:1–16.

Ngugi HK, King SB, Abayo GO, Reddy YV (2002) Prevalence, incidence, and severity of sorghum diseases in western Kenya. Plant Dis 86:65–70.

Ogwumike O (2002) Hemopoietic effect of aqueous extract of the leaf sheath of *Sorghum bicolor* in albino rats. Afr J Biomed Res 5:1–2.

Oladiji A, Jacob T, Yakubu M (2007) Anti-anaemic potentials of aqueous extract of *Sorghum bicolor* (L.) Moench stem bark in rats. J Enthnopharmac 111:651–656.

Oliveira NCF, Lobato AK, Gonçalves-Vidigal MC, Costa RC, Santos Filho BG, Alves GA, Maia WJ, Cruz FJ, Neves HK, Lopes MS (2009) Carbon compounds and chlorophyll contents in sorghum submitted to water deficit during three growth stages. J Food Agri Env 7:588–593.

Olías R, Eljakaoui Z, Li JU, De Morales PA, Marín-manzano MC, Pardo JM, Belver A (2009) The plasma membrane Na^+/H^+ antiporter SOS_1 is essential for salt tolerance in tomato and affects the partitioning of Na^+ between plant organs. Plant Cell Env 32:904–916.

Oono Y, Seki M, Nanjo T, Narusaka M, Fujita M, Satoh R, Satou M, Sakurai T, Ishida J, Akiyama K, Iida K (2003) Monitoring expression profiles of Arabidopsis gene expression during rehydration process after dehydration using ca. 7000 full-length cDNA microarray. Plant J 34:868–887.

Ouzounidou G (1993) Changes in variable chlorophyll fluorescence as a result of Cu-treatment: dose-response relations in Silene and Thlaspi. Photosynthetica 29:455–462.

O'Kennedy MM, Grootboom A, Shewry PR (2006) Harnessing sorghum and millet biotechnology for food and health. J Cereal Sci 44:224–235.

Palavecino P, Martín M, Cecilia P, Georgina C, Domínguez, Pablo DR (2016) Chemical composition and physical properties of sorghum flour prepared from different sorghum hybrids grown in Argentina. Starch-Stärke 68:1055–1064.

Parker C (2009) Observations on the current status of Orobanche and Striga problems worldwide. Pest Manage Sci 65:453–459.

Pawlak-Sprada S, Arasimowicz-Jelonek M, Podgórska M, Deckert J (2011) Activation of phenylpropanoid pathway in legume plants exposed to heavy metals. Part I. Effects of cadmium and lead on phenylalanine ammonia-lyase gene expression, enzyme activity and lignin content. Acta Biochim Pol 58:211–216.

Pendleton BB, Teetes GL, Parker RD (2000) Quantifying Texas sorghum growers' use of IPM for insect pests. Southwestern Entomol 25:39–52.

Pereira IS, da Silva DD, Casela CR, Tardin FD, de Abreu MS (2011) Resistance of parent lines and simple hybrids of sorghum to anthracnose fungus *Colletotrichum sublineolum*. Revista Caatinga 24:46–51.

Peterson GC, Ali AE, Teetes GL, Jones JW, Schaefer K (1989) Grain sorghum resistance to midge by yield loss vs. visual scores. Crop Sci 29:1136–1140.

Prasad PV, Staggenborg SA (2009) Growth and production of sorghum and millets. Soils Plant Growth Crop Prod 2.

Ramya S, Ramesh V, Muralidharan J, Purushothaman MR (2017) Fodder yield and chemical composition of hybrid Napier and multi-cut sorghum fodder at different stages of cutting. IJSR 23:181–185.

Ravet K, Touraine B, Boucherez J, Briat JF, Gaymard F, Cellier F (2009) Ferritins control interaction between iron homeostasis and oxidative stress in Arabidopsis. Plant J 57:400–412.

Reddy RH, Balasamy JK, Somanath A, Ramadoss BR, Venkatesan T, Sundaram GM (2016) Rapid screening assay for precise and reliable estimation of cyanide content in sorghum. Aust J of Crop Sci 10:1388.

Reddy K, Zehr UB (2004) Novel strategies for overcoming pests and diseases in India. In: Proceedings of the 4th International Crop Science Congress on New Directions for a Diverse Planet, Brisbane, Australia

Rehman M, Fahad S, Saleem MH, Hafeez M, Habib ur Rahman M, Liu F, Deng G (2020) Red light optimized physiological traits and enhanced the growth of ramie (*Boehmeria nivea* L.). Photosynthetica 58:922–931.

Reynolds M, Tattaris M, Cossani CM, Ellis M, Yamaguchi-Shinozaki K, Saint Pierre C (2015) Exploring genetic resources to increase adaptation of wheat to climate change. In: Advances in wheat genetics: from genome to field. Springer, Berlin/Heidelberg, Germany, pp 355–368.

Ricachenevsky FK, Menguer PK, Sperotto RA, Williams LE, Fett JP (2013) Roles of plant metal tolerance proteins (MTP) in metal storage and potential use in biofortification strategies. Front Plant Sci 14:144.

Rio D, Rodriguez-Mateos A, Spencer JP, Tognolini M, Borges G, Crozier A (2013) Dietary (poly) phenolics in human health: structures, bioavailability, and evidence of protective effects against chronic diseases. Antioxid Redox Sig 18:1818–1892.

Rodríguez M, Canales E, Borrás-Hidalgo O (2005) Molecular aspects of abiotic stress in plants. Biotecnol Apl 22:1–10.

Rooney WL, Blumenthal J, Bean B, Mullet JE (2007) Designing sorghum as a dedicated bioenergy feedstock. Biofuel Bioprod Biorefin 1:147–157.

Sabir P, Ashraf M, Akram NA (2011) Accession variation for salt tolerance in proso millet (*Panicum miliaceum* L.) using leaf proline content and activities of some key antioxidant enzymes. J Agr Crop Sci 197:340–347.

Sajjad H, Muhammad M, Ashfaq A, Waseem A, Hafiz MH, Mazhar A, Nasir M, Asad A, Hafiz UF, Syeda RS, Fahad S, Depeng W, Wajid N (2019) Using GIS tools to detect the land use/land cover changes during 40 years in Lodhran district of Pakistan. Environ Sci Pollut Res 27:39676–39692. https://doi.org/10.1007/s11356-019-06072-3.

Saleem MH, Fahad S, Adnan M, Mohsin A, Muhammad SR, Muhammad K, Qurban A, Inas AH, Parashuram B, Mubassir A, Reem MH (2020a) Foliar application of gibberellic acid endorsed phytoextraction of copper and alleviates oxidative stress in jute (*Corchorus capsularis* L.) plant grown in highly copper-contaminated soil of China. Environ Sci Pollution Res 27:37121–37133. https://doi.org/10.1007/s11356-020-09764-3.

Saleem MH, Fahad S, Shahid UK, Mairaj D, Abid U, Ayman ELS, Akbar H, Analía L, Lijun L (2020c) Copper-induced oxidative stress, initiation of antioxidants and phytoremediation potential of flax (Linum usitatissimum L.) seedlings grown under the mixing of two different soils of China. Environ Sci Poll Res 27:5211–5221. https://doi.org/10.1007/s11356-019-07264-7.

Saleem MH, Rehman M, Fahad S, Tung SA, Iqbal N, Hassan A, Ayub A, Wahid MA, Shaukat S, Liu L, Deng G (2020b) Leaf gas exchange, oxidative stress, and physiological attributes of rapeseed (*Brassica napus* L.) grown under different light-emitting diodes. Photosynthetica 58:836–845.

Salih AA, Ali IA, Lux A, Luxova M, Cohen Y, Sugimoto Y, Inanaga S (1999) Rooting, water uptake, and xylem structure adaptation to drought of two sorghum cultivars. Crop Sci 39:168–173.

Sanchez AC, Subudhi PK, Rosenow DT, Nguyen HT (2002) Mapping QTLs associated with drought resistance in sorghum (*Sorghum bicolor* L. Moench). Plant Mol Biol 48:713–726.

Satish K, Gutema Z, Grenier C, Rich PJ, Ejeta G (2012) Molecular tagging and validation of microsatellite markers linked to the low germination stimulant gene (lgs) for Striga resistance in sorghum [*Sorghum bicolor* (L.) Moench]. Theor App Genet 124:989–1003.

Saud S, Chen Y, Fahad S, Hussain S, Na L, Xin L, Alhussien SA (2016) Silicate application increases the photosynthesis and its associated metabolic activities in Kentucky bluegrass under drought stress and post-drought recovery. Environ Sci Pollut Res 23:17647–17655. https://doi.org/10.1007/s11356-016-6957-x.

Saud S, Chen Y, Long B, Fahad S, Sadiq A (2013) The different impact on the growth of cool season turf grass under the various conditions on salinity and drought stress. Int J Agric Sci Res 3:77–84.

Saud S, Fahad S, Cui G, Chen Y, Anwar S (2020) Determining nitrogen isotopes discrimination under drought stress on enzymatic activities, nitrogen isotope abundance and water contents of Kentucky bluegrass. Sci Rep 10:6415. https://doi.org/10.1038/s41598-020-63548-w.

Saud S, Fahad S, Yajun C, Ihsan MZ, Hammad HM, Nasim W, Amanullah Jr, Arif M, Alharby H (2017) Effects of nitrogen supply on water stress and recovery mechanisms in kentucky bluegrass plants. Front. Plant Sci 8:983. https://doi.org/10.3389/fpls.2017.00983.

Saud S, Li X, Chen Y, Zhang L, Fahad S, Hussain S, Sadiq A, Chen Y (2014) Silicon application increases drought tolerance of Kentucky bluegrass by improving plant water relations and morph physiological functions. SciWorld J 2014:1–10. https://doi.org/10.1155/2014/368694.

Schmincke K (2003) Medicinal plants for forest conservation and healthcare. Non-wood For Prod 11.

Setter T, Waters I (2003) Review of prospects for germplasm improvement for waterlogging tolerance in wheat, barley and oats. Plant Soil 253:1–34.

Shafi MI, Adnan M, Fahad S, Fazli W, Ahsan K, Zhen Y, Subhan D, Zafar-ul-Hye M, Martin B, Rahul D (2020) Application of single superphosphate with humic acid improves the growth, yield and phosphorus uptake of wheat (*Triticum aestivum* L.) in calcareous soil. Agron 10:1224. https://doi.org/10.3390/agronomy10091224.

Shah SH, Ali S, Hussain Z, Jan SA, Ali GM (2016) Genetic improvement of tomato (*Solanum lycopersicum*) with AtDREB1A dene for cold stress tolerance using optimized agrobacterium-mediated transformation system. Int J Agric Biol 18:471–782.

Shah F, Lixiao N, Kehui C, Tariq S, Wei W, Chang C, Liyang Z, Farhan A, Fahad S, Huang J (2013) Rice grain yield and component responses to near 2 °C of warming. Field Crop Res 157:98–110.

Shao HB, Chu LY, Jaleel CA, Zhao CX (2008) Water-deficit stress-induced anatomical changes in higher plants. Comptes Rendus Bio 331:215–225.

Sharma SS, Dietz KJ (2006) The significance of amino acids and amino acid-derived molecules in plant responses and adaptation to heavy metal stress. J Exp Bot 57:711–726.

Sharma HC, Taneja SL, Rao NK, Rao KP (2003) Evaluation of sorghum germplasm for resistance to insect pests, International Crops Research Institute for the Semi-Arid Tropics, India.

Sharma R, Thakur RP, Senthilvel S, Nayak S, Reddy SV, Rao VP, Varshney RK (2011) Identification and characterization of toxigenic Fusaria associated with sorghum grain mold complex in India. Mycopathol 171:223–230.

Shi H, Ishitani M, Kim C, Zhu JK (2000) The Arabidopsis thaliana salt tolerance gene SOS1 encodes a putative Na+/H+ antiporter. Proc Natl Acad Sci 97:6896.

Shinozaki K, Yamaguchi-Shinozaki K (2000) Molecular responses to dehydration and low temperature: differences and cross-talk between two stress signaling pathways. Curr Opin Plant Bio 3: 217–223.

Singh M, Boora K (2008) Molecular characterization of anthracnose resistance gene in sorghum. Plant Genom Bioinformat 373–388.

Singh V, Nguyen CT, van Oosterom EJ, Chapman SC, Jordan DR, Hammer GL (2015b) Sorghum genotypes differ in high temperature responses for seed set. Field Crop Res 171:32–40.

Singh BU, Padmaja PG, Seetharama N (2004) Biology and management of the sugarcane aphid, *Melanaphis sacchari* (Zehntner) (Homoptera: Aphididae), in sorghum: a review. Crop Prot 23:739–755.

Siripornadulsil S, Traina S, Verma DP, Sayre RT (2002) Molecular mechanisms of proline-mediated tolerance to toxic heavy metals in transgenic microalgae. Plant Cell 14:2837–2847.

Sloderbeck PE, Michaud JP, Whitworth RJ (2008) Sorghum insect management. Kansas State University Extension Publication MF-742.

Song SY, Chen Y, Chen J, Dai XY, Zhang WH (2011) Physiological mechanisms underlying OsNAC5-dependent tolerance of rice plants to abiotic stress. Plant 234:331–345.

Sonobe K, Hattori T, An P, Tsuji W, Eneji AE, Kobayashi S, Kawamura Y, Tanaka K, Inanaga S (2010) Effect of silicon application on sorghum root responses to water stress. J Plant Nut 34:71–82.

Spurthi GS, Patil RK, Puttanavar MS, Ramegowda GK (2009) Effect of intercropping on the incidence of stem borer and armyworm in sorghum. J Entomol Res 33:89–92.

Srivastava S, Pathak AD, Gupta PS, Shrivastava AK, Srivastava AK (2012) Hydrogen peroxide-scavenging enzymes impart tolerance to high temperature induced oxidative stress in sugarcane. J Env Biol 33:657.

Stemler AB, Harlan JR, de Wet JM (1975) Evolutionary history of cultivated sorghums (*Sorghum bicolor* [Linn.] Moench) of Ethiopia. Bull Torrey Bot Club 325–333.

Subhan D, Zafar-ul-Hye M, Fahad S, Saud S, Martin B, Tereza H, Rahul D (2020) Drought stress alleviation by ACC deaminase producing *Achromobacter xylosoxidans* and *Enterobacter cloacae*, with and without timber waste biochar in maize. Sustain 12:6286. https://doi.org/10.3390/su12156286.

Suhendro E, Kunetz C, McDonough C, Rooney L, Waniska R (2000) Cooking characteristics and quality of noodles from food sorghum. Cereal Chem 77:96–100.

Surwenshi A, Chimmad VP, Jalageri BR, Kumar V, Ganapathi M, Nakul HT (2010) Characterization of sorghum genotypes for physiological parameters and yield under receding soil moisture conditions. Res J Agric Sci 1:242–244.

Swami AK, Alam SI, Sengupta N, Sarin R (2011) Differential proteomic analysis of salt stress response in *Sorghum bicolor* leaves. Env Exp Bot 71:321–328.

Swindell WR, Huebner M, Weber AP (2007) Transcriptional profiling of Arabidopsis heat shock proteins and transcription factors reveals extensive overlap between heat and non-heat stress response pathways. BMC Genom 8:125.

Sènou M, Tchogou A, Assogba F, Agossadou A, Dougnon T, Agbangnan D, Loko F (2017) Study of biological tolerance of aqueous extract of *Sorghum bicolor*. J App Biosci 109:10640–10648.

Tadross M, Suarez P, Lotsch A, Hachigonta S, Mdoka M, Unganai L, Lucio F, Kamdonyo D, Muchinda M (2007) Changes in growing-season rainfall characteristics and downscaled scenarios of change over southern Africa: implications for growing maize. IPCC Regional Expert Meeting on Regional Impacts, Adaptation, Vulnerability, and Mitigation, Nadi, Fiji, pp 193–204.

Tari I, Gábor L, Zoltán T, Péter P (2013) Response of sorghum to abiotic stresses: a review. J Agron Crop Sci 199:264–274.

Tariq M, Ahmad S, Fahad S, Abbas G, Hussain S, Fatima Z, Nasim W, Mubeen M, Hamid ur Rehman M, Khan MA, Adnan M (2018) The impact of climate warming and crop management on phenology of sunflower-based cropping systems in Punjab, Pakistan. Agri For Met 256:270–282.

Tavakkoli E, Rengasamy P, McDonald GK (2010) High concentrations of Na+ and Cl–ions in soil solution have simultaneous detrimental effects on growth of faba bean under salinity stress. J Exp Bot 61:4449–4459.

Teige M, Scheikl E, Eulgem T, Dóczi R, Ichimura K, Shinozaki K, Dangl JL, Hirt H (2004a) The MKK2 pathway mediates cold and salt stress signaling in Arabidopsis. Mol Cell 15:141–152.

Teige M, Scheikl E, Eulgem T, Dóczi R, Ichimura K, Shinozaki K, Dangl JL, Hirt H (2004b) Sensing and signalling during plant flooding. Plant Physiol Biochem 42:273–282.

Teixeira EI, de Ruiter J, Ausseil AG, Daigneault A, Johnstone P, Holmes A, Tait A, Ewert F (2018) Adapting crop rotations to climate change in regional impact modelling assessments. Sci Total Environ 616:785–795.

Thakur RP, Reddy BV, Mathur K (2007) Screening techniques for sorghum diseases. Information: Bulletin No. 76. International Crops Research Institute for the Semi-Arid Tropics.

Tingting XI, Peixi SU, Lishan S (2010) Photosynthetic characteristics and water use efficiency of sweet sorghum under different watering regimes. Pak J Bot 42:3981–3994.

Török Z, Goloubinoff P, Horváth I, Tsvetkova NM, Glatz A, Balogh G, Varvasovszki V, Los DA, Vierling E, Crowe JH, Vígh L (2001) Synechocystis HSP17 is an amphitropic protein that stabilizes heat-stressed membranes and binds denatured proteins for subsequent chaperone-mediated refolding. Proc Natl Acad Sci 98:3098–3103.

Uemura M, Steponkus PL (1997) Effect of cold acclimation on membrane lipid composition and freeze-induced membrane destablization. In: Li PH, Chen THH (eds) Plant cold hardiness. Springer, Boston, MA

Uz Zaman Q, Zubair A, Muhammad Y, Muhammad ZI, Abdul K, Fahad S, Safder B, Ramzani PMA, Muhammad N (2017) Zinc biofortification in rice: leveraging agriculture to moderate hidden hunger in developing countries. Arch Agron Soil Sci 64:147–161. https://doi.org/10.1080/03650340.2017.1338343.

Vinebrooke RL, Cottingham K, Norberg, Marten SJI, Dodson SC, Maberly S, Sommer U (2004) Impacts of multiple stressors on biodiversity and ecosystem functioning: the role of species co-tolerance. Oikos 104:451–457.

Viswanathan C, Zhu JK (2002) Molecular genetic analysis of cold–regulated gene transcription. Philos Trans R Soc Lond 357:877–886.

Waheed A (2014) Screening and selection of tomato genotypes/cultivars for drought tolerance using multivariate analysis. Pak J Bot 46:1165–1178.

Wahid F, Fahad S, Subhan D, Adnan M, Zhen Y, Saud S, Manzer HS, Martin B, Tereza H, Rahul D (2020) Sustainable management with mycorrhizae and phosphate solubilizing bacteria for enhanced phosphorus uptake in calcareous soils. Agri 10:334. https://doi.org/10.3390/agriculture10080334.

Wahid A, Gelani S, Ashraf M, Foolad MR (2007) Heat tolerance in plants: an overview. Env Expl Bot 61:199–223.

Wajid N, Ashfaq A, Asad A, Muhammad T, Muhammad A, Muhammad S, Khawar J, Ghulam MS, Syeda RS, Hafiz MH, Muhammad IAR, Muhammad ZH, Muhammad Habib ur R, Veysel T, Fahad S, Suad S, Aziz K, Shahzad A (2017) Radiation efficiency and nitrogen fertilizer impacts on sunflower crop in contrasting environments of Punjab. Pak Environ Sci Pollut Res 25:1822–1836. https://doi.org/10.1007/s11356-017-0592-z.

Wang CQ, Tao WA, Ping MU, Li ZC, Ling YA (2013) Quantitative trait loci for mercury tolerance in rice seedlings. Rice Sci 20:238–242.

Wasaya Allah, Xiying Z, Qin F, Zongzheng Y (2018) Root phenotyping for drought tolerance: a review. Agron 8:241.

Watson A, Ghosh S, Williams MJ, Cuddy WS, Simmonds J, Rey MD, Hatta MA, Hinchliffe A, Steed A, Reynolds D, Adamski NM (2018) Speed breeding is a powerful tool to accelerate crop research and breeding. Nat Plant 4:23.

Wen JQ, Oono K, Imai R (2002) Two novel mitogen-activated protein signaling components, OsMEK1 and OsMAP1, are involved in a moderate low-temperature signaling pathway in rice. Plant Physiol 129:1880–1891.

Williams JP, Khan MU, Mitchell K, Johnson G (1988) The effect of temperature on the level and biosynthesis of unsaturated fatty acids in diacylglycerols of *Brassica napus* leaves. Plant Physiol 87:904–910.

Wong JH, Lau T, Cai N, Singh J, Pedersen JF, Vensel WH, Buchanan BB (2009) Digestibility of protein and starch from sorghum (*Sorghum bicolor*) is linked to biochemical and structural features of grain endosperm. J Cereal Sci 49:73–82.

Wu C, Kehui C, She T, Ganghua L, Shaohua W, Fahad S, Lixiao N,Jianliang H, Shaobing P, Yanfeng D (2020) Intensified pollination and fertilization ameliorate heat injury in rice (*Oryza sativa* L.) during the flowering stage. Field Crops Res 252:107795.
Wu C, Tang S, Li G, Wang S, Fahad S, Ding Y (2019) Roles of phytohormone changes in the grain yield of rice plants exposed to heat: a review. PeerJ 7:e7792. https://doi.org/10.7717/peerj.7792.
Xie T, Su P, Shan L (2010) Photosynthetic characteristics and water use efficiency of sweet sorghum under different watering regimes. Pak J Bot 42:3981–3994.
Xiong L, Wang RG, Mao G, Koczan JM (2006) Identification of drought tolerance determinants by genetic analysis of root response to drought stress and abscisic acid. Plant Phyiol 142:1065–1074.
Yan F, Kuang Y, Ren B, Wang J, Zhang D, Lin H, Yang B, Zhou X, Zhou H (2018) Highly efficient A· T to G· C base editing by Cas9n-guided tRNA adenosine deaminase in rice. Mol Plant 11:631–634.
Yang Z, Zhang Z, Zhang T, Fahad S, Cui K, Nie L, Peng S, Huang J (2017) The effect of season-long temperature increases on rice cultivars grown in the central and southern regions of China. Front Plant Sci 8:1908. https://doi.org/10.3389/fpls.2017.01908.
Yruela I (2005) Copper in plants. Braz J Plant Physiol 17:145–156.
Yu J, Tuinstra M (2001) Genetic analysis of seedling growth under cold temperature stress in grain sorghum seedlings. Crop Sci 41:1438–1443.
Zafar-ul-Hye M, Muhammad N, Subhan D, Fahad S, Rahul D, Mazhar A, Ashfaq AR, Martin B, Ji˘rí H, Zahid HT, Muhammad N (2020a) Alleviation of cadmium adverse effects by improving nutrients uptake in bitter gourd through cadmium tolerant rhizobacteria. Environ 7:54. https://doi.org/10.3390/environments7080054.
Zafar-ul-Hye M, Tahzeeb-ul-Hassan M, Muhammad A, Fahad S,, Martin B, Tereza D, Rahul D, Subhan D (2020b) Potential role of compost mixed biochar with rhizobacteria in mitigating lead toxicity in spinach. Sci Rep 10:12159. https://doi.org/10.1038/s41598-020-69183-9.
Zahida Z, Hafiz FB, Zulfiqar AS, Ghulam MS, Fahad S, Muhammad RA, Hafiz MH, Wajid N, Muhammad S (2017) Effect of water management and silicon on germination, growth, phosphorus and arsenic uptake in rice. Ecotoxicol Environ Saf 144:11–18.
Zegada-Lizarazu W, Monti A (2013) Photosynthetic response of sweet sorghum to drought and re-watering at different growth stages. Physiol Plant 149:56–66.
Zerbini E, Thomas D (2003) Opportunities for improvement of nutritive value in sorghum and pearl millet residues in South Asia through genetic enhancement. Field Crops Res 84:3–15.
Zhang YL, Pennerman KK, Wang H, Yin G (2016b) Characterization of a Sorghum mosaic virus (SrMV) isolate in China. Saudi J Biol Sci 23:237–242.
Zhang F, Wang Y, Yu H, Zhu K, Zhang Z, Zou FLJ (2016a) Effect of excessive soil moisture stress on sweet sorghum: physiological changes and productivity. Pak J Bot 48:1–9.
Zhu Y, Hai-Jun G, Jun-Liang Y (2019) Role of silicon in mediating salt tolerance in plants: a review. Plants 8:147.
Zuo Y, Zhang F (2011) Soil and crop management strategies to prevent iron deficiency in crops. Plant Soil 339:83–95.

7

Plant Chemical Ecology: The Focused Arena for Plant Survival and Productivity

Ansar Hussain[1], Muhammad Ifnan Khan[1], Muhammad Arif[2], Shahid Farooq[3], Madiha Zainab[4], Ali Raza[5], Muhammad Anwar[6], Muhammad Furqan Ashraf[7], Muhammad Ijaz[8], Mona Hussein[9], Ali Noman[10], Shahzadi Mahapara[1], and Ali Bakhsh[1]

[1]Department of Plant Breeding & Genetics, Faculty of Agricultural Sciences, Ghazi University, D. G. Khan, Pakistan

[2]Department of Plant Protection, Faculty of Agricultural Sciences, Ghazi University, D. G. Khan, Pakistan

[3]Department of Agronomy, Faculty of Agricultural Sciences, Ghazi University, D. G. Khan, Pakistan

[4]College of Life Science, Fujian Agriculture and Forestry University, Fuzhou, China

[5]Institute of Tropical Agriculture and Forestry, Hainan University, Haikou, China

[6]College of Life Sciences and Oceanography, Shenzhen University, Shenzhen, China

[7]College of Life Sciences, South China Agricultural University, Guangzhou, China

[8]College of Agriculture, BZU Bahadur Campus, Layyah, Pakistan

[9]Biology Department, Taibah University, Madinah, Saudi Arabia.

[10]Department of Botany, Government College University, Faisalabad, Pakistan

Corresponding Author Ali Noman (PhD) Department of Botany, Government College University, Faisalabad

Ansar Hussain (PhD) Department of Plant Breeding & Genetics, Ghazi University, D. G. Khan, Pakistan

7.1 Introduction

Chemical plant ecology is a multi-faceted field that probes the prevalence, task, and combination of plant-based secondary metabolites (Kuhlisch and Pohnert 2015). The understanding of these secondary metabolites helps in the discovery of the applications of these secondary metabolites in medical science. Therefore, chemical ecology progress is strictly linked to the analytic chemistry evolution (Meinwald and Eisner 2008). Secondary metabolites are a gigantic cluster of complex chemicals composed of non-volatile and volatile compounds (Pichersky et al. 2006); micro molecules like alkaloids and terpenoids; and macro molecules including polysaccharides or peptides. Secondary metabolites don't play a direct role in primary functions of cells, that is, growth, development, or reproduction (Jenke-Kodama et al. 2008). The role of these chemicals has their significance as defence agents to stimulate and transform the concerned processes. The role of secondary metabolites as defence agents has been confirmed experimentally (Dicke and van Loon 2000). Secondary metabolites play a very vital role in pollinator's attraction and in competition with adjacent plants. The origin of diversity in secondary metabolites lies in the evolution of life. These compounds are source of development of adaptations in plants to attain a cutting edge in their environment (Macel et al. 2007).

Secondary metabolites' biosynthesis pathways are derivatives of primary metabolites during evolution by duplication, followed by mutation of copied genes (Pezzini et al. 2018). The polypeptide synthesis and the essential fatty acid synthesis are good examples of the similarity between pathways and display of alike enzymatic activity (Nelson et al. 2009). Several secondary metabolites have been demonstrated this sort of analogy. Nevertheless, there persist some deviations where the evolutionary links are not found (Jenke-Kodama et al. 2008). A few biosynthesis pathways resulted from symbiotic fungi which were the result of lateral genetic transformation in nature (Futuyama 2009). Some special substances are originated for specific ecological niches, due to rapid selection processes. On the other hand, in these very specified structures, there are many more products such as terpenoids or alkaloids synthesized by more classical reactions pathway (Krause et al. 2013). In some species, secondary metabolites are competitively more variable comparing to primary metabolism as they result through independent evolutionary pressures in plants and manipulate enzymatic activity. A huge diversity of chemical structures is produced by miscellaneous secondary metabolic pathways with limited genetic resources (Jenke-Kodama et al. 2008). The vast diversity of secondary metabolites might be the big portion attributed to the greatly unpredictable plant chemistry. The chemical defence approaches play a vital role in plant protection (Cariñanos et al. 2017) from insects, microorganisms, and herbivores. This defence grates detoxification in return, which enables the plants to evolve even more unique and advanced defence mechanism. This process of evolution and co-evolution guides to the joint adaptation that may be specific by pair-wise or diffused (Turcotte et al. 2012). Mutual evolution of species adapting one another's characteristics is pair-wise coevolution. While the adaptation process of single species under the influence of the interaction with several species is diffuse co-evolution, this can lead to evolutionary competition. Adaptation of *Brassicaceae* using sulphur-based glucosinolate–myrosinase defence system is an example of evolutionary defensive, which is significant trait of *Brassicaceae* family (Svergun and Koch 2003). Additionally, various *Brassicaceae* genera have developed unique structures different from other plant families. Conversely, the correlation of phylogenetic and similarity in chemicals proximity does not match completely. Generally, a host shift of associated herbivores is best suited with plant chemistry as compared to plant phylogeny. Overall, phylogenetic and evolutionary record contains valued data on plant–animal interactions (Futuyama 2009).

7.2 Soil Chemical Ecology

Abiotic components of the ecosystem can be influenced by chemicals released by plants, that is, accumulation and availability of inorganic ions. Activities of allelochemicals are subjected to ecological features like nutrients' availability, photoperiod, and moisture deficiency (Inderjit and Nilsen 2003; Legrand et al. 2003). Proteoid is clusters of rootlets developed in *Lupinus albus* in response to poor phosphorus-availability conditions. *L. albus* has proteoid types of roots that excrete citrate, as a result of which citrate is present in higher proportions in soil around proteoid roots in comparison to bulk soil. This is involved in lowering soil pH, mobilizing soluble phosphate, and influencing the solubility of Fe and Mn (Johnson 2007). Phytosiderophores are compounds released by some graminaceous plants under the conditions of Fe and Zn deficiency, which takes part in solubilization of metal ions, mainly Fe. For example, phenolics constitute complexes with nutrient ions (Arnold et al. 2004; Barbehenn et al. 2005). Phenolics compete for anion absorption sites and bind with soluble Al, Fe, and Mn, which bind to phosphate and resultantly increase P availability (Filley et al. 2002; Filley et al. 2006). Phenolic acids, ferulic, p-coumaric, p-hydroxybenzoic, and protocatechuic acids manipulate organic N in soil and accumulation of inorganic ions, including Al, Mn, Fe, and PO_4 (Hood-Nowotny et al. 2010). It is reported that protocatechuic acid may increase solubility of Fe and Al by forming complexes with them (Del Arco et al. 2010). The growth of certain plant species was inhibited by two anthraquinones, physcion and emodin, isolated from roots, rhizomes of *P. sachalinense,* and from allied soils under controlled conditions, and resulted in implication of allelopathy (Vrchotová and Šerá 2008). It has also been reported that availability of certain inorganic ions was influenced by emodin and physcion (Chubar et al.

2005). Various research works imply that influence of *P. sachalinense* on various species may be due to two anthraquinones or their influence on soil inorganic ions. Phenolics and terpenoids have a critical role in nitrification inhibition (Warren et al. 2015; Zacchino et al. 2017) and hence affect the nutrient status of plant community. Phenols, including ferulic acids, myricetin, caffeic, tannins, and tannin-derived compounds, hinder NH_4^+ to NO_2 oxidation by *Nitro-somonas* (Wahman et al. 2005). Terpenoids might have a critical role in nitrification inhibition (Subbarao et al. 2012b). On the other hand, it has been recommended that terpenoids boosted ammonium-N immobilization by soil micro-organisms instead of nitrification inhibition (Cui and Song 2007; Kanerva 2007). Nitrification rate is lowered in succession, may be because of allelochemicals. Some researchers also disagreed, reporting no boost in nitrification through succession (Sun et al. 2016). Polyphenols formed by plants are involved in allelopathy and plant defence; however, this group of chemicals has been revealed to impart key influences on organic and inorganic soil nutrient pools and fluxes (Hättenschwiler and Vitousek 2000). For example, polyphenols produced by litter of leaves of *P. muricata* in soils of northern California's pygmy forest affected the discharge of soluble organic and mineral nitrogen (Ockels et al. 2007). In these forests, the soil pH is low and polyphenols from the forest ground formulate compounds with Al, consequently augmenting metal's toxicity. The mix of 3-carene, pinene, camphene, monoterpenes, myrcene, limonene, and phellandrene inhibits nitrification when these are added into soil (Dietz et al. 2013). These examples demonstrate the possible influence of allelochemicals on soil nutrient reserves. Any effect on soil nutrient reserve may ultimately influence the growth and development of plants in the environs, even though such a system doesn't match the conventional theory of plant to plant allelopathy. Various factors influence the phytotoxicity of phenolic acids such as mineral nutrition, soil pH, and soil type. Climatic factors like heat and intensity of light may also modify allelopathic prospective of weeds' residues including *A. retroflexus* and *S. glauca* on *Zea mays* (Naseem et al. 2009). The residues of these weeds had strong inhibitory effects when maize crop was grown under low temperature. *Ceratiola ericoides*, a perennial shrub, produced inactive dihydrochalcone ceratiolin (Scrivanti et al. 2003). Ceratiolin undergoes transformation under some specific pH, temperature, and light conditions, and resultantly, hydrocinnamic acid was produced, which is a toxic compound (Hewitt and Menges 2008). Microbial degradation of hydrocinnamic acid results in the formation of acetophenone (Scrivanti et al. 2003). Acetophenone and hydrocinnamic acid are suggested to play a crucial role to hinder germination and growth of the *S. scoparium* grass. Therefore, an active allelochemical isn't produced by *C. ericoides* that influence the grass, while biotic and abiotic factors convert an inactive compound (ceratiolin) into active allelochemicals (acetophenone and hydrocinnamic acid), which influence structure of community. Lower light conditions decreased the total phenolic content in the submerged macrophyte *Myriophyllum spicatum*, whereas the light regime has no effect on the amount of tellimagrandin II (Martinou and Youle 2011). Environmental constituents (light, moisture, and temperature), herbivory, ecological factors (e.g. mineral deficiencies), and rivalry for available resource may cause quantitative and qualitative changes in flavonoids (Chaves et al. 2001). The quantity of flavonoid in *Cistus ladanifer* is influenced by ultraviolet light, water availability, and heat (Chaves et al. 2001). The amount of apigenin and kaempferol was relatively high in summer season, and quantity of exudates of apigenin-4'-(O) methylin of *C. ladanifer* was decreased after rainfall during in monsoon period. Climatic factors have critical role in accumulation of allelochemicals or their breakdown substances to bioactive levels.

7.3 Plant–Soil–Microbe Interaction

Previous research work suggests that soil microbial populations can perform a vital role in organizing various plant communities (Bever 2015; Classen et al. 2015). Soil microbial populations possess a variety of plant-symbiotic and plant-antagonistic microorganisms, and plants react to these soil communities in species-specific reactions (Mangan et al. 2010). These species-specific reactions aid soil microbes to manipulate composition of plant communities (Classen et al. 2015; Isbell et al. 2015; Smith-Ramesh and Reynolds 2017). For example, divergence in plant reactions to soil microbe population may affect the relative abundance of plant species (Klironomos 2002; Klironomos and Hart

2002; Mangan et al. 2010), and non-native plant species are less negatively affected by soil microbes often as compared with native plant species (Kulmatiski et al. 2008; Callaway et al. 2011). This may be possible because non-native species have 'escaped' from pathogens present in their local environment (Colautti et al. 2004). Soil microbial organisms can also persuade plant performance indirectly by mediating interactions between plant species (Emam et al. 2014; Ewnetu et al. 2014). Soil microorganisms responsible for species-specific benefits can help their hosts to achieve competitive edge over other plant species, as can occur with mycorrhizal fungi (Bailey et al. 2002; Sigüenza et al. 2006). Mycorrhizae can enhance diversity and can reduce competitive inequalities between species (Johnson et al. 2004; Wagg et al. 2011; Sutherland et al. 2017). Hostile soil microorganisms may also aggravate or deteriorate the effects of competition (Bever 2015; Albornoz et al. 2016), depending on the plant species observed and the degree of their reactions to the soil microorganisms. However, in some studies, it is reported that soil microbial agents have no effect on plant competition (Casper and Castelli 2007), or that competition overwhelms the influence that soil microbes have on plant performance (Crawford and Knight 2017). Environmental hassle and their fluctuations are the most important factors that influence agricultural productions by troubling rhizosphere execution (Singh 2015). A vigorous plant rhizosphere is very important not only in nutrient uptake and water availability to plants, but also provides persistent benefits to the microbe's diversity, which jointly supports plant health. The composition of root exudates primarily depends on plant genotypes that are responsible for recruitment of microbes in the rhizosphere (Shrivastava and Kumar 2015). The composite of beneficial microorganisms' mixture around the root zones has a critical role in provision of improved plant growth-promoting attributes inclusive of rhizobacteria, mycorrhizal fungi, and other microbes (Hamilton et al. 2016). These beneficial microorganisms sustain plant health by interacting with plant roots symbiotically via myriad of mechanisms, including pathogens control, secondary metabolites production, and enhanced resistance against various stresses (Glick 2014; Rashid et al. 2016). Beneficial fungal complexes surrounding the root zones safeguard the plants against different pathogens and hence play their role as potential biocontrol organisms (Del Fabbro et al. 2014). The fungal webs augment P attainment and water accessibility in the circumstance of harsh water insufficiency (Barnawal et al. 2017). Hence, the plant–microbe symbiotic relationships could be a feasible alliance, which could certainly boost the attainment of necessary nutrients and moisture by plants. This resultantly boosts up utilizable forms of N, discharges growth-boosting hormones, enzymes, metabolites, and organic volatiles, and promotes health of plants. Nitrogen is the most important nutrient element essentially required to plants for growth enhancement and fulfilling nutritional desires (Krapp 2015). The N takes parts actively in photosynthesis and chlorophyll biosynthesis in plants (Zimmer et al. 2008). Besides this, N dynamically participates in several biochemical or physiological reactions in plants (Krapp 2015). Nitrogen is suggested as a key nutrient essential in sufficient quantity to enhance sustainable yield and quality of crop (Wang et al. 2013; Sainju et al. 2015). Conversely, the extreme application of agrochemicals in modern cropping systems severely boosts anthropogenic N loading as higher quantities of the N supplied to crops aren't consumed (Yang et al. 2014). Consequently, extreme use of N fertilizers causes economic loss and direct to a cascade of vast-scale ecological impression such as challenging sustainability of ecosystem around the globe through terrestrial and marine eutrophication, acidification, and by generating large hypoxic zones (Galloway et al. 2008; Yang et al. 2014). The vital role of soil microbial organisms in management of soil N stress is well established. Microbial agents-mediated biological N_2 fixation (BNF) adds 180 × 106 t of fixed N per annum worldwide, 80% of which comes from symbiotic relations and remaining from free living or associative systems. N_2 fixation is extremely energy-consuming practice, necessitating approximately 16 mol of adenosine triphosphate (ATP) for reduction of each mole of elemental N_2. It would be helpful if carbon (C) from bacterial resources is aimed at oxidative phosphorylation, which favours ATP synthesis, rather than glycogen production, as later store energy in the form of glycogen (Glick 2014). Bell determined that organic component inputs in rhizosphere from root secretion amend microbial organism diversity that establishes the N uptake of plants (Stefani et al. 2015). The *Rhizobiaceae* family members, that is, *Azorhizobium, Mesorhizobium, Bradyrhizobium, Allorhizobium, Sinorhizobium,* and *Rhizobium,* possibly fixed 80% of biological N in agricultural systems in connection with leguminous plants, which play a crucial role in altering the chemical ecology of environment. In leguminous crops like soybean, pea, peanut, and alfalfa, *Rhizobium* and *Bradyrhizobium* develop symbiotic relationships with plant roots, transform N_2 into ammonia, and make it readily available N source for plant use (Badawi et al. 2011). The plant growth-promoting *rhizobacteria* (PGPR) those exist in inner parts of plant are known as intracellular PGPR (i-PGPR). These PGPR constitutes very novel nodular structure to augment the N_2 fixation rate (Gray and Smith 2005). Endophytes consist of a range of soil microbial bacteria such as *Azorhizobium, Allorhizobium, Mesorhizobium, Bradyrhizobium*, and *Rhizobium* belonging to family Rhizobiaceae, normally occupy the root zone of agricultural plants to form the N-fixing nodules (Zimmer et al. 2016), and encourage plant development and indirectly and directly growth of crop plants. The PGPR available on the rhizoplane or within the apoplast of the root cortex are known as extracellular PGPR (e-PGPR). The e-PGPR consists of *Agro-bacterium, Arthro-bacter, Azoto-bacter, Azo-spirillum, Bacillus, Burkholderia, Caulo-bacter, Chromo-bacterium*, Erwinia, *Flavo-bacterium, Micrococcus, Pseudomonas*, and *Serratia* (Gray and Smith 2005). Frankia and endophytes species of i-PGPR are able to fix N symbiotically in alliance with higher and developed plants (Badawi et al. 2011). It has been suggested that majority of the plants constitute mycorrhizal symbiotic relationships, mostly dicots and monocots (Bressano et al. 2010). The major mycorrhizal groups known to constitute symbiotic relationships with plants are as follows: *arbuscular mycorrhizae, ecto-mycorrhizae, ectendo-mycorrhizae, arbutoid mycorrhizae, ericoid mycorrhizae, monotropoid mycorrhizae*, and *orchid mycor-rhizae* (Bressano et al. 2010). N_2 fixation, very vital source of N responsible for productivity of plant, is initial step in N cycling to the biosphere from the atmosphere (Nadeem et al. 2014). Fungi demonstrate improved N status when inoculated with rhizobia (Barrett et al. 2015), PGPR (Ortiz et al. 2015), or both (Barnawal et al. 2017). Beneficial N_2 fixation is a well-known procedure mediated by bacteria solely, the sole organisms having nitrogenase enzyme, which reduces N_2 to ammonia specifically in symbiotic root nodules (Zimmer et al. 2016). The *Nitrogenase* genes (*nif* genes) are responsible for N_2 fixation and available in symbiotic and free-living systems as well (Boyd et al. 2011).

7.4 Plant Chemical Ecology

The plants possess a layer of a cuticle, whose surface plays a critical role in autecology and physiology, initiating the defence system of plants. This cuticle of plant also serves as the initial contact point to interact with other microbial organisms at almost all trophic stages. The layer of cuticle wrap covers all aerial parts of almost all higher plants, for example fruits, stems, and leaves, while the stems don't have this protective layer because of its secondary growth. An uninterrupted sheet that merely gaps have the stomata pores serves the purpose of gaseous exchange with environment, it is crucial to study the physical and chemical cuticle nature which will help to understand the numerous important biological functioning of cuticle. Contemporary information of cuticle properties enables us to assess the signals on the outer surface of higher plant which are used by microbial organisms or insects (Bernays 1991). Grasshoppers and locusts can differentiate between resistant and susceptible crop plants only by contact between the surface wax and their chemoreceptors. This is because of chemical properties and physical structure of plant cuticle (Jeffree 2006). Several studies have conducted on individual properties of surface characteristics such as ability to wet (Müller and Riederer 2005; Koch and Barthlott 2009) or optical characteristics (Kerstiens 2006). Functions of the cuticle have been studied by various scientists (Müller and Riederer 2005; Jeffree 2006; Kim et al. 2007). The important role of the cuticle in inter-actions with other living organisms was explained previously in relation to microbes (Mendgen and Hahn 2002), herbivorous insects (Eigenbrode 2004; Städler and Reifenrath 2009), and efficiency of predatory insects (Eigenbrode 2004). It is believed that density of cuticle and different functions of cuticle have been underacknowledged, and very less information is available in previous ecological studies. Cuticle contains cuticular waxes in the lipophilic portion. Waxes are composed of esters of long-chain primary alcohols together with long-chain fatty acids. Here, the word 'waxes' is used in a broad sense to comprise of triterpenoids. cyclic compounds like penta-cyclic (Zimmermann et al. 2004). The aliphatic portion of waxes comprises of multi-faceted mix of compounds having hydrocarbon backbone ranging carbon atoms between 21 and >40. Their biosynthesis is dependent on the action of elongates on moieties, derived from long-chain fatty acid synthetase, and additions of oxygen-containing substituents in further reactions and the removal of one carbon atom lead to aliphatic compound, classes present in cuticle waxes (Uchida et al. 2005). There are well-known hypotheses about physiological and ecological occurrence of costs to chemical defence, and these resulted in production of vast literature on plasticity involved in defence, and theory linked to plasticity has facilitated to illuminate the costs (Crispo 2007). Particularly, several theories have been hypothesized on such factors to manipulate provision of capital to chemical defence in plants (Massad et al. 2012; Heath et al. 2014; Smilanich et al. 2016). The most famous among these hypotheses is the hypothesis of optimal defence, and it hypothesizes that plant-defensive techniques are coupled with variation in selective pressure from different plant pathogens, herbivores, and fitness consequences of damage to tissues as well (Heil and McKey 2003). The plant-appearance hypothesis recommends that chemistry of plant and herbivory is decided by predictability of plants in space and time decides (Smilanich et al. 2016). The hypothesis of carbon–nutrient balance speculates that C and N availability determines the production of plant secondary metabolites (Kong et al. 2005). The hypothesis of resource availability focuses on settlement of plants to meagre soils by chemical defence of tissues that are hard to replace (Endara and Coley 2011). The hypothesis of growth–differentiation balance hypothesizes that resources distribution among various physiological reactions is dependent on ecological circumstances and limitations (Glynn et al. 2007). These all hypotheses spotlight on defences as adaptations to biotic and abiotic selective stresses. Though the carbon–nutrient balance and growth–differentiation balance hypotheses are mainly plasticity hypotheses, they centred on ecological reactions of plant chemistry to environmental circumstances. These hypotheses have enthused numerous studies and reviews, but few generalities have appeared across plant taxa or classes of secondary metabolites. For the plasticity hypotheses, resource availability significantly modifies chemical defence for diverse plant taxa and classes of composites (Massad et al. 2012), although the relative effectiveness of any of these hypothesizes is based on the compounds under study – that is growing C:N proportions more positively affects the phenolics as compared with other compounds (Bezemer et al. 2006). A practical aspect for the continual exploration of these diverse hypotheses is to believe deviations in plant defensive composites as incorporated layers, for which every hypothesis presents diverse approach into the allocation and natural variation of secondary compounds, as well as insights into ontogenetic courses, physiological systems, evolutionary origins, and functional consequences (Schuman and Baldwin 2016). Plant secondary metabolites have a vital role in different ecological interactions across various spatial, temporal, and organizational scales. Several useful hypotheses explain about the influences of plant secondary metabolites, accumulated metals (Rascio and Navari-Izzo 2011), or meagre dietary mixes of primary metabolites on herbivores, and their rivals. Hypotheses considering trophic interactions mediated chemically are centred on sequestration, specialization, and population diversity. The conventional viewpoint of tritrophic interactions suggests that insect's herbivores either emit, biotransform, or confiscate plant secondary metabolites. Any of these reactions possibly effect their physiology, dynamics of population, and interactivity with rivals (War et al. 2012). These reactions are usually screened by food span, and confiscation is well-known response associated with specialized group (Dyer et al. 2003). Sequestration may impart pessimistic influence on predators and parasitoids (Ode 2006) or may impart positive influence on parasitoids by means of distraction of herbivore physiological resistance like melanization and encapsulation (Smilanich et al. 2016). According to latest model, it is believed that specialist herbivores are more deadly to predators as compared with generalists, but those specialists are host of numerous and diversified parasitoids (Lampert et al. 2010). Therefore, well-defended plant populations should be distinguished by advanced categories and better reticulate indigenous food networks if there exist abundances of specialist herbivores as compared with generalist herbivores and parasitoids. The problem persists, whether these effects of secondary metabolites on consumers can configure entire arthropod populations. Some researchers have confirmed that entire plant-linked communities are affected by phytochemicals. Such as high volumes of tannin in various genotypes of *Populus* spp. (cottonwood trees) and saponins in *Medicago sativa* plants (alfalfa) were related to enhanced diversity of arthropod (Yang et al. 2009). An additional theory of chemical ecology of trophic reciprocity advocates that plant's secondary compounds are enthusiastic to plants. Because they have crucial role in boosting the secondary consumers' activity, either by smoothing the efficacy of predation and parasitism, or by furnishing chemical signs utilized by predators and parasitoids to detect their prey and hosts (Kessler 2015). To illustrate, after invasion of herbivore on plants, they might generate and release VOCs (volatile organic compounds) for providing location signs of host to the natural rivals of herbivores (Erb et al. 2011). Research studies employing various model systems, inclusive of maize-tomato and tobacco, have demonstrated that in various instances the VOCs excreted by plants could fluctuate depending on the numbers and kinds of herbivores available, thus bestowing specific information about the appropriateness of accessible prey to predators and parasitoids (Kessler 2015). For understanding job of plant chemistry in arbitrating interactions, there exists great prospective in multi-faceted biological communities, principally by quantifying a phytochemical landscape. Combination of all these processes will contribute significantly to fulfil this objective, predominantly genomic and metabolomic methods for analysing statistics from natural systems pooled with segregation and synthesis approach for governing cautious investigations in lab to check the impacts of controlled phytochemical complexes on insect herbivores and predators.

7.5 Modification Chemical Ecology by Human Impact

Humans being are feebly made to comprehend interactions arbitrated chemically because humans realize the globe principally by means of visualization and hearing cues (Yang et al. 2011). In disparity, several oceanic classes don't have eyes and ears. They depend on the chemical signals to observe the surrounding environment, and still the species those have specific organs to observe and listen recon on chemical signals. This phenomenon also happens in humans, but unintentionally. Menstrual cycle in several women is modified by odour of other women (Spencer et al. 2004). Women perceive the body aroma of males having consistent faces (representative of superior genes) more pleasant (Wyart et al. 2007), exotic dancers very much while in estrus as compared to once they aren't (Miller et al. 2007). Consequently, visual species also have strong response to chemical signals.

7.6 Changes in Plant Behaviour and Heavy Metals' Impact

The studies from aquatic, especially freshwater, populations recommend that heavy metals be significant info-disruptors, and comparably same kinds of effects haven't been suggested from terrestrial populations. This may be because of basic variations between aquatic and terrestrial habitation (Boyd 2010). For example, above ground heavy metals in terrestrial populations generally aren't blended in air in higher concentrations, while heavy metals are dissolved in the medium in polluted aquatic populations generally that bathes those organisms. Whereas some heavy metals, including Pb and Hg, have effects on the terrestrial organisms' nervous systems, these can comprise effects on behaviour (Singh et al. 2010), explaining potential effect of heavy metal on the behaviour of terrestrial animals. The lack of extensive investigation leads to scarcity of knowledge about heavy metals' info disruption in terrestrial populations. It is not obvious whether the plants 'perform' in similar way as that of animals in response to heavy metals (Trewavas 2009). Huifang Ma has tried to describe common grounds for both animal and plants 'behaviour' studies (Ma et al. 2016). Viewing the plant root development, it has been confirmed since long time that roots of plants responded to heterogeneity in soil atmosphere (Hodge et al. 2009). The studies on the *Thlaspi caerulescens*, a Zn hyperaccumulator plant, exhibited existence of 'zinco-philic' foraging (Macnair 2003; Oluyemi et al. 2008), the privileged roots propagation in high Zn soil pieces. The data of various researches on plant that hyperaccumulate of other metals (Liu et al. 2013) have revealed analogous cases in root propagation exposing to high-metal soil pieces. In contrary, there may exist some plants' roots to avoid the effect of high metal soil, and can grow well in metal polluted soil. As a result, concentrations of heavy metals in soil may influence patterns and behaviour of root growth of plants as well. Direct response of plant roots and aquatic animals may be similar to soil heavy metals and water-borne heavy metals (Dixit et al. 2015). However, the more delicate info disruption effects of heavy metals that are recognized in aquatic animals are hard to relate any similarity with plants growing in soil containing heavy metals. It is suggested that plant roots may be more responsive to environmental than previous consideration. Particularly, various studies' findings have exhibited that the roots of plants can differentiate among self and non-self-roots (Gruntman and

Novoplansky 2004; Falik et al. 2005). Additionally, it is found those sea rockets roots *Cakile edentula,* var. *lacustris* noticed and respond to conspecifics roots and mainly recommended that they can recognize kin. Discussions have taken place regarding the plants' kin selection claim (Klemens 2008; File et al. 2011) and root-mediated kin identification (Hess and De Kroon 2007), but further experimental work is required to conduct in this field (Hess and De Kroon 2007). Previous work in the area of plant roots of desert plant proposed that *Ambrosia dumosa* and *Larrea tridentate* bushes are able to sense roots hetero-specifics (Callaway and Mahall 2007). The hypothesis of chemical signalling among plant roots generally seems to be accepted (Hodge et al. 2009). Plant communication isn't bounded to the underground dominion. Communication among plants by means of volatile communication has caught the attentions since the 1970s; however, these signals are non-specific and inclined to 'eavesdropping' by surrounding plants as well (Karban et al. 2004). A current study (Karban and Shiojiri 2009) exhibited that individuals of *Artemisia tridentata* (sagebrush) differentiate self-generated and non-self-generated signals. Research work shows that herbivore impairment was considerably less, when the clone of target plants produced volatile signals that is responsible to induce the chemical defence in herbivore (in comparison with volatile signal synthesized by a non-clone plant) (Baldwin et al. 2006). Another pathway is provided by volatile chemical communication by means of which info disruption may be take place. Ability to sense volatile chemical signals by the plant may alter exposure to heavy metals, through the impacts of metal-polluted soils (Engelberth et al. 2004). Stresses caused by heavy metal-contaminated soil exposure may influence a physiology of plant; and therefore, the volatile chemical signalling system has impact on plant growth (Koricheva 2002). Dumping of heavy metal-contaminated soil is an additional opportunity. These dumped heavy metals can influence plants' ecology. Entrance of heavy metal dirt into stomata opening of plant, dew deposition, or metals' dust dissolution in rain may permit heavy metals to infiltrate and enter directly into plant tissues and eventually cells to affect adversely (Greger and Löfstedt 2004). How exposure to heavy metals may manipulate VOCs production or revealing of metal is still poorly understood. The presence of chemical signalling among the plants offers prospect for pollutants of heavy metal to interfere with this signalling pathway, possibly in ways that is analogous to info disruption acknowledged of marine research on animal (Mithöfer et al. 2004). In aquatic systems, there are several ways of signal disruption to occur: heavy metals possibly stop the receptors to received signals, may possibly hinder the reception of receptors, or block signal transmissions (Maksymiec 2007). There aren't such evidences of info disruption to contemporary knowledge, so the question raised, and it can obtain attention of researchers. Heavy metal pollution may interfere plants' interactions with their natural enemies, by using different indirect pathways. Heavy metal pollutants are responsible for plant stresses that amend plant character and derived chemistry as well (Mithöfer et al. 2004). It is suggested that the oxyli-pins production in plants may enhance by heavy metals: such as production of jasmonic acid, which has very vital roles in plant defence response signals (Kessler et al. 2004). Various scientists examined the papers about the plant stress and response hypothesis: the situation that increase stress to plant may cause the changes in plants chemistry that will support the natural of its enemy to performance well against plant defence (Koricheva 2002). The accomplished hypothesis was well established in some cases depends on various factors including stress type to plant and mode of feeding insect. Their review made it crystal clear, but the stress originated by heavy metal toxin is an additional pathway by which can influence the relationships between plant ecological and natural enemies.

7.7 Role of Heavy Metals in Elemental Defences

Atmospheric heavy metals' pollutants may have direct influence on plants by bestowing elemental defence towards natural rivals of plants (Notten et al. 2005). Elemental defence hypothesis was formerly functional, and concentrations of heavy metals' element took place in plants' tissue (Martens and Boyd 2002). These hyperaccumulator plants most probably present in habitation possess naturally high levels of element the plants hoard. For instance, several hyperaccumulators of Ni (Ni hyperaccumulators are the most abundant category of hyperaccumulators (Syam et al. 2016)) happen to serpentine soils, and these soils are mostly Ni rich (Alexander et al. 2007). Investigation of elemental defence has suggested defensive influences in few cases but not in all cases, and numerous elements such as Se, Ni, Cd, and Zn belong to heavy metals (Boyd et al. 2007). Chemosensation is an important process by which microbes and bacteria got information about the environment in which they are exposed (Wiuff et al. 2005). Microbiologists have been eager to discover for the duration of the last 20 years that chemical signalling among bacteria can be widespread and their behaviours advocate resemblance to that of animal behaviours (Diggle et al. 2007). Such as, sensing of quorum is fabrication of bacterial cells' molecules that specify bacterial concentration, and this will encourage the population to behave in the coordinated manner (Diggle et al. 2007; Tait et al. 2009). Discovery of this fact enthused more research efforts into this process, implicated (Waters and Bassler 2005) the prospective uses of these molecules to cure diseases (Larson et al. 2009), and even investigation of social phenomena (Diggle et al. 2007) such as kin selection and cheating (Zhang et al. 2009) in bacteria. Though all microbiologists are not agreed about the phenomenon of behavioural, analogies of eukaryotes (Turovskiy et al. 2007), metals' pollutants, and quorum-sensing do have some direct linkage. For instance, Fe is vital in elemental defence often restraining, source for bacteria that may generate chemicals chelating siderophores to confiscate it from the surrounding atmosphere (Challis 2005). These siderophores may perform various functions such as antibiotics or quorum-sensors (Schertzer et al. 2009). Resistance of a few bacteria to heavy metals is enhanced significantly by developing a bacterial biofilm, and development of bacterial biofilm can be controlled by molecules of quorum-sensing (Sarkar and Chakraborty 2008). One quorum-sensing molecule possesses one heavy metal Zn ion as its integral structural constituents (Hilgers and Ludwig 2001). The research on interference of quorum-sensing focused largely on huge organic molecules (González and Keshavan 2006; Rasmussen and Givskov 2006a), and there are plethora of examples involving the organic compounds that interfere with heavy metals. The gravity of heavy metals involved in quorum-sensing is still under evaluated, but there exist tormenting glimpses about how they operate. For instance, the *Chlamydomonas reinhardtii* (green alga) generates biological compounds that it seems imitate quorum-sensing compounds, synthesized by a number of bacteria (Teplitski et al. 2004). In another study, it is reported that *Delisea pulchra* (red alga) produced biological chemicals *halogenated furanones* those interact with quorum-sensing bacteria, and it may possibly prevent algal biofilm construction on algal thallus surface (Manefield et al. 2002). According to several studies, it is reported that bacterial quorum-sensing is influenced by generation of chemicals by eukaryotes reflecting the interaction between the heavy metals and the chemicals produced (González and Keshavan 2006; Ni et al. 2009).

Scents have crucial roles in the ecological processes and ecological unit. Floral scents are used by pollinators to navigate at short and long spatial scales (McFrederick et al. 2008). Over small distances within 1 m, pollinators, like bees, can use aroma of flower to make the decision whether their visit to flowers will be fruitful or not (Verdonk et al. 2003; McFrederick et al. 2008), while nocturnal moths have proved to the orient to flowers by fragrance of flower alone (Piechulla and Pott 2003). Fragrance also plays important roles in herbivore to plant interactions. These repellents act as direct defences to repel herbivores and behave like protective envelope against predators, parasitoids, and herbivores (Kessler et al. 2004). In predator–prey assays, fragrance performs to designate existence of both predators-prey (Conover et al. 2010) and pheromones as defences against predators. There are many scents-mediated competitive interactions and markings at territorial borders composed of VOCs (Hurst and Beynon 2004). Many insects, particularly moths, utilize scents to announce that they are ready for meeting and give signal to partners; males navigate and sense for lengthy distances of over 100 m to search females for discharging. Sex pheromones are composed of distinctive compounds which are unique mixture of different compounds and receptors (Griffen and Byers 2006; Kurtovic et al. 2007). Aroma is critical in several communal interactions. Animals such as bark beetles use aggregation pheromones to engage population of beetles towards host. Such quorum is essential to conquer the defence barrier of host (Rasmussen and Givskov 2006b). Utilizing pheromones by insect to manipulate the deeds and functioning of adjacent mates is primer source to generate aroma, for succession among the food and colony sources and for identification of nest mate (Vander Meer and Alonso 2002; de Lima Mendonça et al. 2009). In conclusion, even it is still controversial that there are growing proofs that those volatiles can play their role as signals within and among plants (Frost et al. 2008). In same plants,

VOCs from injured parts are exposed to prime and induce defences in uninjured plant parts (Heil and Bueno 2007; Frost et al. 2008). Numerous studies of plant defence have showed that the volatiles discharged of a member of same specie perform to the primary resistance of close by healthy plants (Karban et al. 2004; Heil and Bueno 2007). Chemical signals consist of hydrocarbons, and their reactivity is influenced by the bonds between C atoms (Henze et al. 2008). The hydrocarbons are reactive to NO_3, HO, and O_3. However, NO_3, HO, and O_3 have long inhabited in the lower part of atmosphere, owing to the anthropogenic precursor, and discharge there is the enhancement in the concentrations of NO_3, HO, and O_3 in lower part of atmosphere (Fiore et al. 2002; Chen and Prinn 2006). One-time data reveals that during the early 1990s there was a five-fold raise of O_3 from the preindustrial period (Marenco et al. 2002). In existence of sun light, O_3 is formed by reactions of hydrocarbons, nitrous oxides, and carbon monoxide. Hydrocarbons' quantity remains persistent by photolysis of O_3, together with different compounds including nitrous and nitric acid (Stemmler et al. 2007). In rural atmosphere during day, O_3 and HO shared pronounced pattern. There are trace levels of HO present in the atmosphere, touching only tiny fractions of parts per trillion (Chen and Prinn 2006). The HO higher reactivity poses that it's the key driving factor of atmospheric chemistry regardless of lower concentration of HO as compared to that of NO_3 and O_3. Reaction between nitrogen dioxide and O_3 produced NO_3 that is quickly degenerated by the process of photolysis and hence it is almost absent in the day, except plants under the shady area (McFrederick et al. 2008), and in the night, NO_3 concentrations go up to 20 ppm (Fuentes et al. 2007). Pollutants can respond to hydrocarbons, by two different ways: through HO addition and through abstraction of hydrogen and addition of NO_3 or O_3 (Fuentes et al. 2007). The result of these reactions is loss of quantity and quality of signal. Every VOC group has a selected pace of reactions balanced with all of three pollutants (Evtyugina et al. 2009). Therefore, there will be decline in quantity, according to the constant. Constants' rate indicates the pace of reaction of two compounds. The commencement energy is required for the taking place of reaction and the geometry of array of reacting compounds (Henze et al. 2008). Destruction VOCs' rate depends on both the rate constant of the VOC, to air pollutants, and air pollutants ratio in environment. This reliance on concentration takes place as pollutants required to have a collision, to VOCs for the reaction between two gases to be taken place (Henze et al. 2008). Quality of signal may be impaired in two inter-connected methods: signal specificity loss or variations in ratios of VOCs that compose signals. Interactions of various VOCs with air pollutants (heavy metals) can generate analogous breakdown products and fresh compounds which may be volatile in nature (Fuentes et al. 2007). For instance, the reactions comprising of b-ocimene and b-myrcene together with HO and O_3 result in the production of three compounds including acetone, 4-methyl-3,5-hexadienal, and 4-vinyl4-pentanal (Librando and Tringali 2005; McFrederick et al. 2008). The reaction products like 4-vinyl-4-pentanal and 4-methyl-3, 5-hexadienal are volatile in nature and act as signals. Though the products of reaction are lesser in quantity, give different smell, and are volatilities comparing with parent compounds, it is important to note that the reactions between specific floral, volatiles (e.g. b-ocimene, b-myrcene), and pollutants took place at different reaction rates, which results in the changes in oxidation reaction ratios to generate the compounds that consist of scent-plume and common products as well. So, in the floral scent mixture, the relative concentration, of different volatiles parent, and their break-down products can change as the reactions occur with pollutants. It is very significant because receptors strongly respond to mixture-specific fragrance (Bruce and Pickett 2011). For recipients of scent, this loss of signal quantity and quality may be the reason due to which signals were unrecognizable. This unrecognizable signal situation is very significant, as if recipients are signal specialists and have developed specific receptors to biogenic signal; in this case, the receptors are unable to identify the signal like many insects are unable to catch sex pheromones signals (Kurtovic et al. 2007; De Bruyne and Baker 2008). Chemical signals enlighten consumer exploration and feeding across a striking range of special levels. These signals may not only affect meso-grazer (tiny herbivores usually exist on plants) they eat (Rasher et al. 2013), eating and manoeuvre between hosts over centimetres and meters scales (Coleman et al. 2007), but appear to influence foraging through various sea birds over landscapes covering long distances (Nevitt 2008). General consumers usually react to primary metabolites that hint the presence of food reserves in general (e.g. sugars, amino acids, proteins, respiration products, or metabolic wastes). Specific foodstuff represented may be elucidated by distinctive proportions of primary metabolites (Derby and Steullet 2001). Chemically arbitrated foraging by generalists is usually analysed in the lab, sometimes in flumes and sometimes in the field conditions (Parker et al. 2006; Smee and Weissburg 2006; Nevitt 2008). However, the substances that nod exploration behaviour and motion to a source aren't often investigated. In laboratory experiments or in various field experimentations, peptides have frequently been revealed to attract consumers in customs that recommend they are usually used to get foodstuff or other resources (e.g. settlement sites for barnacle larvae and shells for hermit crabs) (Rittschof and Cohen 2004; Burkepile and Hay 2006). Compatible with essential signals being merged rather than particular molecules within those blends, responses of consumer to certain peptides aren't as vigorous as to natural mixes (Rittschof and Cohen 2004). The role of blends in cuing foraging makes bioassay-guided fractionation and identification of natural forging stimulants difficult. Several aquatic invertebrates wrap their bodies with particular species of anemones, seaweeds, bacteria, or sponges. This chemically mediated behaviour is known as 'clothing.' *Pseudamphithoides incurvaria,* a Caribbean amphipod, builds a dwelling from *Dictyota bartayresii* which is chemically defended seaweed. This seaweed produces diterpene alcohol, pachydictyol-A, that avoids feeding by fishes however motivates dwelling by the amphipod; when in this dwelling, the amphipod is protected against predation by fishes, but the amphipod is rapidly preyed after being isolated from the dwelling or while enforced to construct its dwelling from a seaweed lacking powerful chemical defences (Lane et al. 2009). Similarly, adolescents of *Libinia dubia* (Atlantic decorator crab) selectively exaggerate their crustacean with *Dictyota menstrualis* which is chemically defended seaweed. This seaweed crafts manifold interrelated diterpene alcohols, but dictyol E is single compound, which most robustly dissuades native fishes, a chemical that is used by crab to pick privileged embellishment materials. While tied in the field, crabs adorned with *Dictyota menstrualis* are seldom used, but those adorned with seaweeds lacking chemical defences are generally used. In above examples, the particular chemical that robustly averts feeding by fish is the chemical that stimulates crustacean use of alga. This paradigm advocates that meso-grazer's utilization of chemically toxic seaweeds has been chosen due to the values of seaweeds for predator run off or prevention rather than their straight significance as a food. Actually in case of herbivorous decorator crustacean, the substance responsible for motivation of decorating activities dissuades the crab from feeding on seaweed. In a persuasive example from Antarctic pelagic system (Ichikawa et al. 2012), *Clione antarctica,* a chemically defended pteropod, is seized by amphipod *Hyperiella dilatata*, grasps it on its dorsal side, and becomes sheltered from predation by fish. Native fishes decline the pteropod and decline the amphipod after it grasps the pteropod, but quickly devour the amphipod if the pteropod is detached. The pteropod has polypropionate-derived metabolite pteroeone; hence, it is abandoned by fish.

7.8 Bacterial Chemical Ecology

It is hard to combine these fields due to complexity of severely exploring chemically arbitrating interactions between microorganisms under ordinary conditions. Several oceanic microbial organisms aren't yet cultivable, particularly those in association with particular hosts. In addition, the behaviour of cultured microbes may not match with behaviours in normal multi-species micro-biomes. Microorganisms might exhibit different behaviours merely in reaction to environmental stresses that might be absent in cultivated situation (Moree et al. 2012). For example, the *Phaeocystis globosa,* a phyto-plankton, chemically recognizes conspecifics being assailed, recognizes invader as a copepod or ciliate by means of chemical signals, and modifies its behaviour to reduce vulnerability to that predator (Merchant et al. 2007). Similarly, numerous microbes flourish only in biodiversity of species, which sometimes cannot be multiplied in culture. Such as, microbes' decomposition of complicated hydrocarbons occupies a network of interacting microbial organisms that use their surrounding metabolites after hydrocarbons are consecutively decomposed. Thus, insights from simplify cultural circumstances may not be applicable to complicated natural circumstances (Clay 2014; Lopanik 2014). Regardless of these challenges, a perceptive of micro-biome chemical ecology is rising (Tan et al. 2015). Microbes normally manufacture secondary metabolites that play a vital role in defending their hosts (Lopanik 2014). Foetus of *Palaemon macrodactylus* shrimp are extraordinarily impervious against fungus attack since they are enclosed by bacterium belonging to *Alteromonas sp.* that manufactures 2,3-indolinedione, which represses pathogenic fungi (Gil De Los Santos et al. 2005). Foetuses of those stripped of bacterium by means of antibiotics expire by fungal infection. In comparison, survival rate is higher for mock-treated embryos (without antibiotics treatment), subjected to antibiotics although afterward re-exposed to bacterium or subjected to antibiotics and exposed to isatin only.

Another good example is aquatic bryozoan *Bugulia neritina* (Lopanik 2014). Bioassay-directed fragmentation expresses that *B. neritina* larvae are defended by bryostatins from predators. These chemicals are synthesized by a *Candidatus Endobugula sertula, γ-Proteo-bacterium*, related with the larvae.

For instance, in Papua New Guinea, bright red isopods survive in conspicuous groups on surfaces of reef exposure to fish predators (Rodriguez et al. 2014). The *cyano-bacteria* casing their carapace have red shadow, and fish decline them as foodstuff. Isopods kept under dark conditions for two days sustained a smaller number of *cyano-bacteria* and got edible to fishes. Chemical extricate of typical, *cyano-bacteria*-coated isopods discouraged feeding of fish, advocating that cyano-bacterial metabolites safeguarded the isopods.

Chemical protection of hosts produced by microbial organisms is also recommended for sponges, ascidians, corals, squid egg capsules, fish, seaweeds, and a host of various aquatic organisms (Uria and Piel 2009; Lopanik 2014). Though chemical signals from hosts possibly influence colonization microorganisms (McFall-Ngai et al. 2013).

7.9 Conclusion

Plants have developed various approaches to protect themselves against enemies like herbivores. These defence approaches may be classified into direct and indirect, constitutive, and induced system. The blend of various approaches permits plants to protect them efficiently against opponents. Particularly, the development of advanced techniques made it possible to investigate thoroughly induced indirect defence systems. For instance, research on up-regulated genes. It was of great interest to study correlation and interaction of plants to organisms from upper hierarchical levels. The significance of secondary metabolites like jasmonate and ethylene-linked compounds was recognized and pathways explicated. It is crystal clear now that secondary metabolites have key role in plant defence. On the other hand, plant secondary metabolites may perform other functions like pollination or allelopathy. However, more research work is required in this subject. Several signal transduction pathways are still unclear, while others must be determined yet. The tolerance mechanism, which is mostly unknown yet, also requires elucidation. Ahead of communication inside a plant and among plants, interactions of plants with organisms from other hierarchical levels must be evaluated. Most of the metabolites have been explored in plant leaves; however, only some have been explored in roots. Advance studies, particularly in roots, are compulsory to discover new metabolites with possible worth for medicinal utilization. Additional advancement in these techniques may help in betterment in screenings and enhanced efficacy in identification of plant compounds. Evolution is an everlasting and constant process that will gain researchers' attention for coming years.

REFERENCES

Albornoz FE, Lambers H, Turner BL, Teste FP, Laliberté E (2016) Shifts in symbiotic associations in plants capable of forming multiple root symbioses across a long-term soil chronosequence. Ecol Evol 6:2368–2377.

Alexander EB, Ellis CC, Burke R (2007) A chronosequence of soils and vegetation on serpentine terraces in the Klamath Mountains, USA. Soil Sci 172:565–576.

Arnold T, Appel H, Patel V, Stocum E, Kavalier A, Schultz J (2004) Carbohydrate translocation determines the phenolic content of *Populus foliage*: a test of the sink–source model of plant defense. New Phytol 164:157–164.

Badawi FSF, Biomy A, Desoky A (2011) Peanut plant growth and yield as influenced by co-inoculation with *Bradyrhizobium* and some rhizo-microorganisms under sandy loam soil conditions. Ann Agric Sci 56:17–25.

Bailey VL, Peacock A, Smith JL, Bolton Jr H (2002) Relationships between soil microbial biomass determined by chloroform fumigation–extraction, substrate-induced respiration, and phospholipid fatty acid analysis. Soil Biol Biochem 34:1385–1389.

Baldwin IT, Halitschke R, Paschold A, Von Dahl CC, Preston CA (2006) Volatile signaling in plant-plant interactions: "talking trees" in the genomics era. Science 311:812–815.

Barbehenn R, Cheek S, Gasperut A, Lister E, Maben R (2005) Phenolic compounds in red oak and sugar maple leaves have prooxidant activities in the midgut fluids of *Malacosoma disstria* and *Orgyia leucostigma* caterpillars. J Chem Ecol 31:969–988.

Barnawal D, Bharti N, Pandey SS, Pandey A, Chanotiya CS, Kalra A (2017) Plant growth-promoting rhizobacteria enhance wheat salt and drought stress tolerance by altering endogenous phytohormone levels and TaCTR1/TaDREB2 expression. Physiol Plant 161:502–514.

Barrett LG, Bever JD, Bissett A, Thrall PH (2015) Partner diversity and identity impacts on plant productivity in Acacia–rhizobial interactions. J Ecol 103:130–142.

Bernays EA (1991) Evolution of insect morphology in relation to plants. Philos Trans R Soc Lond 333:257–264.

Bever JD (2015) Preferential allocation, physio-evolutionary feedbacks, and the stability and environmental patterns of mutualism between plants and their root symbionts. New Phytol 205:1503–1514.

Bezemer GL, Bitter JH, Kuipers HP, Oosterbeek H, Holewijn JE, Xu X, Kapteijn F, van Dillen AJ, de Jong KP (2006) Cobalt particle size effects in the Fischer– Tropsch reaction studied with carbon nanofiber supported catalysts. J Am Chem Soc 128:3956–3964.

Boyd RS (2010) Heavy metal pollutants and chemical ecology: exploring new frontiers. J Chem Ecol 36:46–58.

Boyd CA, Boyd CE, Rouse D (2007) Potassium adsorption by bottom soils in ponds for inland culture of marine shrimp in Alabama. J World Aqua Soc 38:85–91.

Boyd ES, Hamilton TL, Peters JW (2011) An alternative path for the evolution of biological nitrogen fixation. Front Microbiol 2:205.

Bressano M, Curetti M, Giachero L, Gil SV, Cabello M, March G, Ducasse DA, Luna CM (2010) Mycorrhizal fungi symbiosis as a strategy against oxidative stress in soybean plants. J Plant Physiol 167:1622–1626.

Bruce TJ, Pickett JA (2011) Perception of plant volatile blends by herbivorous insects–finding the right mix. Phytochem 72:1605–1611.

Burkepile DE, Hay ME (2006) Herbivore vs. nutrient control of marine primary producers: context-dependent effects. Ecology 87:3128–3139.

Callaway RM, Bedmar EJ, Reinhart KO, Silvan CG, Klironomos J (2011) Effects of soil biota from different ranges on Robinia invasion: acquiring mutualists and escaping pathogens. Ecology 92:1027–1035.

Callaway RM, Mahall BE (2007) Plant ecology: family roots. Nature 448:145.

Cariñanos P, Casares-Porcel M, de la Guardia CD, Aira MJ, Belmonte J, Boi M, Elvira-Rendueles B, De Linares C, Fernández-Rodriguez S, Maya-Manzano JM (2017) Assessing allergenicity in urban parks: a nature-based solution to reduce the impact on public health. Environ Res 155:219–227.

Casper BB, Castelli JP (2007) Evaluating plant–soil feedback together with competition in a serpentine grassland. Ecol Lett 10:394–400.

Challis GL (2005) A widely distributed bacterial pathway for siderophore biosynthesis independent of nonribosomal peptide synthetases. Chembiochem 6:601–611.

Chaves N, Sosa T, Escudero J (2001) Plant growth inhibiting flavonoids in exudate of *Cistus ladanifer* and in associated soils. J Chem Ecol 27:623–631.

Chen YH, Prinn RG (2006) Estimation of atmospheric methane emissions between 1996 and 2001 using a three-dimensional global chemical transport model. J Geophys Re Atmos 111.

Chubar N, Kanibolotskyy V, Strelko V, Gallios G, Samanidou V, Shaposhnikova T, Milgrandt V, Zhuravlev I (2005) Adsorption of phosphate ions on novel inorganic ion exchangers. Colloids Surf, A Physicochem Eng Asp 255:55–63.

Classen AT, Sundqvist MK, Henning JA, Newman GS, Moore JA, Cregger MA, Moorhead LC, Patterson CM (2015) Direct and indirect effects of climate change on soil microbial and soil microbial-plant interactions: what lies ahead? Ecosphere 6:1–21.

Clay K (2014) Defensive symbiosis: a microbial perspective. Funct Ecol 28:293–298.

Colautti RI, Ricciardi A, Grigorovich IA, MacIsaac HJ (2004) Is invasion success explained by the enemy release hypothesis? Ecol Lett 7:721–733.

Coleman R, Ramchunder S, Moody A, Foggo A (2007) An enzyme in snail saliva induces herbivore-resistance in a marine alga. Funct Ecol 21:101–106.
Conover MR, Borgo JS, Dritz RE, Dinkins JB, Dahlgren DK (2010) Greater sage-grouse select nest sites to avoid visual predators but not olfactory predators. Condor 112:331–336.
Crawford KM, Knight TM (2017) Competition overwhelms the positive plant–soil feedback generated by an invasive plant. Oecologia 183:211–220.
Crispo E (2007) The Baldwin effect and genetic assimilation: revisiting two mechanisms of evolutionary change mediated by phenotypic plasticity. Evol: Int J Org Evol 61:2469–2479.
Cui X, Song J (2007) Soil NH 4+/NO 3– nitrogen characteristics in primary forests and the adaptability of some coniferous species. Front For Chin 2:1–10.
De Bruyne M, Baker T (2008) Odor detection in insects: volatile codes. J Chem Ecol 34:882–897.
de Lima Mendonça A, Da Silva CE, de Mesquita FLT, da Silva Campos R, Do Nascimento RR, de Azevedo Ximenes ECP, Sant'Ana AEG (2009) Antimicrobial activities of components of the glandular secretions of leaf cutting ants of the genus Atta. Antonie Van Leeuwenhoek 95:295–303.
Del Arco M, Fernandez A, Martin C, Rives V (2010) Solubility and release of fenbufen intercalated in Mg, Al and Mg, Al, Fe layered double hydroxides (LDH): the effect of Eudragit® S 100 covering. J Solid State Chem 183:3002–3009.
Del Fabbro C, Güsewell S, Prati D (2014) Allelopathic effects of three plant invaders on germination of native species: a field study. Biol Invasions 16:1035–1042.
Derby CD, Steullet P (2001) Why do animals have so many receptors? The role of multiple chemosensors in animal perception. Biol Bull 200:211–215.
Dicke M, van Loon JJ (2000) Multitrophic effects of herbivore-induced plant volatiles in an evolutionary context. Entomol Exp Applic 97:237–249.
Dietz M, Machill S, Hoffmann HC, Schmidtke K (2013) Inhibitory effects of *Plantago lanceolata* L. on soil N mineralization. Plant Soil 368:445–458.
Diggle SP, Gardner A, West SA, Griffin AS (2007) Evolutionary theory of bacterial quorum sensing: when is a signal not a signal? Philos Trans R Soc 362:1241–1249.
Dixit R, Malaviya D, Pandiyan K, Singh U, Sahu A, Shukla R, Singh B, Rai J, Sharma P, Lade H (2015) Bioremediation of heavy metals from soil and aquatic environment: an overview of principles and criteria of fundamental processes. Sustainability 7:2189–2212.
Dyer LA, Dodson CD, Stireman J, Tobler M, Smilanich AM, Fincher R, Letourneau DK (2003) Synergistic effects of three Piper amides on generalist and specialist herbivores. J Chem Ecol 29:2499–2514.
Eigenbrode S (2004) The effects of plant epicuticular waxy blooms on attachment and effectiveness of predatory insects. Arthropod Struct Dev 33:91–102.
Emam HE, Mowafi S, Mashaly HM, Rehan M (2014) Production of antibacterial colored viscose fibers using in situ prepared spherical Ag nanoparticles. Carbohydr Polym 110:148–155.
Endara MJ, Coley PD (2011) The resource availability hypothesis revisited: a meta-analysis. Funct Ecol 25:389–398.
Engelberth J, Alborn HT, Schmelz EA, Tumlinson JH (2004) Airborne signals prime plants against insect herbivore attack. Proc Natl Acad Sci 101:1781–1785.
Erb M, Robert CA, Hibbard BE, Turlings TC (2011) Sequence of arrival determines plant-mediated interactions between herbivores. J Ecol 99:7–15.
Evtyugina M, Nunes T, Alves C, Marques M (2009) Photochemical pollution in a rural mountainous area in the northeast of Portugal. Atmos Res 92:151–158.
Ewnetu Y, Lemma W, Birhane N (2014) Synergetic antimicrobial effects of mixtures of Ethiopian honeys and ginger powder extracts on standard and resistant clinical bacteria isolates. Evid Based Complement Alternat Med 2014:562804.
Falik O, Reides P, Gersani M, Novoplansky A (2005) Root navigation by self inhibition. Plant Cell Environ 28:562–569.
File AL, Murphy GP, Dudley SA (2011) Fitness consequences of plants growing with siblings: reconciling kin selection, niche partitioning and competitive ability. Proc R Soc 279:209–218.
Filley T, Cody G, Goodell B, Jellison J, Noser C, Ostrofsky A (2002) Lignin demethylation and polysaccharide decomposition in spruce sapwood degraded by brown rot fungi. Org Geochem 33:111–124.

Filley TR, Nierop KG, Wang Y (2006) The contribution of polyhydroxyl aromatic compounds to tetramethylammonium hydroxide lignin-based proxies. Org Geochem 37:711–727.

Fiore AM, Jacob DJ, Field BD, Streets DG, Fernandes SD, Jang C (2002) Linking ozone pollution and climate change: the case for controlling methane. Geophys Res Lett 29:25–1–25-4.

Frost CJ, Mescher MC, Carlson JE, De Moraes CM (2008) Why do distance limitations exist on plant-plant signaling via airborne volatiles? Plant Signal Behav 3:466–468.

Fuentes JD, Wang D, Bowling DR, Potosnak M, Monson RK, Goliff WS, Stockwell WR (2007) Biogenic hydrocarbon chemistry within and above a mixed deciduous forest. J Atmos Chem 56:165–185.

Futuyama DJ (2009) Evolution. Sinauer Associates, Sunderland, MA, pp 347.

Galloway JN, Townsend AR, Erisman JW, Bekunda M, Cai Z, Freney JR, Martinelli LA, Seitzinger SP, Sutton MA (2008) Transformation of the nitrogen cycle: recent trends, questions, and potential solutions. Science 320:889–892.

Gil De Los Santos J, Storch O, Gil-Turnes C (2005) Bacillus cereus var. toyoii and *Saccharomyces boulardii* increased feed efficiency in broilers infected with *Salmonella enteritidis*. Br Poul Sci 46:494–497.

Glick BR (2014) Bacteria with ACC deaminase can promote plant growth and help to feed the world. Microbiol Res 169:30–39.

Glynn C, Herms DA, Orians CM, Hansen RC, Larsson S (2007) Testing the growth–differentiation balance hypothesis: dynamic responses of willows to nutrient availability. New Phytol 176:623–634.

González JE, Keshavan ND (2006) Messing with bacterial quorum sensing. Microbiol Mol Biol Rev 70:859–875.

Gray E, Smith D (2005) Intracellular and extracellular PGPR: commonalities and distinctions in the plant–bacterium signaling processes. Soil Biol Biochem 37:395–412.

Greger M, Löfstedt M (2004) Comparison of uptake and distribution of cadmium in different cultivars of bread and durum wheat. Crop Sci 44:501–507.

Griffen BD, Byers JE (2006) Intraguild predation reduces redundancy of predator species in multiple predator assemblage. J Anim Ecol 75:959–966.

Gruntman M, Novoplansky A (2004) Physiologically mediated self/non-self discrimination in roots. Proc Natl Acad Sci 101:3863–3867.

Hamilton TL, Bryant DA, Macalady JL (2016) The role of biology in planetary evolution: cyanobacterial primary production in low-oxygen Proterozoic oceans. Environ Microbiol 18:325–340.

Heath JJ, Kessler A, Woebbe E, Cipollini D, Stireman III JO (2014) Exploring plant defense theory in tall goldenrod, *Solidago altissima*. New Phytol 202:1357–1370.

Heil M, Bueno JCS (2007) Within-plant signaling by volatiles leads to induction and priming of an indirect plant defense in nature. Proc Natl Acad Sci 104:5467–5472.

Heil M, McKey D (2003) Protective ant-plant interactions as model systems in ecological and evolutionary research. Ann Rev Ecol Evol System 34:425–553.

Henze D, Seinfeld J, Ng N, Kroll J, Fu T-M, Jacob DJ, Heald C (2008) Global modeling of secondary organic aerosol formation from aromatic hydrocarbons: high-vs. low-yield pathways. Atmos Chem Phys 8:2405–2420.

Hess L, De Kroon H (2007) Effects of rooting volume and nutrient availability as an alternative explanation for root self/non-self discrimination. J Ecol 95:241–251.

Hewitt RE, Menges ES (2008) Allelopathic effects of *Ceratiola ericoides* (Empetraceae) on germination and survival of six Florida scrub species. Plant Ecol 198:47–59.

Hilgers MT, Ludwig ML (2001) Crystal structure of the quorum-sensing protein LuxS reveals a catalytic metal site. Proc Natl Acad Sci 98:11169–11174.

Hodge A, Berta G, Doussan C, Merchan F, Crespi M (2009) Plant root growth, architecture and function. Plant Soil 321:153–187.

Hood-Nowotny R, Umana NH-N, Inselbacher E, Oswald-Lachouani P, Wanek W (2010) Alternative methods for measuring inorganic, organic, and total dissolved nitrogen in soil. Soil Sci Soc Am J 74:1018–1027.

Hurst JL, Beynon RJ (2004) Scent wars: the chemobiology of competitive signalling in mice. Bioessays 26:1288–1298.

Hättenschwiler S, Vitousek PM (2000) The role of polyphenols in terrestrial ecosystem nutrient cycling. Trends Ecol Evol 15:238–243.

Ichikawa N, Sasagawa M, Yamamoto M, Komaki H, Yoshida Y, Yamazaki S, Fujita N (2012) DoBISCUIT: a database of secondary metabolite biosynthetic gene clusters. Nucl Acids Res 41:D408–D414.

Inderjit, Nilsen ET (2003) Bioassays and field studies for allelopathy in terrestrial plants: progress and problems. Crit Rev Plant Sci 22:221–238.
Isbell F, Craven D, Connolly J, Loreau M, Schmid B, Beierkuhnlein C, Bezemer TM, Bonin C, Bruelheide H, De Luca E (2015) Biodiversity increases the resistance of ecosystem productivity to climate extremes. Nature 526:574.
Jeffree CE (2006) The fine structure of the plant cuticle. Biol Plant Cuticle 23:11–125.
Jenke-Kodama H, Müller R, Dittmann E (2008) Evolutionary mechanisms underlying secondary metabolite diversity. In: Petersen F, Amstutz R (eds) Natural compounds as drugs volume I. Springer, Basel, pp 119–140.
Johnson DB (2007) Physiology and ecology of acidophilic microorganisms, physiology and biochemistry of extremophiles. Ame Soc Microbiol 257–270.
Johnson D, Vandenkoornhuyse PJ, Leake JR, Gilbert L, Booth RE, Grime JP, Young JPW, Read DJ (2004) Plant communities affect arbuscular mycorrhizal fungal diversity and community composition in grassland microcosms. New Phytol 161:503–515.
Kanerva S (2007) Plant secondary compounds and soil microbial processes in carbon and nitrogen cycling in relation to tree species. Dissertationes Forestales
Karban AS, Okazaki T, Panhuysen CI, Gallegos T, Potter JJ, Bailey-Wilson JE, Silverberg MS, Duerr RH, Cho JH, Gregersen PK (2004) Functional annotation of a novel NFKB1 promoter polymorphism that increases risk for ulcerative colitis. Human Mol Genet 13:35–45.
Karban R, Shiojiri K (2009) Self-recognition affects plant communication and defense. Ecol Lett 12:502–506.
Kerstiens G (2006) Water transport in plant cuticles: an update. J Exp Bot 57:2493–2499.
Kessler A (2015) The information landscape of plant constitutive and induced secondary metabolite production. Cur Opin Insect Sci 8:47–53.
Kessler A, Halitschke R, Baldwin IT (2004) Silencing the jasmonate cascade: induced plant defenses and insect populations. Science 305:665–668.
Kim KS, Park SH, Jenks MA (2007) Changes in leaf cuticular waxes of sesame (*Sesamum indicum* L.) plants exposed to water deficit. J Plant Physiol 164:1134–1143.
Klemens JA (2008) Kin recognition in plants? Biol Lett 4:67.
Klironomos JN (2002) Feedback with soil biota contributes to plant rarity and invasiveness in communities. Nature 417:67.
Klironomos JN, Hart MM (2002) Colonization of roots by arbuscular mycorrhizal fungi using different sources of inoculum. Mycorrhiza 12:181–184.
Koch K, Barthlott W (2009) Superhydrophobic and superhydrophilic plant surfaces: an inspiration for biomimetic materials. Philos Trans R Soc 367:1487–1509.
Kong AY, Six J, Bryant DC, Denison RF, Van Kessel C (2005) The relationship between carbon input, aggregation, and soil organic carbon stabilization in sustainable cropping systems. Soil Sci Soc Am J 69:1078–1085.
Koricheva J (2002) The carbon-nutrient balance hypothesis is dead; long live the carbon-nutrient balance hypothesis? Oikos 98:537–539.
Krapp A (2015) Plant nitrogen assimilation and its regulation: a complex puzzle with missing pieces. Curr Opin Plant Biol 25:115–122.
Krause D, Nagaraja T, Wright A, Callaway T (2013) Board-invited review: rumen microbiology: leading the way in microbial ecology. J Anim Sci 91:331–341.
Kuhlisch C, Pohnert G (2015) Metabolomics in chemical ecology. Nat Prod Rep 32:937–955.
Kulmatiski A, Beard KH, Stevens JR, Cobbold SM (2008) Plant–soil feedbacks: a meta-analytical review. Ecol Lett 11:980–992.
Kurtovic A, Widmer A, Dickson BJ (2007) A single class of olfactory neurons mediates behavioural responses to a Drosophila sex pheromone. Nature 446:542.
Lampert EC, Dyer LA, Bowers MD (2010) Caterpillar chemical defense and parasitoid success: Cotesia congregata parasitism of Ceratomia catalpae. J Chem Ecol 36:992–998.
Lane AL, Nyadong L, Galhena AS, Shearer TL, Stout EP, Parry RM, Kwasnik M, Wang MD, Hay ME, Fernandez FM (2009) Desorption electrospray ionization mass spectrometry reveals surface-mediated antifungal chemical defense of a tropical seaweed. Proc Natl Acad Sci 106:7314–7319.
Larson NI, Story MT, Nelson MC (2009) Neighborhood environments: disparities in access to healthy foods in the US. Am J Prevent Med 36:74–81. e10.

Legrand C, Rengefors K, Fistarol GO, Graneli E (2003) Allelopathy in phytoplankton-biochemical, ecological and evolutionary aspects. Phycologia 42:406–419.
Librando V, Tringali G (2005) Atmospheric fate of OH initiated oxidation of terpenes. Reaction mechanism of α-pinene degradation and secondary organic aerosol formation. J Environ Manage 75:275–282.
Liu X, Song Q, Tang Y, Li W, Xu J, Wu J, Wang F, Brookes PC (2013) Human health risk assessment of heavy metals in soil–vegetable system: a multi-medium analysis. Sci Total Environ 463:530–540.
Lopanik NB (2014) Chemical defensive symbioses in the marine environment. Funct Ecol 28:328–340.
Ma H, Evangelou MW, Tandy S, Felderer B, Vontobel P, Schulin R (2016) Effect of uneven root allocation on shoot Zn uptake by crop plants growing on soil with heterogeneous P and Zn distribution. Environ Exp Bot 130:226–232.
Macel M, Lawson CS, Mortimer SR, Šmilauerova M, Bischoff A, Crémieux L, Doležal J, Edwards AR, Lanta V, Bezemer TM (2007) Climate vs. soil factors in local adaptation of two common plant species. Ecology 88:424–433.
Macnair MR (2003) The hyperaccumulation of metals by plants. Adv Bot Res 40:63–105.
Maksymiec W (2007) Signaling responses in plants to heavy metal stress. Acta Physiol Plant 29:177.
Manefield M, Rasmussen TB, Henzter M, Andersen JB, Steinberg P, Kjelleberg S, Givskov M (2002) Halogenated furanones inhibit quorum sensing through accelerated LuxR turnover. Microbiol 148:1119–1127.
Mangan SA, Schnitzer SA, Herre EA, Mack KM, Valencia MC, Sanchez EI, Bever JD (2010) Negative plant–soil feedback predicts tree-species relative abundance in a tropical forest. Nature 466:752.
Marenco F, di Sarra A, De Luisi J (2002) Methodology for determining aerosol optical depth from Brewer 300–320-nm ozone measurements. Appl Opt 41:1805–1814.
Martens SN, Boyd RS (2002) The defensive role of Ni hyperaccumulation by plants: a field experiment. Am J Bot 89:998–1003.
Martinou J-C, Youle RJ (2011) Mitochondria in apoptosis: Bcl-2 family members and mitochondrial dynamics. Dev Cell 21:92–101.
Massad TJ, Dyer LA, Vega G (2012) Costs of defense and a test of the carbon-nutrient balance and growth-differentiation balance hypotheses for two co-occurring classes of plant defense. PLoS One 7:e47554.
McFall-Ngai M, Hadfield MG, Bosch TC, Carey HV, Domazet-Lošo T, Douglas AE, Dubilier N, Eberl G, Fukami T, Gilbert SF (2013) Animals in a bacterial world, a new imperative for the life sciences. Proc Natl Acad Sci 110:3229–3236.
McFrederick QS, Kathilankal JC, Fuentes JD (2008) Air pollution modifies floral scent trails. Atmos Environ 42:2336–2348.
Meinwald J, Eisner T (2008) Chemical ecology in retrospect and prospect. Proc Natl Acad Sci 105:4539–4540.
Mendgen K, Hahn M (2002) Plant infection and the establishment of fungal biotrophy. Trends Plant Sci 7:352–356.
Merchant SS, Prochnik SE, Vallon O, Harris EH, Karpowicz SJ, Witman GB, Terry A, Salamov A, Fritz-Laylin LK, Maréchal-Drouard L (2007) The Chlamydomonas genome reveals the evolution of key animal and plant functions. Science 318:245–250.
Miller G, Tybur JM, Jordan BD (2007) Ovulatory cycle effects on tip earnings by lap dancers: economic evidence for human estrus? Evol Human Behav 28:375–381.
Mithöfer A, Schulze B, Boland W (2004) Biotic and heavy metal stress response in plants: evidence for common signals. FEBS Lett 566:1–5.
Moree WJ, Phelan VV, Wu C-H, Bandeira N, Cornett DS, Duggan BM, Dorrestein PC (2012) Interkingdom metabolic transformations captured by microbial imaging mass spectrometry. Proc Natl Acad Sci 109:13811–13816.
Müller C, Riederer M (2005) Plant surface properties in chemical ecology. J Chem Ecol 31:2621–2651.
Nadeem SM, Ahmad M, Zahir ZA, Javaid A, Ashraf M (2014) The role of mycorrhizae and plant growth promoting rhizobacteria (PGPR) in improving crop productivity under stressful environments. Biotechnol Adv 32:429–448.
Naseem M, Aslam M, Ansar M, Azhar M (2009) Allelopathic effects of sunflower water extract on weed control and wheat productivity. Pak J Weed Sci Res 15:107–116.
Nelson E, Mendoza G, Regetz J, Polasky S, Tallis H, Cameron D, Chan KM, Daily GC, Goldstein J,

Kareiva PM (2009) Modeling multiple ecosystem services, biodiversity conservation, commodity production, and tradeoffs at landscape scales. Front Ecol Environ 7:4–11.
Nevitt GA (2008) Sensory ecology on the high seas: the odor world of the procellariiform seabirds. J Exp Biol 211:1706–1713.
Ni N, Li M, Wang J, Wang B (2009) Inhibitors and antagonists of bacterial quorum sensing. Med Res Rev 29:65–124.
Notten M, Oosthoek A, Rozema J, Aerts R (2005) Heavy metal concentrations in a soil–plant–snail food chain along a terrestrial soil pollution gradient. Environ Pollut 138:178–190.
Ockels FS, Eyles A, McPherson BA, Wood DL, Bonello P (2007) Phenolic chemistry of coast live oak response to *Phytophthora ramorum* infection. J Chem Ecol 33:1721–1732.
Ode PJ (2006) Plant chemistry and natural enemy fitness: effects on herbivore and natural enemy interactions. Annu Rev Entomol 51:163–185.
Oluyemi E, Feuyit G, Oyekunle J, Ogunfowokan A (2008) Seasonal variations in heavy metal concentrations in soil and some selected crops at a landfill in Nigeria. Afr J Environ Sci Tech 2:089–096.
Ortiz N, Armada E, Duque E, Roldán A, Azcón R (2015) Contribution of arbuscular mycorrhizal fungi and/or bacteria to enhancing plant drought tolerance under natural soil conditions: effectiveness of autochthonous or allochthonous strains. J Plant Physiol 174:87–96.
Parker JD, Burkepile DE, Hay ME (2006) Opposing effects of native and exotic herbivores on plant invasions. Science 311:1459–1461.
Pezzini S, Van Delft M, Schoop LM, Lotsch B, Carrington A, Katsnelson M, Hussey N, Wiedmann S (2018) Unconventional mass enhancement around the Dirac nodal loop in ZrSiS. Nat Phys 14:178.
Pichersky E, Noel JP, Dudareva N (2006) Biosynthesis of plant volatiles: nature's diversity and ingenuity. Science 311:808–811.
Piechulla B, Pott MB (2003) Plant scents—mediators of inter-and intraorganismic communication. Plant 217:687–689.
Rascio N, Navari-Izzo F (2011) Heavy metal hyperaccumulating plants: how and why do they do it? And what makes them so interesting? Plant Sci 180:169–181.
Rasher DB, Hoey AS, Hay ME (2013) Consumer diversity interacts with prey defenses to drive ecosystem function. Ecology 94:1347–1358.
Rashid MI, Mujawar LH, Shahzad T, Almeelbi T, Ismail IM, Oves M (2016) Bacteria and fungi can contribute to nutrients bioavailability and aggregate formation in degraded soils. Microbiol Res 183:26–41.
Rasmussen TB, Givskov M (2006a) Quorum sensing inhibitors: a bargain of effects. Microbiol 152:895–904.
Rasmussen TB, Givskov M (2006b) Quorum-sensing inhibitors as anti-pathogenic drugs. Int J Med Microbiol 296:149–161.
Rittschof D, Cohen JH (2004) Crustacean peptide and peptide-like pheromones and kairomones. Peptides 25:1503–1516.
Rodriguez AB, Fodrie FJ, Ridge JT, Lindquist NL, Theuerkauf EJ, Coleman SE, Grabowski JH, Brodeur MC, Gittman RK, Keller DA (2014) Oyster reefs can outpace sea-level rise. Nat Climat Chang 4:493.
Sainju UM, Allen BA, Caesar-TonThat T, Lenssen AW (2015) Dryland soil carbon and nitrogen after 30 years of tillage and cropping sequence combination. Agron J 107:1822–1830.
Sarkar S, Chakraborty R (2008) Quorum sensing in metal tolerance of *Acinetobacter junii* BB1A is associated with biofilm production. FEMS Microbiol Lett 282:160–165.
Schertzer JW, Boulette ML, Whiteley M (2009) More than a signal: non-signaling properties of quorum sensing molecules. Trends Microbiol 17:189–195.
Schuman MC, Baldwin IT (2016) The layers of plant responses to insect herbivores. Annu Rev Entomol 61:373–394.
Scrivanti LR, Zunino MP, Zygadlo JA (2003) Tagetes minuta and Schinus areira essential oils as allelopathic agents. Biochem System Ecol 31:563–572.
Shrivastava P, Kumar R (2015) Soil salinity: a serious environmental issue and plant growth promoting bacteria as one of the tools for its alleviation. Saudi J Biol Sci 22:123–131.
Sigüenza C, Crowley DE, Allen EB (2006) Soil microorganisms of a native shrub and exotic grasses along a nitrogen deposition gradient in southern California. Appl Soil Ecol 32:13–26.
Singh A (2015) Soil salinization and waterlogging: a threat to environment and agricultural sustainability. Ecol Indic 57:128–130.
Singh A, Sharma RK, Agrawal M, Marshall FM (2010) Health risk assessment of heavy metals via dietary

intake of foodstuffs from the wastewater irrigated site of a dry tropical area of India. Food Chem Toxicol 48:611–619.
Smee DL, Weissburg MJ (2006) Hard clams (*Mercenaria mercenaria*) evaluate predation risk using chemical signals from predators and injured conspecifics. J Chem Ecol 32:605–619.
Smilanich AM, Fincher RM, Dyer LA (2016) Does plant apparency matter? 30 years of data provide limited support but reveal clear patterns of the effects of plant chemistry on herbivores. New Phytol 210:1044–1057.
Smith-Ramesh LM, Reynolds HL (2017) The next frontier of plant–soil feedback research: unraveling context dependence across biotic and abiotic gradients. J Veg Sci 28:484–494.
Spencer NA, McClintock MK, Sellergren SA, Bullivant S, Jacob S, Mennella JA (2004) Social chemosignals from breastfeeding women increase sexual motivation. Horm Behav 46:362–370.
Stefani FO, Bell TH, Marchand C, Ivan E, El Yassimi A, St-Arnaud M, Hijri M (2015) Culture-dependent and-independent methods capture different microbial community fractions in hydrocarbon-contaminated soils. PLoS One 10:e0128272.
Stemmler K, Ndour M, Elshorbany Y, Kleffmann J, D'anna B, George C, Bohn B, Ammann M (2007) Light induced conversion of nitrogen dioxide into nitrous acid on submicron humic acid aerosol. Atmos Chem Phys 7:4237–4248.
Städler E, Reifenrath K (2009) Glucosinolates on the leaf surface perceived by insect herbivores: review of ambiguous results and new investigations. Phytochem Rev 8:207–225.
Subbarao G, Sahrawat KL, Nakahara K, Rao IM, Ishitani M, Hash C, Kishii M, Bonnett D, Berry W, Lata J-C (2012b) A paradigm shift towards low-nitrifying production systems: the role of biological nitrification inhibition (BNI). Ann Bot 112:297–316.
Sun L, Lu Y, Yu F, Kronzucker HJ, Shi W (2016) Biological nitrification inhibition by rice root exudates and its relationship with nitrogen-use efficiency. New Phytol 212:646–656.
Sutherland WJ, Barnard P, Broad S, Clout M, Connor B, Côté IM, Dicks LV, Doran H, Entwistle AC, Fleishman E (2017) A 2017 horizon scan of emerging issues for global conservation and biological diversity. Trends Ecol Evol 32:31–40.
Svergun DI, Koch MH (2003) Small-angle scattering studies of biological macromolecules in solution. Rep Prog Phys 66:1735.
Syam N, Wardiyati T, Maghfoer MD, Handayanto E, Ibrahim B, Muchdar A (2016) Effect of accumulator plants on growth and nickel accumulation of soybean on metal-contaminated soil. Agric Agric Sci Proc 9:13–19.
Tait K, Williamson H, Atkinson S, Williams P, Cámara M, Joint I (2009) Turnover of quorum sensing signal molecules modulates cross-kingdom signalling. Environ Microbiol 11:1792–1802.
Tan J, Zuniga C, Zengler K (2015) Unraveling interactions in microbial communities-from co-cultures to microbiomes. J Microbiol 53:295–305.
Teplitski M, Chen H, Rajamani S, Gao M, Merighi M, Sayre RT, Robinson JB, Rolfe BG, Bauer WD (2004) Chlamydomonas reinhardtii secretes compounds that mimic bacterial signals and interfere with quorum sensing regulation in bacteria. Plant Physiol 134:137–146.
Trewavas A (2009) What is plant behaviour? Plant Cell Environ 32:606–616.
Turcotte MM, Corrin MS, Johnson MT (2012) Adaptive evolution in ecological communities. PLoS Biol 10:e1001332.
Turovskiy Y, Kashtanov D, Paskhover B, Chikindas ML (2007) Quorum sensing: fact, fiction, and everything in between. Adv Appl Microbiol 62:191–234.
Uchida Y, Ohshima T, Sasaki Y, Suzuki H, Yanai S, Yamashita N, Nakamura F, Takei K, Ihara Y, Mikoshiba K (2005) Semaphorin3A signalling is mediated via sequential Cdk5 and GSK3β phosphorylation of CRMP2: implication of common phosphorylating mechanism underlying axon guidance and Alzheimer's disease. Genes Cells 10:165–179.
Uria A, Piel J (2009) Cultivation-independent approaches to investigate the chemistry of marine symbiotic bacteria. Phytochem Rev 8:401–414.
Vander Meer RK, Alonso LE (2002) Queen primer pheromone affects conspecific fire ant (Solenopsis invicta) aggression. Behav Ecol Sociobiol 51:122–130.
Verdonk JC, De Vos CR, Verhoeven HA, Haring MA, van Tunen AJ, Schuurink RC (2003) Regulation of floral scent production in petunia revealed by targeted metabolomics. Phytochem 62:997–1008.

Vrchotová N, Šerá B (2008) Allelopathic properties of knotweed rhizome extracts. Plant Soil Environ 54:301–303.
Wagg C, Jansa J, Schmid B, van der Heijden MG (2011) Belowground biodiversity effects of plant symbionts support aboveground productivity. Ecol Lett 14:1001–1009.
Wahman DG, Katz LE, Speitel GE (2005) Cometabolism of trihalomethanes by *Nitrosomonas europaea*. Appl Environ Microbiol 71:7980–7986.
Wang J, Liu W-Z, Dang T-H, Sainju UM (2013) Nitrogen fertilization effect on soil water and wheat yield in the Chinese Loess Plateau. Agron J 105:143–149.
War AR, Paulraj MG, Ahmad T, Buhroo AA, Hussain B, Ignacimuthu S, Sharma HC (2012) Mechanisms of plant defense against insect herbivores. Plant Signal Behav 7:1306–1320.
Warren RL, Keeling CI, Yuen MMS, Raymond A, Taylor GA, Vandervalk BP, Mohamadi H, Paulino D, Chiu R, Jackman SD (2015) Improved white spruce (*Picea glauca*) genome assemblies and annotation of large gene families of conifer terpenoid and phenolic defense metabolism. Plant J 83:189–212.
Waters CM, Bassler BL (2005) Quorum sensing: cell-to-cell communication in bacteria. Annu Rev Cell Dev Biol 21:319–346.
Wiuff C, Zappala R, Regoes R, Garner K, Baquero F, Levin B (2005) Phenotypic tolerance: antibiotic enrichment of noninherited resistance in bacterial populations. Antimicrob Agents Chemother 49:1483–1494.
Wyart C, Webster WW, Chen JH, Wilson SR, McClary A, Khan RM, Sobel N (2007) Smelling a single component of male sweat alters levels of cortisol in women. J Neurosci 27:1261–1265.
Yang L, Jelsbak L, Marvig RL, Damkiær S, Workman CT, Rau MH, Hansen SK, Folkesson A, Johansen HK, Ciofu O (2011) Evolutionary dynamics of bacteria in a human host environment. Proc Natl Acad Sci 108:7481–7486.
Yang J, Martinson TE, Liu RH (2009) Phytochemical profiles and antioxidant activities of wine grapes. Food Chem 116:332–339.
Yang Z, Xu M, Liu Y, He F, Gao F, Su Y, Wei H, Zhang Y (2014) Nitrogen-doped, carbon-rich, highly photoluminescent carbon dots from ammonium citrate. Nanoscale 6:1890–1895.
Zacchino SA, Butassi E, Di Liberto M, Raimondi M, Postigo A, Sortino M (2017) Plant phenolics and terpenoids as adjuvants of antibacterial and antifungal drugs. Phytomedicine 37:27–48.
Zhang Y, Zhu Y, Zhu Y, Li Y (2009) The importance of engineering physiological functionality into microbes. Trends Biotechnol 27:664–672.
Zimmer S, Messmer M, Haase T, Piepho H-P, Mindermann A, Schulz H, Habekuß A, Ordon F, Wilbois K-P, Heß J (2016) Effects of soybean variety and Bradyrhizobium strains on yield, protein content and biological nitrogen fixation under cool growing conditions in Germany. Eur J Agron 72:38–46.
Zimmer K, Meyer C, Gebauer G (2008) The ectomycorrhizal specialist orchid *Corallorhiza trifida* is a partial myco-heterotroph. New Phytol 178:395–400.
Zimmermann R, Strauss JG, Haemmerle G, Schoiswohl G, Birner-Gruenberger R, Riederer M, Lass A, Neuberger G, Eisenhaber F, Hermetter A (2004) Fat mobilization in adipose tissue is promoted by adipose triglyceride lipase. Science 306:1383–1386.

8

Plant—Microbe Interaction under Climate Change

Abdel Rahman M. Al-Tawaha[1], Ezz Al-DeinAl-Ramamneh[2], Muhammad Aasim, Canan Sevinc[3], Abdel Razzaq Al-Tawaha[4], Yousef M Abu-Zaitoon[1], Mohammad Alhawatema[5], Ali M. Qaisi[6], Amanullah[7], Imran[7], Nusrat Jahan[8], Moh'd Ali Wedyan[9], Munir Turk[10], Abdur Rauf[11], Shah Khalid[7], Mohd Abas Shah[12], Rokayya Sami[13], Devarajan Thangadurai[14], Jeyabalan Sangeetha[15], Wafa'a A. Al-Taisan[16], and Shah Fahad[17]

[1]*Department of Biological Sciences, Al Hussein Bin Talal University, P.O. Box 20, Ma'an, Jordan*

[2]*Department of Agricultural Sciences, AL-Shouback University College, Al-Balqa Applied University, AL-Shouback, Ma'an, Jordan*

[3]*Necmettin Erbakan University, Faculty of Science, Department of Biotechnology, Konya, Turkey*

[4]*Department of Crop Science, Faculty of Agriculture, Universiti Putra Malaysia, 43400 Serdang, Selangor, Malaysia*

[5]*Department of applied biological science, Faculty of Science, Tafila Technical University, Tafila, Jordan*

[6]*Department of Pharmaceutical Sciences, School of Pharmacy, University of Jordan, Amman, Jordan*

[7]*Agronomy Department, The University of Agriculture, Peshawar, Pakistan*

[8]*Department of Botany, Faculty of Life Sciences Aligarh Muslim University, Aligarh 202002, UP, India*

[9]*Department of Biology and Biotechnology, The Hashemite University, Zarqa, Jordan*

[10]*Department of Plant Production, Jordan University of Science and Technology, Irbid, Jordan*

[11]*Department of Chemistry, University of Swabi, Anbar, Khyber Pakhtunkhwa, Pakistan*

[12]*ICAR-Central Potato Research Station, Jalandhar, Punjab 144 003, India*

[13]*Department of Nutrition and Food Science, Taif University, Taif, Al-huwayah 888, Kingdom of Saudi Arabia*

[14]*Department of Botany, Karnatak University, Dharwad 580003, Karnataka, India*

[15]*Department of Environmental Science, Central University of Kerala, Kasaragod, 671316, Kerala, India*

[16]*Department of Biology, College of Science, Imam Abdulrahman Bin Faisal University, P.O. Box 1982, Dammam, Saudi Arabia*

[17]*Department of Agronomy, University of Haripur, Haripur, Khyber Pakhtunkhwa, Pakistan*

8.1 Plant Natural Systems

Plants coexist with several microbial communities in their vicinity. This situation resulted in complex interactions that can explain much of the observed ecosystem functions. Study of plant rhizosphere indicated the mediation of microbes in partitioning soil resources and the impact of plant genome, on the other hand, in shaping soil microbial communities (Ho et al. 2017).

8.1.1 Host–microbe ecology: factors affecting rhizosphere community

8.1.2 Plant metabolites and root exudates

Plants synthesize and accumulate compounds, namely metabolites and root exudates, which function in various ways (Paiva 2000; Turk and Tawaha 2002a,b; Turk and Tawaha 2003a,b; Turk et al. 2003c; Turk et al. 2005; Al-Tawaha and Odat 2010; Ho et al. 2017; Al-Zoubi and Tawaha 2019). These natural products can defend plants against pathogens, attract beneficial plant growth-promoting rhizobacteria onto roots, or adversely attract harmful soil microbes (Benizri et al. 2001; Hadacek 2002; Ren et al. 2016).

i. Sugars

Carbon is a major element in the universe, and its presence in soil indicated soil health conditions. Roots are important plant organs involved in the absorption of raw materials from the soil and, as a sink, import the synthesized food transported by the phloem from aerial parts (Hennion et al. 2019).

The release of carbon components by living roots, a process called rhizodeposition, is a major step in the terrestrial carbon flow cycle (Derrien et al. 2004). By this process, roots release about 10 to 40% of assimilated C as exudates (Hennion et al. 2019). A tracing of ^{13}C-labelled carbon and ^{13}C carbohydrates in wheat seedlings exposed to $^{13}CO_2$ pulse labelling showed the presence of polymerized glucose, soluble glucose, and fructose in the rhizodeposits, and that neutral sugar counts for 40% of the root-derived carbon. However, the decreased contribution of labelled sugars to^{13}C in soil and the increased microbial sugars over time might partially indicate metabolization of root-released sugars by micro-organisms (Derrien et al. 2004). Studying tissue culture-derived tobacco plants showed that the higher the root content of carbohydrates, the larger and more diverse are the microbial communities (Tilston et al. 2008). Sugars are the primary source of energy necessary for all eukaryotes to perform their physiological processes. In fact, the amount of carbohydrates in soil influences the prosperity of microorganisms. A portion of the released sugars, within the rhizosphere, are utilized by beneficial microorganisms, while another portion can be diverted by parasitic microbes (Hennion et al. 2019).

Arbuscular mycorrhiza–symbiotic plant interaction is a prominent well-documented phenomenon (Kiers et al. 2011; Garcia et al. 2016). Plant and fungus exchange materials, where plant provides photosynthetic products and in return receives nutrients as phosphorus and nitrogen from the fungus (Hennion et al. 2019). The regulation of such an interaction occurs via the control of plant and fungal transport, and plants show various expression levels of sucrose transport genes in response to inoculation with arbuscular mycorrhiza (Doidy et al. 2012). The soil pathogen–plant relation is of a different nature than mentioned with the mycorrhiza, while the plant limits available apoplastic sugars to the parasite and thus restricts pathogen proliferation, and the pathogen in turns divert sugars by interference with its transport in host plants (Ruan 2014; Zhao et al. 2018).

ii. Organic acids and amino acids

The dynamics of soil microbial communities can be influenced by their competition for precursors of amino acids, oligopeptides, and proteins. This is clearly understood by the quick uptake of amino acid molecules by heterotrophic microbes once they are available (Hobbie and Hobbie 2012). As a matter of fact, the preference for amino acids by microbes contributes to the versatility of the competition between plants and microbes for raw materials. Although microbes compete better than plants for soil nutrients, it was found for example that while low molecular weight amino acid, glycine, represented a non-favourable poor substrate for microbial growth, it was a favourable amino acid with a rapid uptake by the sedge *Kobresia myosuroides* (Lipson et al. 1999). Therefore, the variation in microbial preferences for amino acids provides a specialized niche for plants (Lipson et al. 1999). Away from competition for nutrients, plants and microbes were found to act although independently but in a synergistic way to ameliorate rhizosphere properties. The availability of minerals like Fe in soil can be hindered sometimes, resulting in low solubility of its crystalline form (Colombo et al. 2014). However, owing to the importance of such elements in the life processes, plants exudates, for example, flavonoids, phyto siderophores, and organic acids, interact with siderophores released by microbes, and as a result, soluble Fe complexes form that are more available for plants (Colombo et al. 2014). The nature of formed organic acids in root exudates varied with different plant species and growth conditions. Rhizosphere of corn plants cultured on sand with various media (nutrient media or biosolid-treated plant irrigated with nutrient media without trace minerals) contained more organic acids when biosolid-supplemented media, and as a result, plants were vigorous with higher biomass yield than control treatment (Koo et al. 2006). Organic acids formed under the conditions of study included lactic, acetic, butyric, oxalic, tartaric, maleic, succinic, valeric, glutaric, pyruvic, propionic, lactic, acetic, and butyric acids. Results indicated that nutrient medium used and plants could have provided the substrate for bacterial growth as evident by formation of much of the organic acids as the fermentation products of bacterial growth (Koo et al. 2006). Studying the rhizosphere of durum wheat for low molecular weight organic acids using GC techniques resulted in the separation of several organic acids: succinic, acetic, butyric, malic, propionic, tartaric, malonic, fumaric, citric, and trans-aconitic acids respectively (Szmigielska et al. 1997).

iii. Antimicrobial compounds

Plants responded to external stimuli as the detrimental effects of microbial invasion by secretion of a wide range of secondary metabolites that have one of their effects as antimicrobial compounds (Sudha and Ravishankar 2002). These low molecular weight molecules can be produced in plants prior to the occurrence of a biotic stress (Phytoanticipins) or could be induced in plants as part of a defensive mechanism that is not observed in healthy plants (phytoalexins) (VanEtten et al. 1994). During development of plant resistance to microbes or pests, events that lead to biosynthesis of plant hormones or plant hormonal-like substances such as ethylene, auxins, abscisic acid, jasmonic acid, and salicylic acidic are evident (Pusztahelyi et al. 2016). Examples of the presence of exudates in roots of plants and their use to ameliorate the detrimental effects of microbes are many in literature. Plant root exudates as signalling molecules were secreted by barley in the framework of a defensive mechanism in plants under fusarium attack (Lanoue et al. 2010). Barley roots initiated *de novo* biosynthesis oft-cinnamic acid-phenolic compounds with antimicrobial activity. The effects of intercropping of rice and watermelon were evaluated on the susceptibility of watermelon to the fungus *Fusarium oxysporum* f. sp *niveum* (FON) (Ren et al. 2016). The results indicated the presence of the phenolic acid *p*-coumaric acid in rice root exudates which was found to inhibit the spore germination of FON that caused wilt disease in watermelon.

8.2 Artificial Systems (Microbial-Mediated Transformation of Plants)

As the mechanisms underlying *Agrobacterium*-induced tumours in plants have been unravelled, scientists frequently attempted to manipulate this organism to use it as a vehicle to transfer genes of interest into plants and to efficiently complement crop breeding efforts (Sharma et al. 2005)

8.2.1 Agrobacterium Strains and Host Range

Transformation requires a suitable technique for introducing foreign genes into competent plant cells, which then should be subjected to selection while on growth media to regenerate into complete transgenic plants at a reasonable rate (Datta and Datta 2002). The most common system in use is *Agrobacterium*-mediated transformation. Different strains of *Agrobacterium tumefaciens* were used for transformation as GV3101, LBA4404, C58, GV3101, EHA105, and AGL1 (Chen et al. 2010; Li et al. 2013). Other species of bacteria such as plant-associated symbiotic bacteria were genetically engineered to provide an alternative to *Agrobacterium*-mediated genetic transformation (Broothaerts et al. 2005). Whereas *Agrobacterium*-mediated transformation has been reported to be an efficient method in dicot molecular breeding, the recalcitrant nature of monocots to this technique is well documented (Birch 1997; Nester 2015). Understanding of the factors influencing transient and stable gene expression has increased the host range of *Agrobacterium*-mediated genetic transformation. This includes the use of compounds, for example, thiol (Olhoft et al. 2001), carborundum (Somleva et al. 2002), and surfactants and methods like sonication, heat, and vacuum infiltration (Sivanandhan et al. 2015; Clement et al. 2016)

8.2.2 Molecular Bases of *Agrobacterium*-Mediated Genetic Transformation

The effects of *Agrobacterium* on plants have been studied over the last 100 years, and a clear understanding of the physiological and molecular basics of *Agrobacterium*-induced plant infections is recently provided (de la Riva et al. 1998; Sharma et al. 2005; Chabaud et al. 2007; Nester 2015).

The start of *Agrobacterium* story was with the observation of the crown galls that developed on ornamental and fruit trees (Smith and Townsend 1907), which were found to be started by the presence of this bacterium, but continue in its absence, indicating a unique nature of the effects this bacterium has on plant cells (White and Braun 1941; Braun 1958). Since then many laboratories have intensified their efforts to understand plant infection caused by *Agrobacterium.* The presence of a megaplasmid, called tumour-inducing Ti, was confirmed in virulent strains of *Agrobacterium*, but absent in the avirulent ones (Zaenen et al. 1974; Van Larabeke et al. 1975; Watson et al. 1975). Experiments were pursued, and it was revealed that the part transferred is part of Ti plasmid and is indeed DNA which was called T-DNA for transfer DNA (Chilton et al. 1977). Mutagenesis of the bacterium genome led to the identification of virulence genes, in particular *vir* genes which were thought to be responsible for the transfer of T-DNA to the host plants (Nester 2015). Once inside the plant, the T-DNA is transferred to the host nucleus where it integrates into plant chromosomes and encodes the enzymes of cytokinin and auxin synthesis that plays a role in tumour development (Akiyoshi et al. 1984; Klee et al. 1984). During this process and to attain a feed source, the bacterium encodes the synthesis of other products, namely opines (Nester 2015). Finally, according to the resistant of the host cells to the bacterial effects, the fate of transformation is determined. For genetically modified crops and to attain the desired effects, the so-called disarmed strains have been genetically engineered, where the genes encoding tumours are removed and replaced by the desired genes (Kiyokawa et al. 2009). For increasing the efficiency in plant transformation, vectors were designed such that a helper plasmid containing virulence genes was set apart from T-DNA region onto two replicons within the same bacterium, a system known as T-DNA binary vectors (Lee and Gelvin 2008).

8.2.3 Genetically Modified Crops

Plants could be in principle transformed from unrelated plants, animals, and microbes to increase its gene pool, a process that was not possible earlier (Movsesyan et al. 2001; Nila et al. 2006; Li et al. 2013). Plant biotechnology via *Agrobacterium* utilizes tissue culture techniques and the genetically modified bacterium to produce new products. *Agrobacterium* has been engineered with genes of interest, for example, for disease, herbicide as well as genes for stress tolerance, for example, for salt or drought tolerance (Chai et al. 2003; Toguri et al. 2003; Zhang and Wang 2015; Li et al. 2018). Transformation was also attempted for reduced ethylene sensitivity and compact growth in ornamentals (Christensen et al. 2008; Sanikhani et al. 2009).

8.3 Plant–Microbes Interaction and Plant Immune System

The interaction between plants and beneficial plant microbe helps plants to enhance resistance against biotic or abiotic stresses and elevated crop yield with low inputs (Reid 2011). The examples are bacterial (Oldroyd et al. 2011) and fungal (Smith and Smith 2011) symbiosis with plants, quorum sensing, and biofilm forming (Ramey et al. 2004) which affects the plants' transcriptome and proteome

(Mathesius 2009), bacterial enzymes and genes' necessary plant growth promotion, and plant-induced systemic resistance (ISR) by beneficial endophytes (Kloepper and Ryu 2006). In order to decoding the relationship between plants and microbes, modern biotechnological and molecular biology tools/techniques (genome sequencing, metagenomics, proteomics, transcriptomics, etc.) are valuable.

The interaction between plant and *Agrobacterium tumefaciens* is biotrophic with altered morphology or physiology of infected host plants rather than showing disease symptoms in plants (Pitzschke 2013). *Agrobacterium*–plant interaction revealed the activation of virulence genes due to plant-derived signalling molecules (Subramoni et al. 2014) like acidic signals (Wu et al. 2012), plant-based phenolic, and rhizosphere sugars (Nair et al. 2011; Hu et al. 2013). Various factors like host defence mechanisms (MAPKs, hormones, defence gene expression) and recalcitrance nature of plant control the genetic transformation efficiency (Zhao et al. 2011; Pitzschke 2013). It is evident that recognition, response, and regulation of plant-based chemical signals are necessary for the breaking of plants defence.

The immune system of plants responding pathogens is based on different chemical, structural, and protein-based defence mechanism dependent mainly on efficient signalling mechanisms and innate plant immunity (Ausubel 2005). The first step is the recognition of pathogens' entry into the plant cells. When pathogens break the primary defence, it resulted in elicitation of two major plant immune response known as (i) MAMPs/PAMPs (microbial/pathogen-linked molecular patterns) prompted immunity (MTI/PTI), also known as horizontal or basal immunity and (ii) ETI (effector-triggered immunity), also known as vertical immunity or *R*-gene-based, which recognized the avirulence (AVR) proteins (van der Hoorna and Kamoun 2009) in the pathogen. Other immune responses of plants include gene silencing and SAR (systemic acquired response) (Padmanabhan et al. 2009; Sahu et al. 2012). Apart from that, *trans*-generational immune memory phenomena in plants highlight the transfer of immune memory to the successive generation with effective adaptations in next generation against specific stress, also known as acquired immune power (Molinier et al. 2006; Jaskiewicz et al. 2011; Muthamilarasan and Prasad 2013).

Injection of bacterial virulence effector proteins into the plant cells occurs through hypersensitive response and pathogenicity (Hrp) type-III secretion system (T3SS) (Nans et al. 2015) to grow inside the cell and to localize disease lesions and for elicitation of hypersensitive response (HR) in non-hosts. Cytoplasmic chaperones and T3SS proteins are responsible for the delivery of effectors to pass though cell wall or plasma membrane of plants.

The colonization of *Trichoderma* spp. in the root zone tends to increase plant growth and development, enhanced resistance and nutrient uptakes, and plant protection by different responses like SAR or rhizobacterium-induced systemic resistance against invaded pathogens (Harman et al. 2004). *Piriformospora indica* is an endosymbiont that colonizes in root zone of higher plants and beneficial for them in many ways ranging from plant growth and development to resistance against stresses (Unnikumar et al. 2013). Certain studies revealed the *DREB2A*, *CBL1*, and *RD29A* genes responsible for abiotic stresses (Ansari et al. 2013) and osmoprotectants (Waller et al. 2005; Trivedi et al. 2013) from *P. indica* colonization. A study by Zuccaro et al. (2011) revealed certain gene numbers and their impact during colonization which recorded 579 genes (36–48 hpi) for prepenetration, 397 genes (3 dpi) for early colonization), and 641 genes (5 dpi) were recognized as singly regulated in comparison to fungus-free roots (Zuccaro et al. 2011).

The roles of plant hormonal signals like jasmonic acid, ethylene, and methyl jasmonate have significant impact on plant immune system by colonization of *P. indica* in the root zone of host plant (Stein et al. 2008). Some studies reflect the role of JA-based signals (*VSP*, *PDF1.2*, *LOX2*) plus ethylene (*ERF1*) signals for upregulation of genes against powdery mildew (Waller et al. 2005; Unnikumar et al. 2013) in the *P. indica*-treated plants. Whereas role of IAA and ethylene on biotrophic symbiosis as a compatibility factor has been highlighted by Hilbert et al. (2012). Sanders et al. (2002) and Ramakrishna et al. (2016) reported augmented intracellular Ca^{2+} pool as a second messenger for signalling pathways in plants. Akum et al. (2015) reported the colonization of *P. indica* by the suppression of salicylate-mediated basal resistance response governed by PIIN_08944.

The advancements in next-generation sequencing (NGS) genomics tools like dual RNA-sequencing (RNA-seq) of both pathogens and plants transcriptomes and non-coding RNAs (ncRNAs) have been used for the regulation of plant defence responses and also for better understanding of the system

(Westermann et al. 2012; Meyer et al. 2016; Westermann et al. 2016) by detecting deferentially expressed genes and changes occur during transcriptional regulatory events (Westermann et al. 2012; Mardis 2013). Whereas non-coding RNAs like phasiRNAs (phased, secondary, small interfering RNAs), lincRNAs (long intergenic non-coding RNAs), and microRNAs play significant role in PAMP immunity mechanism and effector-triggered defence responses and give knowhow about the impact of epigenetics on regulating the plant genes (Lenandias-Briere et al. 2016; Meyer et al. 2016) against pathogens. Although researchers are unable to open out the mysteries about plant immune system, some mysteries like *Avr* genes identification, root's immune mechanisms, molecular mechanisms involved in pathogen's colonization, cellular activity regulation, gene expression, and signalling mechanisms are still unveiled and need further work and advancement in post-genomic era technologies.

8.4 Plant–Microbes Interaction and Genomics

The interaction between microbes and their respective surroundings (environment and plant) can be endophytic or epiphytic especially with plants which lead to positive, negative, or neutral impact on growth and development (Newton et al. 2010). However, there is a possibility of pathogenic effect which may cause serious infections (Strange and Scott 2005) and significant yield loss and posing global food security threat. Therefore, specific signalling pathways that govern the mutual interactions or plants' response to disease are highly significant (Imam et al. 2015a,b). The negative impact of global warming on plant defence system (Schenk et al. 2008) and establishment of effective disease management (Madden and Wheelis 2003) enforces to recognize the plant–pathogen synergy. In recent years, application of biotechnological and molecular biology tools and technologies like genomics data analyses, metabolomics, metagenomics, NGS, proteomics, or transcriptomics is playing vital role to study and understand the plant–pathogens interactions and associations.

The three major areas of plant–microbes interaction with established signalling network mechanism are (i) plants–mycorrhizae symbiosis (Smith and Smith 2011), (ii) N_2 fixation (Oldroyd et al. 2011), and (iii) pathogenesis (Wirthmueller et al. 2013). However, modern disease management strategies include resistance genes or QTLs counter to pathogens (Akhon and Machray 2009) or exploiting the plants biotic/abiotic bearings. The limitations associated with biotic advent include intimately interaction of plants and microbes, plant-based physiological changes, and plant needs against pathogens protection or resistance without affecting their own growth.

The onset of genome-based studies compassionates the pathogens and their mutualistic plants' association (Guttman et al. 2014). The effector genes are associated with repeat-rich regions or accessory chromosomes (ACs) of the genome. The examples are *Magnaporthe oryzae* derived *AvrPita*, *SIX* genes from *Fusarium oxysporum,* PEP cluster from *Nectria haematococca* (Rodriguez et al. 2008), ToxA from *Pyrenophora tritici-repentis*, Tox3 and Tox1 from *Parastagonospora nodorum* (Liu et al. 2012), and necrosis-inducing proteins like ZtNIP1 and ZtNIP2 from *Zymoseptoria tritici* (Ben M'barek et al. 2015). Although in-depth RNA-seq revealed the identification of several 'effector-like' genes or secondary metabolite clusters (*Z. tritici*), no effector gene is associated with virulence (Kellner et al. 2014; Rudd et al. 2015) except *Mg3LysM* (Marshall et al. 2011) unveiled earlier by close homology to another effector gene *Ecp6* (Marshall et al. 2011). Whereas three small secreted proteins (SSPs) responsible for virulence were reported by Poppe et al. (2015). Pathogen-genome projects using bioinformatics revealed the identification of T3SS-pathway targeting signals, effectors, and T3SS-Hrp regulon promoters (Alfano and Collmer 2004) associated with mobile genetic elements, for horizontal gene transfer (Arnold et al. 2003a,b).

The advancement of NGS technologies with sequencing or re-sequencing of plants and pathogens is leading to genomics and post-genomics era to use such technologies for broader disease resistance management spectra against different pathogenic microorganisms under circumvent climatic conditions (Knief 2014). Apart from that, proteomics with the aid of bioinformatics and computational biology enables to elucidate plant–pathogen interactions by different proteins at specific time and conditions (Imam et al. 2014). Proteome-level study is based on certain molecules, engaged for intervene peculiar cellular processes (Kav et al. 2007) but facing huge challenge of involving both plant and microbes proteins (Mathesius 2009). Relative knowledge about proteins involved in plant growth and development, controlling physiological and biochemical pathways, is highly significant for genomics and proteomics studies for the revelation of unique genes for crop improvement programs (Scherp et al. 2011; Jayaraman et al. 2012).

8.5 Genome-Scale Modelling at Metabolic Level

The recent advances in the field of biotechnology enable the researchers to understand the molecular basis of plant pathogen systems, signalling pathways, and plant metabolism. The understanding of phytosystem can be divided into three stages: (i) transcriptomic (CDNA microarrays, SuperSAGE gene expression) based initial stage (Matsumura et al. 2003), (ii) NGS based on transcriptional profiling (RNA-seq) and proteomic technologies (MS/MS, 2D gels, iTRAQ) (Jones et al. 2004), and (iii) post-genomic era based on sequences to phenotypes (Huitema et al. 2004).

The entire genome sequences of bacteria and plant open the window to understand the plant–pathogen interaction in the post-genomic era based on annotation, modelling, sequencing, and reconstruction (Seaver et al. 2012). The Genome-scale modelling ('GEMing') is based on metabolic capabilities like designing of rational metabolic engineering strategies, directionality, elucidation of metabolism, physiology, gene–protein–reaction associations, metabolic reaction steps, localization, reaction compilation, and reversibility of reactions (Kauffman et al. 2003; Murabito et al. 2009; Liu et al. 2010; Senger 2010; Seaver et al. 2012). Due to advancement in metabolomics, sequencing, and proteomics, this GEMing model has been employed for different microorganisms, animals, and plants (Baumler et al. 2011; Orth et al. 2011; Rolfsson et al. 2011).

The metabolic reconstruction model known as genome-scale reconstruction model (GSRM) is used for analysing and interpreting the reactions and cell's metabolite concentrations under distinct conditions and time. GSRM focused on unified networks rather than focusing on genes or pathways with the aid of variable computational techniques and is highly significant for complete untangle of complex phenotype–genotype accord in the cells (Oberhardt et al. 2009).

8.6 Molecular Techniques and Soil Microorganisms

Techniques used for the assay of microbial biomass, diversity, and catabolic activities can be classified into three groups (Rincon-Florez et al. 2013). These techniques include (i) PCR-based, (ii) Non-PCR-based techniques, and (iii) high-throughput sequencing technologies (Table 8.1). Some of the given techniques were also employed for genomic studies.

Microarrays offer the assay of large data volumes for the detection of gene expression or array sequences (Murray et al. 2001; Chang et al. 2008) by employing variable microarrays techniques like (i) community genome arrays, (ii) functional gene arrays (GeoChip arrays), (iii) metagenomic arrays, (iv) open reading frame arrays, and (v) phylogenetic oligonucleotide arrays.

These techniques are based on considering microbial populations (16 S rRNA genes by PhyloChip microarrays) (Russo et al. 2012) or organic carbon (Davinic et al. 2012).

Proteome and metaproteome analysis attributes towards microbial biodiversity and functions in relation to plant nutrition, nutrient recycling, and molecular dialog (Badri et al. 2009; Xiong et al. 2013; Renella et al. 2014). The Integrated Microbial Genome (IMG) system developed for analysis, annotation, and dissemination of microbial genome and metagenome datasets (Markowitz et al. 2014) by the US Department of Energy Joint Genome Institute (DOE JGI) offers the review and analysis of genomes annotations (structural and functional) in a comparative context.

Different available IMG datamarts are IMG/W (http://img.jgi.doe.gov/w) for openly accessible genomes, IMG/ER (http://img.jgi.doe.gov/er) for expert review of genome annotations, and IMG/EDU (http://img.jgi.doe.gov/edu) for education or training. Apart from that, the Community Cyber

TABLE 8.1
Molecular Techniques Used for Microbial Assay

Technique	Reference
Group I: PCR-based techniques	
DDGE/TGGE (denaturing gradient gel electrophoresis /temperature gradient gel electrophoresis)	Kirk et al. (2004)
	Dowd et al. (2008)
	Cleary et al. (2012)
T-RFLP (Terminal Restriction Fragment Length Polymorphism)	Kakavas et al. (2008)
SSCP (Single-Strand Conformation Polymorphism)	Nocker et al. (2007)
ARISA/RISA (Automated Ribosomal Intergenic Spacer Analysis/Ribosomal RNA intergenic spacer analysis)	Kirk et al. (2004)
LH-PCR (Length-Heterogeneity PCR)	Okubo and Sugiyama (2009);
	Mills et al. (2007)
RAPD (Random Amplified Polymorphic DNA)	Newbury and Ford-Lloyd (1993)
	Fritsch and Rieseberg (1996);
ARDRA (Amplified Ribosomal DNA Restriction Analysis)	Nocker et al. (2007)
q-PCR (Real-time quantitative PCR)	Smith and Osborn (2009)
Group II: Non-PCR-based techniques	
CFE (Chloroform Fumigation-Extraction)	Ocio and Brookes (1990)
	Alessi et al. (2011)
PLFA (Phospholipid Fatty Acid Analysis)	Zhang et al. (2007)
FDA (Fluorescein Diacetate)	Adam and Duncan (2001)
SIP (Stable-Isotope Probing)	Dumont and Murrell (2005)
DNA arrays	Li and Liu (2003)
FISH (Fluorescence *In Situ* hybridization)	Moter and Go¨bel (2000)
Group III: High-throughput sequencing technologies	
454 Genome Sequencer	Mardis (2008)
Illumina genome analyser	Loman et al. (2012)
SOLiD (Applied Biosystems Sequencing by Oligonucleotide Ligation and Detection)	Magi et al. (2010)
	Shokralla et al. (2012)
PGM (Ion Personal Genome Machine)	Egan et al. (2012)
	Loman et al. (2012)
HeliScope Single Molecule Sequence	Magi et al. (2010)
	Shokralla et al. (2012)
SMRT DNA Sequencer	

infrastructure for Advanced Microbial Ecology Research and Analysis (CAMERA, http://camera.calit2.net/) database also offers a single system for deposition, location, analysis, visualization, and data sharing about microbial biology using advanced web-based analysis portal (Sun et al. 2011).

Microbial metagenomes are dependent on direct DNA extraction for the systemization of metagenomic libraries for decoding the molecular processes about biodiversity, nutrient uptake, and transport processes. A work by Fierer et al. (2012) on metagenomes of different grasses revealed the abundance of osmoregulation and dormancy-linked genes compared to low number of genes linked with nutrient cycle and catabolism in desert communities.

REFERENCES

Adam G, Duncan H (2001) Development of a sensitive and rapid method for the measurement of total microbial activity using fluorescein diacetate (FDA) in a range of soils. Soil Biol Biochem 33:943–951

Akhon MAY, Machray GC (2009) Biotech crops: technologies, achievements and prospects. Euphytica 166:47–59.

Akiyoshi DE, Klee H, Amasino RM et al (1984) T-DNA of Agrobacterium encodes an enzyme of cytokinin biosynthesis. Proc Natl Acad Sci USA 81:5994–5998. https://doi.org/10.1073/pnas.81.19.5994.

Akum FN, Steinbrenner J, Biedenkopf D et al (2015) The *Piriformospora indica* effector PIIN_08944 promotes the mutualistic Sebacinalean symbiosis. Front Plant Sci 6:906. https://doi.org/10.3389/ fpls.2015 .00906.

Al-Tawaha AM, Odat N (2010). Use of sorghum and maize allelopathic properties to inhibit germation and growth of wild barley (*Hordeum spontaneum*). Not Bot Horti Agrobot Cluj Napoca 38:124–127.

Al-Zoubi OM, Tawaha AM (2019) Allelopathic effect of beetroot (*Beta vulgaris* L) on germination and growth *Zea mays* and *Vigna umbellate*. Int J Bot Stud 4:47–51.

Alessi DS, Walsh DM, Fein JB et al (2011) Uncertainties in determining microbial biomass C using the chloroform fumigation–extraction method. Chem Geol 280:58–64.

Ansari MW, Bains G, Shukla, A et al (2013) Low temperature stress ethylene and not Fusarium might be responsible for mango malformation. Plant Physiol Biochem 69:34–38.

Arnold AE, Mejía LC, Kyllo D et al (2003a) Fungal endophytes limit pathogen damage in a tropical tree. Proc Natl Acad Sci USA 100:15649–15654.

Arnold DL, Pitman A, Jackson RW et al (2003b) Pathogenicity and other genomic islands in plant pathogenic bacteria. Mol Plant Pathol 4:407–420.

Ausubel FM (2005) Are innate immune signaling pathways in plants and animals conserved? Nat Immunol 6:973–979. https://doi.org/10.1038/ni1253.

Badri DV, Weir TL, van der Lelie D et al (2009) Rhizosphere chemical dialogues: plant–microbe interactions. Curr Opin Biotechnol 20:642–650.

Baumler DJ, Peplinski RG, Reed JL et al (2011) The evolution of metabolic networks of E. coli. BMC Syst Biol 5:182.

Ben M'barek S, Cordewener JHG, Tabib Ghaffary SM et al (2015) FPLC and liquid-chromatography mass spectrometry identify candidate necrosis-inducing proteins from culture filtrates of the fungal wheat pathogen *Zymoseptoria tritici*. Fungal Genet Biol 79:54–62.

Benizri E, Baudoin E, Guckert A (2001) Root colonization by inoculated plant growth-promoting rhizobacteria. Biocontrol Sci Tech 11:557–574.

Birch RG (1997) Plant transformation: problems and strategies for practical application. Ann Rev Plant Biol 48:297–326.

Braun AC (1958) Aphysiological basis for autonomous growth of the crown-gall tumor cell. Proc Natl Acad Sci USA 44:344–349. https://doi.org/10.1073/pnas.44.4.344.

Broothaerts W, Mitchell HJ, Weir B et al (2005) Gene transfer to plants by diverse species of bacteria. Nature 10:629–633.

Chabaud M, Rater P, De Sousa Araújo S et al (2007) Agrobacterium tumefaciens-mediated transformation and in vitro plant regeneration of *M. truncatula*. In: *Medicago truncatula* handbook. Nobel Research Institute, Oslo

Chai ML, Wang BL, Kim JY et al (2003) Agrobacterium-mediated transformation of herbicide resistance in creeping bentgrass and colonial bentgrass. J Zhejiang Univ Sci 4:346–351.

Chang HW, Sung Y, Kim KH et al (2008) Development of microbial genome-probing microarrays using digital multiple displacement amplification of uncultivated microbial single cells. Environ Sci Technol 42:6058–6064.

Chen X, Equi R, Baxter H et al (2010) A high-throughput transient gene expression system for switchgrass (*Panicum virgatum* L.) seedlings. Biotechnol Biofuels 3:9. http://www.biotechnologyforbiofuels.com/content/3/1/9.

Chilton MD, Drummond MH, Merlo DJ et al (1977) Stable incorporation of plasmid DNA into higher plant cells: the molecular basis of crown gall tumorigenesis. Cell 11:263–271.

Christensen B, Sriskandarajah S, Serek M et al (2008) Transformation of *Kalanchoe blossfeldiana* with rol-genes is useful in molecular breeding towards compact growth. Plant Cell Rep 27:1485–1495.

Cleary DFR, Smalla K, Mendonca-Hagler LCS et al (2012) Assessment of variation in bacterial composition among microhabitats in a mangrove environment using DGGE fingerprints and barcoded pyrosequencing. PLoS One 7:e29380.

Clement WKF, Lai KS, Wong MY et al (2016) Heat and hydrolytic enzymes treatment improved the Agrobacterium-mediated transformation of recalcitrant indica rice (*Oryza sativa* L.). Plant Cell Tissue Organ Cult 125:183–190.

Colombo C, Palumbo G, He JZ et al (2014) Review on iron availability in soil: interaction of Fe minerals, plants, and microbes. J Soils Sediments 14:538–548.

Datta K, Datta SK (2002) Plant transformation. In Gilmartin PM, Bowler C (eds) Molecular plant biology: a practical approach, Vol. 1. Oxford University Press, UK, pp 13–32.

Davinic M, Fultz LM, Acosta-Martinez V, Calderon FJ et al (2012) Pyrosequencing and mid-infrared spectroscopy reveal distinct aggregate stratification of soil bacterial communities and organic matter composition. Soil Biol Biochem 46:63–72.

de la Riva GA, González-Cabrera J, Vázquez-Padrón R et al (1998) *Agrobacterium tumefaciens*: a natural tool for plant transformation. Electron J Biotechnol 1:118–133.

Derrien D, Marol C et al (2004) The dynamics of neutral sugars in the rhizosphere of wheat. An approach by 13C pulse-labelling and GC/C/IRMS. Plant Soil 267:243–253.

Doidy J, van Tuinen D, Lamotte O et al (2012) The medicago truncatula sucrose transporter family. Characterization and implication of key members in carbon partitioning towards arbuscular mycorrhizal fungi. Mol Plant 5:1346–1358.

Dowd SE, Sun Y, Secor PR et al (2008) Survey of bacterial diversity in chronic wounds using pyrosequencing, DGGE, and full ribosome shotgun sequencing. BMC Microbiol 8:43.

Dumont MG, Murrell JC (2005) Stable isotope probing-linking microbial identity to function. Nat Rev Microbiol 3:499–504.

Egan AN, Schlueter J et al (2012) Applications of next generation sequencing in plant biology. Am J Bot 99:175–185.

Fierer N, Leff JW, Adams BJ et al (2012) Crossbiome metagenomic analyses of soil microbial communities and their functional attributes. Proc Natl Acad Sci USA 109:21390–21395.

Fritsch P, Rieseberg LH (1996) The use of random amplified polymorphic DNA (RAPD) in conservation genetics. In: Smith TB, Wayne RK (eds) Molecular genetics approaches in conservation. Oxford University Press, Oxford, pp 54–73.

Garcia K, Doidy J, Zimmermann SD et al (2016) Take a trip through the plant and fungal transportome of mycorrhiza. Trends Plant Sci 21:937–950.

Guttman DS, McHardy AC, Schulze-Lefert P (2014) Microbial genome-enabled insights into plant-microorganism interactions. Nat Rev Genet 15:797–813.

Hadacek F (2002) Secondary metabolites as plant traits: current assessment and future perspectives. Crit Rev Plant Sci 21:273–322.

Harman GE, Howell CR, Viterbo A et al (2004) Trichoderma species- opportunistic, avirulent plant symbionts. Nat Rev Microbiol 2:43–56.

Hennion N, Durand M, Vriet C et al (2019) Sugars en route to the roots. Transport, metabolism and storage within plant roots and towards microorganisms of the rhizosphere. Physiol Plant 165:44–57.

Hilbert M, Voll LM, Ding Y et al (2012) Indole derivative production by the root endophyte *Piriformospora indica* is not required for growth promotion but for biotrophic colonization of barley roots. New Phytol 196:520–534.

Ho Ying-Ning, Mathew D, Huang Chieh-Chen (2017) Plant-microbe ecology: interactions of plants and symbiotic microbial communities. In: Zubaida Y (ed) Plant ecology: traditional approaches to recent trends. IntechOpen. https://doi.org/10.5772/intechopen.69088.

Hobbie JE, Hobbie EA (2012) Amino acid cycling in plankton and soil microbes studied with radioisotopes: measured amino acids in soil do not reflect bioavailability. Biogeochemistry 107:339–360.

Hu X, Zhao J, Degrado WF, Binns AN (2013) *Agrobacterium tumefaciens* recognizes its host environment using ChvE to bind diverse plant sugars as virulence signals. Proc Natl Acad Sci USA 110:678–683. https://doi.org/10.1073/pnas.1215033110.

Huitema E, Bos JI, Tian M, Win J et al (2004) Linking sequence to phenotype in Phytophthora-plant interactions. Trends Microbiol 12:193–200. https://doi.org/10.1016/j.tim.2004.02.008.

Imam J, Alam S, Mandal NP et al (2015a) Molecular identification and virulence analysis of Avr genes in rice blast pathogen, *Magnaporthe oryzae* from Eastern India. Euphytica 206:21–31.
Imam J, Mandal NP, Variar M, Shukla, P (2015b) Advances in molecular mechanism toward understanding plant-microbe interaction: a study of *M. oryzae* Vs rice. In: Shukla P (ed) Frontier discoveries and innovations in interdisciplinary microbiology. Springer, New Delhi, pp 79–96.
Imam J, Nitin M, Toppo NN et al. (2014) A comprehensive overview on application of bioinformatics and computational statistics in rice genomics towards an amalgamated approach for improving acquaintance base. In: Kishor PBK, Bandyopadhyay R, Suravajhala P (eds) Agricultural bioinformatics. Springer, Berlin.
Jaskiewicz M, Conrath U, Peterhänsel C (2011) Chromatin modification acts as a memory for systemic acquired resistance in the plant stress response. EMBO Rep 12:50–55. https://doi.org/10.1038/embor.2010.186.
Jayaraman D, Forshey KL, Grimsrud PA, Ane JM (2012) Leveraging proteomics to understand plant–microbe interactions. Front Plant Sci 3:44. https://doi.org/10.3389/fpls.2012.00044.
Jones AM, Thomas V, Truman B et al (2004) Specific changes in the Arabidopsis proteome in response to bacterial challenge: differentiating basal and R-gene mediated resistance. Phytochemistry 65:1805–1816.
Kakavas VK, Plageras P, Vlachos TA et al (2008) PCR-SSCP: a method for the molecular analysis of genetic diseases. Mol Biotechnol 38:155–163.
Kauffman KJ, Prakash P, Edwards JS (2003) Advances in flux balance analysis. Curr Opin Biotechnol 14:491–496. https://doi.org/10.1016/j.copbio.2003.08.001.
Kav NNV, Srivastava S, Yajima W, Sharma N (2007) Application of proteomics to investigate plant-microbe interactions. Curr Proteom 4:28–43. https://doi.org/10.2174/157016407781387357.
Kellner R, Bhattacharyya A, Poppe S et al (2014) Expression profiling of the wheat pathogen *Zymoseptoria tritici* reveals genomic patterns of transcription and host-specific regulatory programs. Genome Biol Evol 6:1353–1365.
Kiers ET, Duhamel M, Beesetty Y et al (2011) Reciprocal rewards stabilize cooperation in the mycorrhizal symbiosis. Science 333:880–882.
Kirk JL, Beaudette LA, Hart M, et al (2004) Methods of studying soil microbial diversity. J Microbiol Methods 58:169–188.
Kiyokawa K, Yamamoto S, Sakuma K et al (2009) Construction of disarmed Ti plasmids transferable between *Escherichia coli* and Agrobacterium species. Appl Environ Microbiol 75:1845–1851.
Klee H, Montoya A, Horodyski F, et al (1984) Nucleotide sequences of the tms genes of the pTiA6NC octopineTi- plasmid: two gene products involved in plant tumorigenesis. Proc Natl Acad Sci USA 81:1728–1732. https://doi.org/10.1073/pnas.81.6.1728.
Kloepper J, Ryu CM (2006) Bacterial endophytes as elicitors of induced systemic resistance. Soil Biol 9:33–52. https://doi.org/10.1007/3-540-33526-9_3.
Knief C (2014) Analysis of plant microbe interactions in the era of next generation sequencing technologies. Front Plant Sci 5:216.
Koo BJ, Chang AC, Crowley De, Page Al L (2006) Characterization of organic acids recovered from rhizosphere of corn grown on biosolids-treated medium. Commun Soil Sci Plant Anal 37:871–887. https://doi.org/10.1080/00103620600564158.
Lanoue A, Burlat V, Schurr U, Röse USR (2010) Induced root-secreted phenolic compounds as a belowground plant defense. Plant Signal Behav 5:1037–1038. https://doi.org/10.4161/psb.5.8.12337.
Lee LY, Gelvin SB (2008) T-DNA binary vectors and systems. Plant Physiol 146:325–332.
Lenandias-Briere C, Moreau J, Hartmann C, Crespi M (2016) Noncoding RNAs, emerging regulators in root endosymbiosis. Mol Plant-Microb Interact 29:170–180. https://doi.org/10.1094/MPMI-10-15-0240-FI.
Li CW, Lee SH, Chan MT (2013) Utilization of the plant methionine sulfoxide reductase B genes as selectable markers in Arabidopsis and tomato transformation. Plant Cell Tissue Organ Cult 113:555–563.
Li ESY, Liu WT (2003) DNA microarray technology in microbial ecology studies-principle, applications and current limitations. Microbes Environ 18:175–187.
Li J, Phan TT, Li YR et al (2018) Isolation, transformation and overexpression of sugarcane SoP5CS Gene for drought tolerance improvement. Sugar Technol 20:464–473.
Lipson DA, Raab TK, Schmidt SK, Monson RK (1999) Variation in competitive abilities of plants and microbes for specific amino acids. Biol Fertil Soil 29:257–261.

Liu LM, Agren R, Bordel S, Nielsen J (2010) Use of genome-scale metabolic models for understanding microbial physiology. FEBS Lett 584:2556–2564. https://doi.org/10.1016/j.febslet.2010.04.052.

Liu Z, Zhang Z, Faris JD et al (2012) The cysteine rich necrotrophic effector SnTox1 produced by *Stagonospora nodorum* triggers susceptibility of wheat lines harbouring Snn1. PLoS Pathog 8:e1002467.

Loman NJ, Constantinidou C, Chan JZM, et al (2012) High-throughput bacterial genome sequencing: an embarrassment of choice, a world of opportunity. Nat Rev Microbiol 10:599–606.

Madden LV, Wheelis M (2003) The threat of plant pathogens as weapons against U.S. crops. Annu Rev Phytopathol 41:155–176. https://doi.org/10.1146/annurev.phyto.41.121902.102839.

Magi A, Benelli M, Gozzini A, et al (2010) Bioinformatics for next generation sequencing data. Genes 1:294–307.

Mardis ER (2008) Next-generation DNA sequencing methods. Annu Rev Genom Human Genet 9:387–402.

Mardis ER (2013) Next-generation sequencing platforms. Annu Rev Anal Chem 6:287–303. https://doi.org/10.1146/annurev-anchem-062012-092628.

Markowitz VM, Chen IMA, Palaniappan K et al (2014) IMG 4 version of the integrated microbial genomes comparative analysis system. Nucleic Acids Res 42:560–567.

Marshall R, Kombrink A, Motteram J, et al (2011) Analysis of two in planta expressed LysM effector homologs from the fungus *Mycosphaerella graminicola* reveals novel functional properties and varying contributions to virulence on wheat. Plant Physiol 156:756–769.

Mathesius U (2009) Comparative proteomic studies of root-microbe interactions. J Proteom 72:353–366. https://doi.org/10.1016/j.jprot.2008.12.006.

Matsumura H, Reich S, Ito A, et al (2003) Gene expression analysis of plant host-pathogen interactions by SuperSAGE. Proc Natl Acad Sci USA 100:15718–15723. https://doi.org/10.1073/pnas.2536670100.

Meyer AFE, Shuey LS, Naidoo S et al (2016) Dual RNA-sequencing of *Eucalyptus nitens* during *Phytophthora cinnamomi* challenge reveals pathogen and host factors influencing compatibility. Front Plant Sci 7:191. https://doi.org/10.3389/fpls.2016.00191.

Mills DEK, Entry JA, Gillevet PM, Mathee K (2007) Assessing microbial community diversity using amplicon length heterogeneity polymerase chain reaction. Soil Sci Soc Am J 71:572–578.

Molinier J, Ries G, Zipfel C, Hohn B (2006) Transgeneration memory of stress in plants. Nature 442:1046–1049. https://doi.org/10.1038/nature05022.

Moter A, Go¨bel UB (2000) Fluorescence in situ hybridization (FISH) for direct visualization of microorganisms. J Microbiol Methods 41:85–112.

Movsesyan NR, Alizade K, Musiychuk KA et al (2001) Tobacco plants expressing bacterial genes for thermostable glucanases. Russ J Genet 37:610–616. https://doi.org/10.1023/A:1016613005650.

Murabito E, Simeonidis E, Smallbone K, Swinton J (2009) Capturing the essence of a metabolic network: a flux balance analysis approach. J Theor Biol 260:445–452. https://doi.org/10.1016/j.jtbi.2009.06.013.

Murray AE, Lies D, Li G et al (2001) DNA/DNA hybridization to microarrays reveals gene-specific differences between closely related microbial genomes. Proc Natl Acad Sci USA 98:9853–9858.

Muthamilarasan M, Prasad M (2013) Plant innate immunity: an updated insight into defense mechanism. J Biosci 38:433–449. https://doi.org/10.1007/s12038-013-9302-2.

Nair GR, Lai X, Wise AA, et al (2011) The integrity of the periplasmic domain of the VirA sensor kinase is critical for optimal coordination of the virulence signal response in *Agrobacterium tumefaciens*. J Bacteriol 193:1436–1448. https://doi.org/10.1128/JB.01227-10.

Nans A, Kudryashev M, Saibil HR, Hayward RD (2015) Structure of a bacterial type III secretion system in contact with a host membrane in situ. Nat Commun 6:10114.

Nester EW (2015) Agrobacterium: nature's genetic engineer. Front Plant Sci 5:730. https://doi.org/10.3389/fpls.2014.00730.

Newbury HJ, Ford-Lloyd BV (1993) The use of RAPD for assessing variation in plants. Plant Growth Regul 12:43–51.

Newton AC, Fitt BDL, Atkins SD et al (2010) Pathogenesis, parasitism and mutualism in the trophic space of microbe-plant interactions. Trends Microbiol 18:365–373. https://doi.org/10.1016/j.tim.2010.06.002.

Nila AG, Sandalio LM, López MG et al (2006) Expression of a peroxisome proliferator-activated receptor gene (xPPARα) from *Xenopus laevis* in tobacco (*Nicotiana tabacum*) plants. Plant 224:569–581.

Nocker A, Burr M, Camper AK (2007) Genotypic microbial community profiling: a critical technical review. Microb Ecol 54:276–289.

Oberhardt MA, Palsson BØ, Papin JA (2009) Applications of genome-scale metabolic reconstructions. Mol Syst Biol 5:320.

Ocio JA, Brookes PC (1990) An evaluation of methods for measuring the microbial biomass in soils following recent additions of wheat straw and the characterization of the biomass that develops. Soil Biol Biochem 22:685–694.

Okubo A, Sugiyama S (2009) Comparison of molecular fingerprinting methods for analysis of soil microbial community structure. Ecol Res 24:1399–1405.

Oldroyd EDG, Murray JD, Poole PS, Downie JA (2011) The rules of engagement in the legume-rhizobial symbiosis. Annu Rev Genet 45:119–144. https://doi.org/10.1146/annurev-genet-110410-132549.

Olhoft P, Lin K, Galbraith J et al (2001) The role of thiol compounds in increasing Agrobacterium-mediated transformation of soybean cotyledonary-node cells. Plant Cell Rep 20:731–737.

Orth JD, Conrad TM, Na J, Nielsen N, Somers D et al (2011) A comprehensive genome-scale reconstruction of *E. coli* metabolism-2011. Mol Syst Biol 7:1–9.

Padmanabhan C, Zhang X, Jin H (2009) Host small RNAs are big contributors to plant innate immunity. Curr Opin Plant Biol 12:465–472. https://doi.org/10.1016/j.pbi.2009.06.005.

Paiva NL (2000) An introduction to the biosynthesis of chemicals used in plant-microbe communication. J Plant Growth Regul 19:131–143.

Pitzschke A (2013) Agrobacterium tumefaciens and plant defense-transformation success hangs by a thread. Front Plant Sci 4:519. https://doi.org/10.3389/fpls.201. 00519

Poppe S, Dorsheimer L, Happel P, Stukenbrock EH (2015) Rapidly evolving genes are key players in host specialization and virulence of the fungal wheat pathogen Zymoseptoria tritici (*Mycosphaerella graminicola*). PLoS Pathog 11:e1005055.

Pusztahelyi T, Holb IJ, Pócsi I (2016) Plant fungal interactions: special secondary metabolites of biotrophic, necrotrophic, and other specific interactions. In: Mérillon J-M, Ramawat KG (eds) Fungal metabolites, Springer International Publishing, Cham, pp 1–58. https://doi.org/10.1007/978-3-319-19456-1_39-2.

Ramakrishna A, Gill SS, Sharma KK et al (2016) Indoleamines (serotonin and mel- atonin) and calcium-mediated signaling in plants. In: Ravishankar GA, Ramakrishna A (eds) Serotonin and melatonin: their functional role in plants, food, phytomedicine, and human health. Taylor and Francis, Boca Raton, FL, pp 85–95. ISBN: 978-1-4987-3905-4 (Ch. 7)

Ramey BE, Koutsoudis M, von Bodman SB, Fuqua C (2004) Biofilm formation in plant–microbe associations. Curr Opin Microbiol 7:602–609. https://doi.org/10.1016/j.mib.2004.10.014.

Reid A (2011) Microbes helping to improve crop productivity. Microbe 6:435–439.

Ren L, Huo H, Zhang F et al (2016) The components of rice and watermelon root exudates and their effects on pathogenic fungus and watermelon defense. Plant Signal Behav 11:6. https://doi.org/10.1080/15592324.2016.1187357.

Renella G, Ogunseitan O, Giagnoni L, Arenella M (2014) Environmental proteomics: a long march in the pedosphere. Soil Biol Biochem 69:34–37.

Rincon-Florez VA, Carvalhais LC, Schenk PM (2013) Culture independent molecular tools for soil and rhizosphere microbiology. Diversity 5:581–612.

Rodriguez RJ, Henson J, Van Volkenburgh E et al (2008) Stress tolerance in plants via habitat-adapted symbiosis. Int Soc Microb Ecol 2:404–416.

Rolfsson O, Palsson BO, Thiele I (2011) The human metabolic reconstruction Recon 1 directs hypothesis of novel human metabolic functions. BMC Syst Biol 5:155. https://doi.org/10.1186/1752-0509-5-155.

Ruan YL (2014) Sucrose metabolism: gateway to diverse carbon use and sugar signaling. Ann Rev Plant Biol 65:33–67.

Rudd J, Kanyuka K, Hassani-Pak K et al (2015) Transcriptome and metabo- lite profiling the infection cycle of *Zymoseptoria tritici* on wheat (*Triticum aestivum*) reveals a biphasic interaction with plant immunity involving differential pathogen chromosomal contributions, and a variation on the hemib- iotrophic lifestyle definition. Plant Physiol 167:1158–1185.

Russo SE, Legge R, Weber KA et al (2012) Bacterial community structure of contrasting soils underlying Bornean rain forests: inferences from microarray and next-generation sequencing methods. Soil Biol Biochem 55:48–59.

Sahu PP, Puranik S, Khan M, Prasad M (2012) Recent advances in tomato functional genomics: utilization of VIGS. Protoplasma 249:1017–1027. https://doi.org/10.1007/s00709-012-0421-7.

Sanders D, Pelloux J, Brownlee C, Harper JF (2002) Calcium at the crossroads of signaling. Plant Cell 14:401–417.
Sanikhani M, Mibus H, Stummann BM, Serek M (2009) Transformation of *Kalanchoe blossfeldiana* for ethylene insensitivity. Acta Hortic 847:103–108. https://doi.org/10.17660/actahortic.2009.847.12.
Schenk PM, McGrath KC, Lorito M (2008) Plant-microbe and plant- insect interactions meet common grounds. New Phytol 179:251–255. https://doi.org/10.1111/j.1469-8137.2008.02538.x.
Scherp P, Ku G, Coleman L, Kheterpal I (2011) Gel-based and gel-free proteomic technologies. Methods Mol Biol 702:163–190. https://doi.org/10.1007/978-1-61737-960-4_13.
Seaver SMD, Henry CS, Hanson AD (2012) Frontiers in metabolic reconstruction and modeling of plant genomes. J Exp Bot 63:2247–2258. https://doi.org/10.1093/jxb/err371.
Senger RS (2010) Biofuel production improvement with genome-scale models: the role of cell composition. Biotechnol J 5:671–685. https://doi.org/10.1002/biot.201000007.
Sharma KK, Bhatnagar-Mathur P, Thorpe TA (2005). Genetic transformation technology: status and problems. In Vitro Cell Dev Biol Plant 41:102–112.
Shokralla S, Spall JL, Gibson JF, Hajibabaei M (2012) Next generation sequencing technologies for environmental DNA research. Mol Ecol 21:1794–1805.
Sivanandhan G, Dev GK, Theboral J et al (2015) Sonication, vacuum infiltration and thiol compounds enhance the Agrobacterium-mediated transformation frequency of *Withania somnifera* (L.) Dunal. PLoS One 10:e0124693.
Smith CJ, Osborn AM (2009) Advantages and limitations of quantitative PCR (qPCR)-based approaches in microbial ecology. FEMS Microbiol Ecol 67:6–20.
Smith SE, Smith FA (2011) Roles of arbuscular mycorrhizas in plant nutrition and growth: new paradigms from cellular to ecosystem scales. Annu Rev Plant Biol 62:227–250. https://doi.org/10.1146/annurev-arplant-042110-103846.
Smith EF, Townsend CO (1907) A plant tumor of bacterial origin. Science 25:671–673. https://doi.org/10.1126/science.25.643.671.
Somleva MN, Tomaszewski Z, Conger BV (2002) Agrobacterium-mediated genetic transformation of switchgrass. Crop Sci 42:2080–2087.
Stein E, Molitor A, Kogel KH, Waller F (2008) Systemic resistance in Arabidopsis conferred by the mycorrhizal fungus *Piriformospora indica* requires jasmonic acid signaling and the cytoplasmic function of NPR1. Plant Cell Physiol 49:1747–1751.
Strange RN, Scott PR (2005) Plant disease: a threat to global food security. Annu Rev Phytopathol 43:83–116. https://doi.org/10.1146/annurev.phyto.43.113004.133839.
Subramoni S, Nathoo N, Klimov E, Yuan Z (2014) *Agrobacterium tumefaciens* responses to plant-derived signaling molecules. Front Plant Sci 5:322. https://doi.org/10.3389/fpls.2014.00322.
Sudha G, Ravishankar GA (2002) Involvement and interaction of various signaling compounds on the plant metabolic events during defense response, resistance to stress factors, formation of secondary metabolites and their molecular aspects. Plant Cell Tissue Organ Cult 71:181–212.
Sun S, Chen J, Li W, et al (2011) Community cyberinfrastructure for advanced microbial ecology research and analysis: the CAMERA resource. Nucleic Acids Res 39:546–551.
Szmigielska AM, Van Rees KCJ, Cieslinski G, Huang PM (1997) Comparison of liquid and gas chromatography for analysis of low molecular weight organic acids in rhizosphere soil. Commun Soil Sci Plant Anal 28:99–111. https://doi.org/10.1080/00103629709369775.
Tilston EI, Halpin C, Hopkins DW (2008) Tissue culture propagation alters plant-microbe interactions in tobacco rhizosphere. Biol Fertility Soil 44:897–901.
Toguri T, Ogawa T, Kakitani M, et al (2003) Agrobacterium-mediated transformation of chrysanthemum (*Dendranthema grandiflora*) plants with a disease resistance gene (pac1). Plant Biotechnol 20:121–127.
Trivedi DK, Bhatt H, Pal R et al (2013) Sequence specific 1H, 13C and 15N NMR assignments of cyclophilin A like protein from *Piriformospora indica* involved in salt tolerance. Biomol NMR Assign 7:175–178.
Turk MA, Lee KD, Tawaha AM (2005) Inhibitory effects of aqueous extracts of black mustard on germination and growth of radish. Res J Agric Biol Sci 1:227–231.
Turk MA, Shatnawi MK, Tawaha AM (2003c) Inhibitory effects of aqueous extracts of black mustard on germination and growth of alfalfa. Weed Biol Manage 3:37–40.
Turk MA, Tawaha AM (2002a) Inhibitory effects of aqueous extracts from black mustard (B*rassica nigra* L.) on germination and growth of wheat. Pak J Biol Sci 5:278–280.

Turk MA, Tawaha AM (2002b). Inhibitory effects of aqueous extracts of black mustard on germination and growth of lentil. Pak J Agron 1:28–30.

Turk MA, Tawaha AM (2003a) Allelopathic effect of black mustard (*Brassica nigra* L.) on germination and growth of wild oat (*Avena fatua* L.). Crop Prot 22:673.

Turk MA, Tawaha, AM (2003b) Allelopathic effects of black mustard (*Brassica nigra*) on germination and growth of wild barley (*Hordeum spontaneum*). J Agron Crop Sci 189:298–303.

Unnikumar KR, Sowjanya SK, Varma A (2013) *Piriformospora indica*: a versatile root endophytic symbiont. Symbiosis 60:107–113.

van der Hoorna RAL, Kamoun S (2009) From guard to decoy: a new model for perception of plant pathogen effectors. Plant Cell 20:2009–2017.

Van Larabeke N, Schell J, Schilperoort RA et al (1975) Acquisition of tumor-inducing ability by non- oncogenic agrobacteria as a result of plasmid transfer. Nature 255:742–743. https://doi.org/10.1038/255742a0.

VanEtten HD, Mansfield JW, Bailey JA, Farmer EE (1994). Two classes of plant antibiotics: phytoalexins versus "phytoanticipins". Plant Cell 6:1191.

Waller F, Achatz B, Baltruschat H et al (2005) The endophytic fungus *Piriformospora indica* reprograms barley to salt-stress tolerance, disease resistance, and higher yield. Proc Natl Acad Sci USA 102:13386–13391.

Watson B, Currier TC, Gordon MP, et al (1975) Plasmid required for virulence of *Agrobacterium tumefaciens*. J Bacteriol 123:255–264.

Westermann AJ, Forstner KU, Amman F, et al (2016) Dual RNA-seq unveils noncoding RNA functions in host-pathogen interactions. Nature 529:496–501. https://doi.org/10.1038/nature16547.

Westermann AJ, Gorski SA, Vogel J (2012) Dual RNA-seq of pathogen and host. Nat Rev Microbiol 10:618–630. https://doi.org/10.1038/nrmicro2852.

White PR, Braun AC (1941) Crown gall production by bacteria-free tumor tissues. Science 94:239–241. https://doi.org/10.1126/science.94.2436.239.

Wirthmueller L, Maqbool A, Banfield MJ(2013) On the front line: structural insights into plant–pathogen interactions Nat Rev Microbiol 11:761–776.

Wu CF, Lin JS, Shaw GC, Lai EM (2012) Acid-induced type VI secretion system is regulated by ExoR-ChvG/ChvI signaling cascade in *Agrobacterium tumefaciens*. PLoS Pathog 8:e1002938. https://doi.org/10.1371/journal.ppat.1002938.

Xiong H, Shen H, Zhang L et al (2013) Comparative proteomic analysis for assessment of the ecological significance of maize and peanut intercropping. J Proteom 78:447–460.

Zaenen I, Van Larabeke N, Teuchy H, et al (1974) Supercoiled circular DNA in crown gall inducing Agrobacterium strains. J Mol Biol 86:109–127. https://doi.org/10.1016/S0022-2836(74)80011-2.

Zhang QC, Wang GH, Yao HY (2007) Phospholipid fatty acid patterns of microbial communities in paddy soil under different fertilizer treatments. J Environ Sci 19:55–59.

Zhang WJ, Wang T (2015) Enhanced salt tolerance of alfalfa (*Medicago sativa*) by rstB gene transformation. Plant Sci 234:110–118.

Zhao F, Chen L, Perl A, et al (2011) Proteomic changes in grape embryogenic callus in response to *Agrobacterium tumefaciens*- mediated transformation. Plant Sci 18:485–495. https://doi.org/10.1016/j.plantsci.2011.07.016.

Zhao D, You Y, Fan H, et al (2018) The role of sugar transporter genes during early infection by root-knot nematodesI nternational. J Mol Sci 19:302.

Zuccaro A, Lahrmann U, Güldener G et al (2011) Endophytic life strategies decoded by genome and transcriptome analyses of the mutualistic root symbiont *Piriformospora indica*. PLoS Pathog 7:e1002290.

9

Mapping QTLs for Abiotic Stress

Ali Fuat Gökçe, Usman Khalid Chaudhry, and Muhammad Daniyal Junaid
Department of Agricultural Genetic Engineering, Ayhan Şahenk Faculty of Agricultural Sciences and Technologies, Niğde Ömer Halisdemir University, Niğde, Turkey

Overview

Crop production is seriously affected by various abiotic stresses, and this issue is increasingly severe in many parts of world. Regions where temperatures are high are more prone to stresses; salt stress and drought stress have increased desertification worldwide as they have severely affected 10% of total world's arable land. Consequently, yields of crops have been reduced up to 50% (Bray et al. 2000; Saud et al. 2013; Shah et al. 2013; Saud et al. 2014; Hafiz et al. 2016; Saud et al. 2016; Aziz et al. 2017a,b; Habib ur Rahman et al. 2017; Kamaran et al. 2017; Saud et al. 2017; Uz Zaman et al. 2017; Wajid et al. 2017; Yang et al. 2017; Zahida et al. 2017; Adnan et al. 2018a,b; Akram et al. 2018a,b; Depeng et al. 2018; Hafiz et al. 2018; Tariq et al. 2018; Adnan et al. 2019; Ahmad et al. 2019; Baseer et al. 2019; Hafiz et al. 2019; Muhammad et al. 2019; Sajjad et al. 2019; Wu et al. 2019; Adnan et al. 2020; Farhana et al. 2020; Farhat et al. 2020; Hafiz et al. 2020a,b; Hussain et al. 2020; Ilyas et al. 2020; Jan et al. 2019; Mubeen et al. 2020; Rehman et al. 2020; Saleem et al. 2020a,b,c; Saud et al. 2020; Shafi et al. 2020; Subhan et al. 2020; Wahid et al. 2020; Wu et al. 2020; Zafar-ul-Hye et al. 2020a,b). So, it is important for plant breeders under these conditions to develop such plant cultivars which can provide better yields and contend such abiotic conditions. However, until now the work in this aspect is relatively slow because physiological basis of stress tolerance has not been yet clear (Cuartero et al. 2006; Munns and Tester 2008; Fahad and Bano 2012; Fahad et al. 2013; Fahad et al. 2014a,b; Fahad et al. 2015a,b; Fahad et al. 2016a,b,c,d; Fahad et al. 2017; Fahad et al. 2018; Fahad et al. 2019a,b; Hesham and Fahad 2020). It is known that major abiotic stress tolerance is quantitative or has been associated with multiple genes. The loci controlling these traits are called quantitative trait loci (QTLs). In this method, parents showing extreme phenotypes are crossed to produce progenies, and then they are screened using appropriate marker (RFLP, RAPD, AFLP, and SNPs). Construction of genetic maps is done, and via statistical software, makers are linked with specific trait (Hussain 2006). When markers are linked with the trait, then it became easier to transfer trait of interest into well-adapted cultivars through marker-assisted breeding approaches. Thus, identification of loci on chromosomes linked to abiotic stresses enables plant breeders to develop stress-tolerant cultivars.

9.1 Abiotic Stress

Climate change has direct influence to aggravate the abiotic stress conditions. Major abiotic stress includes drought, salinity, heat, and cold stress (Fahad and Bano 2012; Fahad et al. 2013; Fahad et al. 2014a,b; Fahad et al. 2015a,b; Fahad et al. 2016a,b,c,d; Fahad et al. 2017; Fahad et al. 2018; Fahad et al.

2019a,b; Hesham and Fahad 2020). Currently, drought and salt stress are widespread global issues, and both of the stresses are predicted to cause severe devastating effect to more than 50% arable land of the world by the year 2050 (Jamil et al. 2011; Kasim et al. 2013). Global earth temperature has been increased up to 0.9°C in the past 70 years, and it is increasing every year due to global warming (Fahad and Bano 2012; Fahad et al. 2013; Rohde et al. 2013; Fahad et al. 2014a,b; Deng et al. 2015; Fahad et al. 2015a,b; Fahad et al. 2016a,b,c,d; Fahad et al. 2017; Fahad et al. 2018; Fahad et al. 2019a,b; Hesham and Fahad 2020). Cold stress is another major constraint for plant development especially in hilly regions. Plants of temperate region are generally tolerant to cold stress. However, their tolerance level varies with a range of chilling (0–15 °C) and freezing (<0 °C) temperatures. Plants as being unable to move to favourable growth conditions suffer from the changing climatic conditions and suffer from growth and yield losses due to aforementioned abiotic stresses. Drought is a situation of non-availability of sufficient water and change in rainfall pattern due to climate change (Fahad et al. 2017). Drought disturbs the plant–water relation as there is least/no water available for uptake by plants (Aksoy et al. 2015). Salinity is the presence of excessive amount of soluble salts in soil. Salinity is divided into two types, primary salinization is caused by natural phenomenon with the presence of high salt contents in underground water, whereas secondary salinization is more concerned with farmer's negligence with an improper application irrigation of poor-quality water and poor soil drainage (Murtaza et al. 2017). All these stresses interfere with the normal functioning of plant, resulting in a stunted plant growth (Ashraf and Foolad 2007; Wahid et al. 2007; Farooq et al. 2012). It disturbs the nutrient uptake through roots, and numerous physiological changes occur in response to stress conditions (Tuna et al. 2007; Shahzad et al. 2016). Abiotic stress disrupts photosynthetic apparatus with least photosynthesis and higher production of reactive oxygen species in plants which causes oxidative stress (Anjum et al. 2011). Plants' exposure to various abiotic stresses is illustrated in Figures 9.1 and 9.2.

9.1.1 Agriculture Affected by Various Abiotic Stresses

Abiotic stress causes enormous production losses globally (Bray et al. 2000). Currently, individual effect of major abiotic stresses has been explored to observe their damages to crops; however, sometimes a crop suffering from one stress might be under the pressure of other stress as well in the field (Chaves et al. 2009). It is observed from drought-stricken areas that plants suffer from combination of

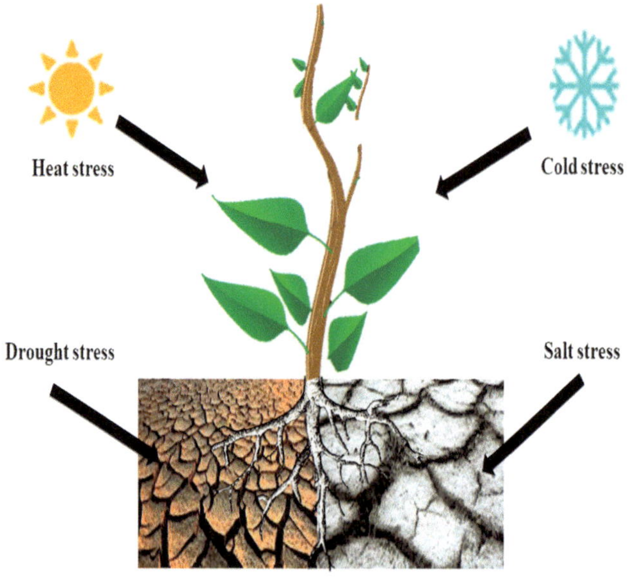

FIGURE 9.1 Plants Exposed to Multifarious Abiotic Stresses Due to Climate Change.

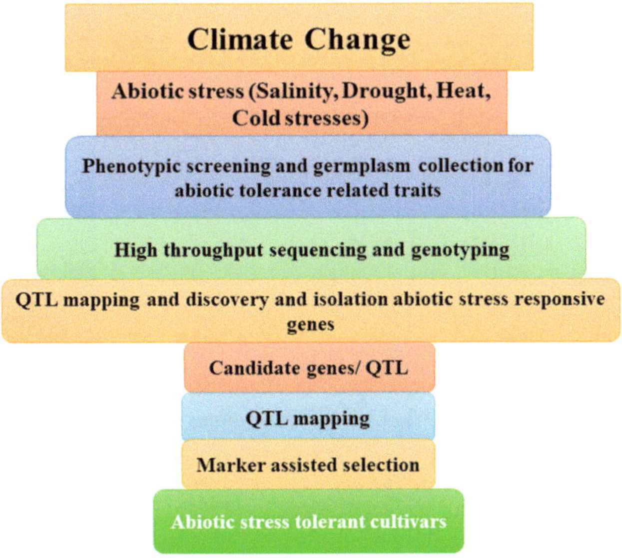

FIGURE 9.2 QTL Mapping for Abiotic Stress Tolerance in Crops.

different stresses, that is, salt and heat stress, or with existence of one stress, it triggers other stresses (Moffat 2002; Rizhsky et al. 2004; Suzuki et al. 2014). Plants are grown under natural conditions and sometimes suffer from adverse effects of two or more than two stresses which might be combination of drought and salinity, heat and salinity, and drought with extreme temperature waves in numerous regions of agricultural field worldwide (Suzuki et al. 2014). Drought and salt stress are most common paradigm that exists together in the field (Moffat 2002). All these environmental adversities significantly affect plant growth and development, resulting in lower yield of crop.

9.2 QTL Mapping

Currently, impervious goal is to identify the genes contributing in variations of traits in different species. Numerous traits in plants are influenced by number of genes, that is, quantitative in nature (Stuber et al. 1992). Basically, there are two approaches to identify function of genes: forward genetics and reverse genetics. Forward genetics is aimed to find genetic base of a specific trait or phenotype, while reverse genetics includes searching and predicting phenotypes that can derive from specific genetic sequence (Pekosz et al. 1999). Functional genomics has aided breeding programmes for abiotic stresses in order to associate phenotypic variations with the genotypic changes, and for this purpose, QTL mapping has paved path in unravelling function of quantitative traits (El-Soda et al. 2015). QTLs are the DNA stretches which are closely associated to the desired trait. They can be located using molecular techniques, for example, PCR for mapping the genomic regions linked to quantitative traits. QTLs govern quantitative characters which control abiotic stress tolerance. The term QTL was coined by Gelderman (1975).

QTL mapping is a forward genetics approach, mainly used for the identification of position of genes on genomic sequences. Historically, first ever genetic map was published by TH Morgan and Alfred Sturtevant in 1911. They exhibited linked genes on a fruit fly chromosome (Mohan et al. 1997). First successful report of QTL mapping in plants was done approximately three decades ago (Bernardo 2008). QTL mapping is now widely used in the field of agriculture for different purposes. It is a reliable source to mark genomic regions for desired trait. Currently, resistance and tolerance in plants to various environmental adversities caused due to climate change are required by minimizing the hazard to our natural environment. In this respect, study of quantitative trait loci allows us to estimate locus number, location, effect, and interaction (Michelmore 1995). QTLs can be divided into two categories: main effect QTLs (M-QTLs) and epistatic QTLs (E-QTLs), and it depends upon the presence of epistasis. The main principle of genetic mapping is based on the segregation of genes in the progeny. After the identification of QTLs of interest, molecular markers can be used for genotyping. The main benefit of this process is that favourable genes can be achieved without doing evaluations in field; thereby, it is less laborious and time-saving approach (Fritsche-Neto et al. 2012).

9.2.1 QTL Mapping Methods and Software

Statistical method is used to estimate genetic loci on chromosomes and mapping for QTL. Various methods and software involved for QTL mapping are described in Table 9.1, and the most commonly used software (Table 9.2), Map manager, MapMaker/QTL, WinQTL Cart, and R/qtl, have been developed and employed for QTL mapping (Soller et al. 1976; Lander and Botstein 1987; Manly et al. 2001; Broman et al. 2003; Wang 2007). Utilization of appropriate statistical software is essential to recognize and dissect quantitative traits that are associated to a trait. It is much easier to transfer required traits into well-adapted cultivars when QTLs are linked to specific traits like salinity, drought, heat, and cold stresses. For this purpose, use of molecular markers is an important strategy as it allows to dissect quantitative traits into single genetic components helping plant breeders to select and breed plants that are resistant (Hussain 2006). Development of linkage maps using various markers or association studies to identify QTLs for economic traits is a good strategy (Araus et al. 2003).

9.2.2 Overview of QTL Mapping Procedure for Abiotic Stresses

In QTL mapping procedure, the first is to choose desirable parents, which contain trait of interests, then to cross the distinct parents and segregating progenies are obtained for a specific trait, and then these

TABLE 9.1

QTL Mapping Methods, Their Advantages, and Disadvantages

QTL Mapping Methods	Advantages	Disadvantages	References
Single marker analysis	Easy to use with least computer intensive	Does not give information and either markers are linked with one QTL or more than one	Soller et al. (1976)
Interval mapping	Estimation of position of QTL	If more than two QTLs are located on a chromosome, then QTL mapping might give biased results and QTL may be mapped to wrong positions	Lander and Botstein (1987)
Composite interval mapping	More effective for estimation and detection of QTL compared to interval mapping	Does not detect epistatic QTLs, irrational selection of cofactors	Jansen and Stam (1994)
Inclusive composite interval mapping	Quicker than composite interval mapping, detects epistasis and dominance	Only restricted to bi-parental mapping population	Li et al. (2007)
Multiple-interval mapping	Evaluate interaction of QTLs	Method involves intensive computational, results differ with selection of model	Kao et al. (1999)
Multiple QTL mapping	Enhanced handling of missing data, huge genomics datasets, more statistical power	Computationally intensive	Arends et al. (2010)

TABLE 9.2
Software for QTL Mapping and Their Operating Platforms

Software Name	Operating System	References
MapQTL	Windows	Manly et al. (2001)
MapMaker/QTL	Windows	Manly et al. (2001)
Ici Mapping	Windows	Li et al. (20070
MultiQTL	Unix/Linux	Manly et al. (2001)
Qgene	Unix/Linux	Joehanes and Nelson (2008)
QTLseqr	Unix/Linux	Mansfeld and Grumet (2018)
FlexQTL	Unix/Linux	Mora et al. (2017)
WinQTL Cart	Unix/Linux, Windows	Wang (2007)

progenies are screened phenotypically. DNA is then extracted, analysis is done using adequate marker, and then data is subjected to appropriate statistical software as mentioned earlier.

9.2.3 Objectives and Applications of QTL Mapping

Major objectives of QTL mapping are investigating number of genes controlling a trait; identify the location of genes that are influencing trait of interest; and knowing the effect of genes causing variation in specific trait. It is also an aim of QTL mapping to find out which type of gene action is present, that is, additive or dominant, and type of alleles associated to positive effects (Khan 2015).

Main applications of QTL mapping are as follows:

- Efficiency of Marker-Assisted Selection (MAS) can be enhanced using QTL alleles (Johnson et al. 2000; Wilcox et al. 2001).
- A comprehensive and detailed understanding of complex genetic traits can be achieved using QTL mapping (Edwards et al. 1987; Bradshaw 1996).
- The location of genes on genetic map whose function is known can be identified (Neale and Savolainen 2004).

QTLs aided in understanding of genetic architecture of crop to enhance coveted traits (Bo et al. 2015). This approach has been exploited for improving various plants traits that are under the direct and indirect influence of abiotic stresses (Welcker et al. 2006).

9.3 QTLs for Salinity Tolerance

Agricultural soils are affected with salinity stress due to climate change and various anthropogenic factors (excessive application of fertilizers, improper application, or poor quality of irrigation and drainage). All these factors contribute to excessive accumulation of salts in soil root zones. It is already known that salinity is controlled by cluster of genes (Morton et al. 2019). So, it is important to identify those genes/group of genes that are controlling salinity stress. In order to identify these regions/gene sequences, QTL mapping is a well-adapted approach for dissecting regions responsible for conferring tolerance in plants against salinity (Munns 2005; Iqbal et al. 2015; Islam et al. 2019). Conventional plant breeding approaches lag behind for selection of salt-tolerant plants because it is polygenic in nature and it is highly affected by environment and exhibit low heritability (Gregorio 1997). However, DNA markers have been successfully exploited for selecting polygenic traits. Molecular markers have numerous benefits such as no genotypic or environmental effect and no epistatic effect with ease of selection (Nadeem et al. 2018). In this regard, QTL mapping is a ground-breaking approach for physical localization and identification of molecular markers tightly associated with salt tolerant QTL. Major hurdle is the parental specificity and screening protocols varying with severity of stress conditions for exploiting marker-assisted selection through the populations of interest (Ashraf and Foolad 2013; Hussain et al. 2017). Phenotypic population is essential for QTL mapping, so development of population for salinity tolerance is a huge problem because every genotype exhibits differential response with the increment in salt contents (4, 8, 12 dSm^{-1}) with different growth phases of the cultivar. This problem can be overcome by following standard protocols for achieving reliable results of phenotyping (Richards 1996; Flowers 2004). QTLs identified for salinity tolerance are given in Table 9.3.

9.3.1 QTL Mapping for Root Architecture/Role against Salinity Stress

Root is the base of the plants. It not only provides anchorage and but also supports plants rather than it absorbs nutrients and transports them to the upper plant parts. The most crucial role during salt stress is selection of essential ions under high pressure of Na^+ ions around its vicinity for uptake by plants and helps plants to adjust themselves with the changing environmental conditions. It is the fact that roots have been extensively studied to cope with salinity stress. Root architecture of the halophytic as well as halo phobic was explored by exploiting contemporary approaches to decipher their response. QTL mapping is one of the approaches successfully used in discerning regions responsible for various root characteristics such as number of roots, weight of root, root angle, volume, and length of root (Reynolds et al. 2007; Manschadi et al. 2008; Wasson et al. 2012; Christopher et al. 2013). Instead of root characteristics and selectivity for sodium (Na^+) ion uptake by roots from saline soil, it is very difficult task to restrict Na^+ entry and enhance uptake of its competitor (K+) with the aid of Na^+ permeable transporters (Horie and Schroeder 2004).

9.3.2 QTL Mapping for Ion Homeostasis in Plants

QTL mapping aided in identifying region for higher uptake of K^+ and limiting the uptake of Na^+ because they compete for binding sites due to physiochemical similarity (Xu et al. 2012). Higher concentration of K^+ is correlated with higher salinity tolerance (Chen et al. 2007). A main QTL chosen is SALTOL, which is responsible for more than 40% of salt uptake by plants (Bonilla et al. 2002). This QTL (SALTOL) is favourable for salinity tolerance conferring promising results at seedling stage; it is chiefly involved in maintaining Na^+/K^+ homeostasis (Thomson et al. 2010). It also supported the fact that sodium transporter (SKC1) gene is located in SALTOL QTL region that is chiefly involved for inhibiting entry of sodium from roots and later transportation to leaves (Haq et al. 2010). SKC1 gene also encodes for high-affinity potassium transporter (HKT) which is selectively responsible for unloading Na^+ to protect plant parts. QTL mapping also dissected regions for genes involved in exclusion of Na^+ from leaves, which encodes for HKT transporters types. The function of this transporter is to take excessive Na^+ from shoots to phloem which is ultimately transferred to roots and eliminated from plant. It also unloads excess amount of Na^+ from xylem to protect aerial parts of plant especially leaves (Ren et al. 2005; Huang et al. 2006; Zhang et al. 2018).

9.3.3 QTL Mapping to Alleviate Osmotic/Oxidative Stress

The entry of sodium in plants causes osmotic and oxidative stress. Although oxidative stress is a secondary effect of salinity, plant metabolism is negatively influenced by salinity stress while plants adapt themselves by regulating various mechanisms including exclusion of ions and osmotic adjustments (Kao 2015). Salt tolerance of crop is the agglomerate of several processes such as ion compartmentation/balance, higher K+ content contrary to Na+ contents in plants. Salt stress also forces plant for stomatal closure, and in this way, disturbance in oxygen and carbon dioxide occurs in plant cells. This surplus oxygen in plant cells later results in development of (ROS) reactive oxygen species which are the main agent of causing oxidative stress. So, the production of ROS species is directly proportional to severity of salt stress as the stress increases it moves towards harmful levels (Xiong and Zhu 2002). Moreover, higher accumulation of ROS can damage nucleic acids, lipids, and protein via oxidation. It is critical to alleviate the formation of reactive oxygen species (Hussain et al. 2018). Numerous ROS

TABLE 9.3

QTLs Identified for Different Abiotic Stresses

QTLs for	Traits	References
Salinity Tolerance		
*SalTol*1-1	Salt tolerance	Islam et al. (2011)
*SalTol*8-1		
*SalTol*10-1		
Saltol	Potassium absorption	Gregorio (1997)
qSDM-5	Seedling dry matter	Prasad et al. (2000)
Q_{NaK}	Na^+:K^+ discrimination	Flowers et al. (2000)
qNa⁺/K⁺LR3-1	Na^+/K^+ in leaf at reproductive stage	Ammar (2004)
qSDS-1	Seedling survival	Lin et al. (2004)
qSKC-1	Shoot K^+ concentration	Ren et al. (2005)
qST3	Shoot K^+ concentration	Lee et al. (2006)
qLB-11	Controlling leaf bronzing	Takehisa et al. (2006)
qCHL2	Leaf chlorophyll contents	Thomson et al. (2010)
qRKC1	Shoot Na^+-K^+ ratio	
qSGEM-6	Seed germination (%)	Prasad et al. (2000)
qSV-6	Seedling vigour	
*qGV*1-1	Germination vigour	Wang et al. (2014)
qGP7-1	Germination percentage	
qRAS-1	Salinity tolerance	Ren et al. (2010)
QSkc-2B	Shoot potassium concentration	Xu et al. (2012)
QRkc-5A1	Root potassium concentration	
Qstgss-3A.1	Salt tolerance index in germination	Ma et al. (2007)
Q.Na2A	Shoot sodium concentration	Genc et al. (2010)
Q.chl2D	Chlorophyll content	
*qaox-s*2.1	Antioxidant activity	Frary et al. (2010)
*qpox-s*1.1	glutathione peroxidase activity	
qMDA-1a	Malondialdehyde content	Jiang et al. (2009)
qNAK-6	Na^+/K^+ ratio	Sabouri et al. (2009)
*LeaNaKCs*l.1	Leaf Na^+/K^+ concentration ratio	Arraouadi et al. (2012)
LRsl.8	Length of roots	Arraouadi et al. (2011)
Lasl.6	Leaf area	
*QRMO.TxFr.*2H	Relative water content	Fan et al. (2015)
*QPC-S.TxFr.*3H	Proline content	
qRL6.1	Root length	Bizimana et al. (2017)
qSL1.1	Shoot length	
qSFW4.1	Shoot fresh weight	
qpht3-1	Plant height	Mahender et al. (2015)
qsl2-1	Stem length	
qRL-1	Root length	
qPH3s	Plant height	Xue et al. (2009)
qGY6s	Grain yield	
qSTR-1, *qSTR*-5	Salt tolerance rating	Ming-zhe et al. (2005)
qDWS-8	Dry matter weight shoots	
*QSO.TxNn.*2H	Superoxide anion	Gill et al. (2019)
*QHP.TxNn.*2H	Hydrogen peroxide	

(*Continued*)

TABLE 9.3 (Continued)

QTLs for	Traits	References
QFv2H	Chlorophyll fluorescence	Siahsar and Narouei (2010)
QWSC2H	Water-soluble carbohydrate	
Drought tolerance		
*qbrt*11d	Basal root thickness	Li et al. (2005)
*qrn*1b	Root numbers	
*brt*11d	Basal root thickness	
*rn*1b	Root numbers	
$QTRL_{IV\text{-}52.11}$	Tap root length	Idrissi et al. (2016)
$QSRL_{IV\text{-}61.63}$	Specific root length	
$QTRL_{IX\text{-}111.15}$	Tap root diameter	
$QTRL_{VII\text{-}21.94}$	Root surface area	
QRA.qgw–2A	Root angle	Christopher et al. (2013)
qGR–1	Germination rate	Mardani et al. (2013)
QPn2AC	Net Photosynthetic rate	Malik et al. (2015)
QCMSa2AC	Cell membrane stability	
QtChl2.1	Chlorophyll content	Guo et al. (2008)
QCMS1.1	Cell membrane stability	Tripathy et al. (2000)
FlrF	Flowering time	Kebede et al. (2001)
qtlOP–2	Osmotic Potential	Saeed et al. (2011)
qtlRLWC1	Relative water contents	Amjid et al. (2015)
Qwue.S42IL-4H	Water use efficiency	Honsdorf et al. (2014)
qRSFW1	Shoot fresh weight	Liu et al. (2011)
QPh1	Plant height	Xu et al. (2005)
qtl 12.1	Grain Yield	Bernier et al. (2009)
qGY-6	Grain Yield	Xu et al. (2018)
*qGYD*1.1	Grain Yield	Solis et al. (2018)
QDT.TxFr.2H	Drought tolerance	Fan et al. (2015)
Heat tolerance		
QHtctd.bhu-7B	Canopy temperature depression	Paliwal et al. (2012)
Qls-dm.bhu-7D	Days to maturity	
*qPH*1.1	Plant height	Ye et al. (2012)
QHtscc.ksu-6A	Chlorophyll contents	Talukder et al. (2014)
QHtpmd.ksu-2B	Plasma membrane damage	
QDtf.tam-4B	Days to flowering	Mason et al. (2010)
Qmat.tam-2D	Days to maturity	Mason et al. (2011)
QTdl.tam-3B	Temperature depression of flag leaf	
QFll.tam-1B	Flag leaf length	
QFlw.tam-2D	Flag leaf width	
QMrs.ksu-4A	Maximum rate of senescence	Vijayalakshmi et al. (2010)
Qspad.ksu-7B	chlorophyll content	
Q.Ctgf.aww-1B	Canopy temperature	Bennett et al. (2012)
rlc1.1	Root length under heat stress	Kilasi et al. (2018)
slht3.1	Shoot length under heat stress	
QHst.cph-3B.1	Maximum quantum efficiency of photosystem	Sharma et al. (2017)

TABLE 9.3 (Continued)

QTLs for	Traits	References
Cold tolerance		
qCTS-1-a	Cold tolerance at seedling stage	Lou et al. (2007)
qCTS-1-c	Cold tolerance at seedling stage	Andaya and Mackill (2003)
Ctb1	Cold tolerance	Saito et al. (2001)
qSCT-1	Seedling cold tolerance	Qian et al. (2000)
qCTS12	Seedling cold tolerance	Andaya and Tai (2006)
qCTB8	Cold tolerance at booting stage	Kuroki et al. (2007)
QFr-H1	Frost tolerance	Reinheimer et al. (2004)
qLTG-3-1	Low temperature germination	Fujino et al. (2004)
qSCT-3	Cold tolerance at seedling stage	Zhi-Hong et al. (2005)
qCTB-8	Cold tolerance for spikelet fertility	Wainaina et al. (2018)
qPL-1	Cold tolerance for panicle length	
qSNP-1	Cold tolerance for spikelet number	
qPN-7-1	Cold tolerance for panicle number	
QYIELD.usw-6A	Grain yield	Fowler et al. (2016)

scavenging enzymes are involved to counteract the production of ROS species in response to salinity and confer salinity tolerance (Shalata et al. 2001). Thereby QTLs responsible for enhancing the anti-oxidant activity were identified against salt stress, and several QTLs have been identified for providing pathways in alleviating oxidative stress (Frary et al. 2010). Malondialdehyde (MDA) is an indicator of stress severity as its concentration increases with the severity of stress (Abbasi et al. 2016). QTLs for MDA have been also mapped deciphering the behaviour of plant under salinity stress (Jiang et al. 2009). Proline is an osmotic protectant; its main role is to maintain osmotic pressure, regulating balance of reactive oxygen species, plant, and embryo development. Proline also exhibited ionic homeostasis by controlling transport of ions across the cellular membranes. There is an intense increase in its levels under salinity stress to overcome osmotic and oxidative stress. Moreover, it also stimulates stress adaptation signalling networks against high salt contents in the plant cells (Ozden et al. 2009; Lehmann et al. 2010). QTLs for proline contents were mapped conferring higher concentration to cope with salinity stress conditions; likewise with the increase in proline content, it was correlated for greater plant height (Fan et al. 2015).

9.3.4 QTL Mapping for Seed Germination and Seedling Emergence under Salt Stress

Higher concentration of salts in soil impedes the initial growth phases of crops such as poor seedling emergence and stand establishment which ultimately results in poor crop yield. The work has been done to map QTL associated with robust seedling emergence under saline conditions (Mano and Takeda 1996). Several QTLs for higher germination, seedling emergence, and shoot weight were mapped for conferring tolerance at initial sensitive growth phases of plant (Mano and Takeda 1997; Wang et al. 2014; Batool et al. 2018).

9.3.5 QTL Mapping for Phenotypic and Yield Traits

Salinity stress influences plant morphology, leading to devastating yield losses. Plants under salt stress exhibit shorter height, a smaller number of leaves, stunted root length, and less number of fruits, and all these agronomic traits result in lesser crop yield. However, QTL mapping has been exploited for locating regions for controlling agronomic traits under salinity stress. Thereby, identification of these regions unravels genes responsible for rendering improved crop characteristics and higher yield. Furthermore, molecular markers can be designed for tracing genes in a segregating population for breeding of salt-tolerant cultivars (Xue et al. 2009). Salt stress causes senescence and early flowering in some crop plants. In some crops, stay green is the important indicator of stress tolerance. QTLs controlling these characteristics of plants were mapped which is essential growth stage for final higher yield of crop. Improvement by achieving higher yield with the use of genetic tool is the main task against such environmental pressure on current agriculture. In order to make high yield crops that can withstand the changing adverse climatic conditions, several QTLs for numerous crops have been dissected with variation in yield under salinity stress (Eleuch et al. 2008; Xue et al. 2009).

9.4 QTL Mapping against Drought Stress

Drought is an increasing problem throughout the world, and drought-affected areas have been doubled in past 30–40 years (FAO 2013). Among all the stresses that affect crop yield and hinder its growth, most alarming is drought, and it is also most resistant to breeders' efforts (Tuberosa and Salvi 2006). Drought is a condition when there is no adequate water supply and soil moisture availability required for normal plant growth (Reynolds et al. 2007). Drought tolerance can be described as the crops' ability to produce constant yield, under environmental adversity (Cattivelli et al. 2002). Drought is a serious menace to crop productivity. However, at molecular level, these conditions are more complex under drought stress; it is a polygenic quantitative trait. There are three major strategies that plant uses against water-deficit conditions: Escape, it allows plants to complete its growth cycle before intense deficiency of water, and this process mainly occurs in short cycle crops. Secondly, avoidance is a strategy in which a plant decreases transpiration and enhances water absorption. Lastly, tolerance is a mechanism that is the ability of a plant to tolerate a stress without altering its growth stages (Tardieu 2005). To find out adequate process and to adapt against drought, the plant behaviour under drought must be known, and this is not easy as plant responses are complex when it faces drought (Jones 2004). For breeding purpose, selection of superior plants that are stress tolerant is a challenge because there is interaction of environment and genotype (G × E). There is gap about the drought tolerance concept, so it is troublesome to characterize drought tolerance traits either phenotypic or physiological. It is evident through various studies that multiple genes control water stress tolerant traits. As it is known that the regions of chromosomes that control quantitative traits are known as QTLs (quantitative trait loci). QTLs for various traits causing tolerance to drought stress have been identified using molecular markers (Hussain 2006). For the development of drought stress-related QTLs, plant breeders have extensively studied genetic loci (Kalladan et al. 2013). It is the main objective of present and future studies to identify QTLs responsible for the drought tolerance along with yield (Maccaferri et al. 2008). Marker-assisted selection is more applicable and efficient when markers are tightly bonded to loci related to stress tolerance.

9.4.1 QTLs Linked with Root Traits for Drought Stress Tolerance

Roots are the sole source of water for plants to provide water necessary for every growth phase. Roots confer unprecedented benefits to crops especially during limited moisture availability in rhizosphere and soil pockets due to extreme climatic conditions. Under limited soil water/drought stress signals, roots try to approach towards soil moisture for plant uptake. Roots play integral role to alleviate mild drought stress conditions and protect from desiccation (Passioura 2012). In this scenario, root length is crucial for capturing available moisture with the onset of initial drought signals. QTL mapping is one the most important approaches to identify regions on chromosomes for improved root characteristics to cope with drought stress conditions and seek water deep in the soil (Price et al. 2002). Therefore, various root traits have been mapped by utilizing QTL mapping which proved to be beneficial for breeding drought tolerant crops (Christopher et al. 2013). Studies showed that large root systems in crops perform well under the water scarce conditions, so development of better root framework can enhance crop yields (Manschadi et al. 2008). Genotypes that have deep and branched root system have ability to extract water from soil reserves. Under drought conditions, root structure has great importance as it can take water and nutrients from deeper soils (Uga et al. 2015). QTL for roots traits against different levels of drought stress throughout the course of plant development was identified, and it was inferred basal root thickness and deeper root system is efficient in coping adverse water stress conditions (Li et al. 2005). Furthermore, QTLs were also identified for the root penetration ability and root growth in soil. QTLs responsible for robustness of root growth such as maximum root length, thickness, and deep root weight with ratio to shoot have been successfully identified to efficiently overcome drought stress (Nguyen et al. 2004).

9.4.2 QTLs Responsible for Physiological Characteristics

Drought has severe influence on physiological processes of crops; therefore, it is essential for tolerant crops to adapt themselves and protect its physiological changes occurring due to water-deficit conditions. It directly destroys the chlorophyll contents, and production of reactive oxygen species is high, and all these factors contribute to diminished photosynthesis rate of plants (Peltzer et al. 2002). As it is known that poor photosynthesis causes poor decreased growth and yield of plant (Ashraf and Foolad 2007). QTLs have been widely identified in different crops for water-deficit conditions (Castonguay and Markhart 1992; Malik et al. 2015). QTLs for numerous physiological characteristics such as photosynthetic rate, cell membrane stability, electrolyte leakage, osmotic adjustment, water-soluble carbohydrates, and relative water contents were identified subjected to drought stress (Malik et al. 2015). Plant closes its stomata during drought stress, wilting of leaves, and cell dehydration, and ultimately loss in turgor hinders photosynthetic activity of plant (Xu et al. 2010). Level of chlorophyll fluorescence is important character for evaluating drought tolerance (Li et al. 2006; Roostaei et al. 2011). In this regard, QTLs were mapped for several plant species for drought stress. The presence of QTLs on chromosomal regions associated with chlorophyll fluorescence emphasized that their expression is genetically controlled (Yang et al. 2007). Numerous QTLs for cell membrane stability are important indicators for the plant which can resist drought stress (Elshafei et al. 2013). Proline production increases under drought stress, and it provides signals during antioxidative defence mechanism and confers tolerance to plants against oxidative stress; therefore, QTL for enhanced proline production has been mapped in various species for drought tolerance (Hayat et al. 2012; Sayed et al. 2012).

9.4.3 QTLs for Yield Traits

Drought is a major reason of crop losses throughout the world. It affects the product in different ways like decreasing grain number, grain size, grain filling, immature flowering, alteration of flower timings, etc., ultimately causing lower yields. QTLs that are associated with yield traits have been located by various researchers, and it has been reported that QTLs can potentially provide the results. Identification of QTLs help us to get better biomass, high number of florets, ears ultimately increasing harvest index (Quarrie et al. 2006). QTL mapping has played important role in investigation of grain components and the chromosomal region, which control them under water scarce conditions (Bennett et al. 2012). QTL mapping for the improvement of yield traits under water-deficient conditions has been helping plant breeders for the development of resistant cultivars (Bernier et al. 2009). Generally, under drought stress, grain yield is affected during reproductive stage of plant. Grain yield-related QTLs have been mapped under drought conditions (Sandhu et al. 2014). For the development of cultivars with good grain yield under drought stress, QTL identification is a quick and reliable technique.

9.5 QTL Mapping for Heat Tolerance

Climate change has direct influence on global warming due to which the temperature of the earth is increasing steadily. Plants being sessile are under direct influence of change in temperature. Slight increase in temperature from optimum requirement of plant is sensed as heat stress by plant. Heat stress disturbs the homeostasis of plant and damages various cellular components with retardation in growth and development leading to death (Mittler 2006). QTLs identified for heat tolerance are given in Table 9.3.

9.5.1 QTL Mapping to Cope with Heat Stress during Seedling Emergence

It disrupts the initial growth phases of seed germination, growth stages, and reproduction (Mittler and Blumwald 2010; Hasanuzzaman et al. 2013). Poor seedling emergence and decreased germination percentage are the obvious effect of heat stress with raise in soil temperature not suitable for ideal seed germination (Piramila et al. 2012). QTL mapping for seed germination characteristics has been mapped to cope rapidly increasing temperature stress (Bettey et al. 2000). Response of plant against heat tolerance depends upon severity and duration of temperature.

9.5.2 QTL Mapping for Plant Morphology against Heat Stress

Transient change in temperature causes malfunctioning of plant's physiological and biochemical machinery with aberrant phenotype (Ahuja et al. 2010). Plants under heat stress exhibit apparent morphological signs. The most obvious ones are sunburn and scorching of aerial plant parts, leaf abscission, and senescence (Zhou et al. 2014). Damage to plant morphology also occurs with onset of heat wave such as drying and rolling of leaves. Stunted shoot growth with higher number of leaves falling can be observed due to high temperature stress. It has detrimental effect on plants' flowering, which is crucial to ensure higher productivity in numerous crops. In grain crops, it harms tillers formation with a drastic reduction in grain formation. Senescence is a complicated trait difficult to quantify with a progression relevant to severity of climatic conditions. However, QTLs associated with have been mapped to ensure robust growth and control of premature ageing of plant parts (Vijayalakshmi et al. 2010). Leaf rolling trait linked to QTLs was also mapped by Tahmasebi et al. (2016). Likewise, other abiotic stresses also interfere with the physiological changes in plants.

9.5.3 QTL Mapping for Plant Physiology against Heat Stress

Photosynthesis is highly sensitive to heat shocks. Thylakoid membrane in chloroplast is the site of damage because it is directly involved in capturing sunlight for carrying out light reaction of photosynthesis with high temperature, and it causes detrimental losses to photosynthetic apparatus. Photosynthesis is further aggravated due to closure of stomata and limiting gas exchange during heat stress. QTL mapping was exploited to decipher chromosomal regions associated for damage of thylakoid membrane, plasma membrane, and reduction in chlorophyll contents during heat stress. Moreover, membrane damages are closely associated with chlorophyll during long exposure to thermal stress and might be the possibility that they are under the same genetic control (Talukder et al. 2014). High temperature stress also triggers oxidative stress by accumulation of ROS species. The main sites of ROS production are photosystems (I and II) in chloroplast, although it is also generated in other plant organelles such as mitochondria and peroxisomes (Soliman et al. 2011). Heat stress damages photosystems of plant which absorbs photons intensity under such high temperature pressure and generates ROS species. Maximum quantum efficiency (Fv/Fm) of photosystem is directly linked with quantum yield of photosynthesis; therefore, QTL was identified for quantum efficiency for improving photosynthesis against heat stress (Sharma et al. 2017). It is well reported that thermal stress induces electron leakage from the thylakoid membrane which disrupts photosynthetic light reaction and enhances oxidative stress. High temperature due to thermal stress is the indicator of decreased antioxidant enzyme activities which ultimately increases MDA content. QTL aided in improving photosynthetic rate and maintaining normal physiological function to alleviate heat stress (Cao et al. 2003).

9.5.4 QTL Mapping for Flowering against Heat Stress

Plant reproductive stage is also susceptible to heat spells, and it is the indispensable phase that with a minor increase in temperature during flowering causes loss of whole grain crop. Moreover, flower abortion, production of no flower, or flowers may not produce fruit and seed with under heat stress (Sato et al. 2006; Lobell et al. 2011). It also disturbs anthesis with severe reduction in anther dehiscence and pollen fertility rate (Hurkman et al. 2009; Ahamed et al. 2010). On the other hand, fertility of flower can be maintained significantly by using QTL linked to that trait as already mapped under heat stress (Zhang et al. 2008; Ye et al. 2012). QTLs for various morphological traits especially responsible for flowering time and reproductive growth stages were mapped to elucidate potential genetic control during these growth phases from heat stress (Mason et al. 2010; Mason et al. 2011).

9.5.5 QTL Mapping for Improving Yield against Heat Stress

High temperature affects the final grain filling and yield of crops. It is a serious threat for food security. It is so terrible during maturity that minute raise in temperature leads towards reduction in crop yields (Savin et al. 1997; Warland et al. 2006; Ali et al. 2018). The average increase of seasonal temperature of 1°C declined grain yield of cereals by 4–10% (Wang et al. 2012). Grains were protected from the adverse effect of high temperature stress by exploiting QTL technique (Tabata et al. 2007). Heat susceptibility index (HIS) is the best parameter for selection of heat-tolerant cultivars. The QTLs responsible for multifarious agronomic traits (grain weight, yield, and grain filling duration) with respect to HSI have been mapped successfully (Paliwal et al. 2012).

9.6 Cold Stress

According to rough estimates, approximately 2/3rd of global landmass is affected by lower temperatures (under freezing point), there are such areas which have –20 °C temperatures, and they constitute half of the earth landmass (Larcher 2001). Many efforts have been done and are still in progress to improve cold tolerance in plants. It is difficult to achieve frost hardening in plants via conventional breeding (Kasuga et al. 1999) because cold tolerance is not a trait controlled by single gene, but it is a product of multiple genes (Fowler and Thomashow 2002). Cold stress syndrome includes fluidity of biomembranes, accumulation of low molecular weight, synthesis and high molecular weight cryoprotectants, and increased potential to face oxidative stress. In most of plants, major damage caused due to low temperature is loss of plasma membrane integrity (Steponkus et al. 1998). Frost has different effects on different types of plant tissues, generally meristematic tissues are less frost hardy (Sakai and Larcher 1987). Plant attributes, especially leaf area, N contents, and greenness, are largely affected by chilling effect (Jompuk et al. 2005). Cold temperatures either affect the cells by dehydrating them by freezing inner water or by lower temperatures. Subtropical and tropical plants are highly influenced by the lower temperatures, because they are not adapted to freezing temperatures, and this damage is called 'chilling,' which results into biomembranes' work loss by lowering fluidity and inactivation of membrane ion pumps. Extent of damage to a plant due to lower temperatures depends on different factors, that is, developmental stage, severity of frost, duration of frost, cooling rates, and ice formation inside cell. Genetic potential along with environmental conditions of a plant controls its sensitivity to frost and frost hardiness, and it also changes with time. The inheritance of cold stress is poorly understood; to identify or locate genomic regions controlling chilling tolerance, QTL analysis is a fruitful tool as it also helps in understanding complex associated traits.

9.6.1 QTL for Root Traits

Root conductivity can also be used for mapping QTLs against cold stress. Biological mechanisms such as transcriptional regulations and posttranscriptional modifications can help in controlling cold tolerance (Sanghera et al. 2011). Introduction of cold tolerance genes by exploiting gene pool of a crop can enhance the production (Xie et al. 2012). QTLs for cold tolerance related to booting stage and cold tolerance have been identified (Shinada et al. 2014). The diameter and length of the root are impervious for coping with cold stress; thereby, the QTLs associated with these traits were dissected for conferring cold stress tolerance (Hund et al. 2004).

9.6.2 QTLs for Seed Germination

Cold stress hinders germination ability, resulting into decreased seedling emergence. The obvious symptoms of cold stress at seedling stage includes yellowing of leaves, withering, and low tillering, and these may also cause pollen sterility (Suzuki et al. 2008). Seedling survival is an important character for developing cold-tolerant cultivars. However, QTLs have been reported for improving germination ability, emergence, and seedling vigour of crops (Knoll et al. 2008). Low temperature influences germination ability of seed which is major hurdle for robust stand establishment especially at high altitudes. QTL mapping also played role in this regard and markers have been developed linked to the desired germinability trait which can be successfully used for improving germination of crop against cold stress (Fujino et al. 2004).

9.6.3 QTLs Related to Physiological Traits

In agricultural crops, cold stress imparts biochemical and physiological mechanisms. Cold stress also interferes with physiological processes like other biotic stresses such as disruption of chlorophyll contents which resultantly yield in lower photosynthesis. In plants, photosynthesis is the main and most important mechanism through which they acquire their energy. Any kind of stress during this process may retard or hinder plant growth. Different QTLs under cold conditions which are related to the proper functioning of photosynthesis apparatus have been identified (Fracheboud et al. 2004; Jompuk et al. 2005). Studies have shown that the control of photosynthetic units and their functioning is under different genetic control. QTLs studies revealed that greenness of leaves is directly associated with the photosynthetic activity. ROs accumulation is also high due to oxidative stress and with higher MDA contents due to cold stress, these all mechanism cause impairment of metabolism (Xie et al. 2009). Contrarily, proline, an important amino acid for coping stress in plants, stabilizes protein and maintains functions of cells (Ashraf and Foolad, 2007). QTLs associated with physiological traits for coping cold stress were reported by Kim and Tai (2011).

9.6.4 QTL Mapping for Reproductive Characteristics against Cold Stress

Production of viable seeds is an ongoing process of plants from one generation to another, and it is compulsory for a plant to have receptive and functional reproductive organs. Cold temperatures can cause male sterility in plants and can lower the productivity of important crops. QTLs have been identified which can control sterility at reproductive stage (Koumoto et al. 2016). At early growth stages, seedling impairment occurs due to sterility caused by low temperatures (Andaya and Mackill 2003). Plant suffers hindrance in seedling growth, stunted height, and poor grain growth because cold stress damages photosynthetic ability (Suh et al. 2010). Severe cold temperatures cause imbalance in plant metabolic activities, especially seed setting is disturbed in various crop species, and QTLs associated with seed setting under severe cold have been identified (Suh et al. 2010). Similar QTLs have also been identified for double haploid populations (Takeuchi et al. 2001). Total proportion of fertile spikelets on the plant is an important indicator that creates knowledge for plants' ability to tolerate cold at reproductive stage, particularly at anthesis (Saito et al. 1995).

9.6.5 QTL for Plant Growth Characteristics under Cold Stress

Cold stress causes stunted plant growth which ultimately results in poor yield. Phenology of the crop and its yield characteristics related to the genetically determined traits like number of leaves, shoot length, plant height and photoperiod must be met adequately for acquiring proper plant yield. Significant differences have been found through QTLs between mapping populations and their parents. Primary yield components include tiller number per plant, per spike number of seeds, and seed size, and these traits are the result of genotypic and environmental interaction. Enormous QTLs are identified for phenological traits in controlled environment. For number of days to anthesis and number of leaves, QTLs have been identified (Fowler et al. 2016). Large number of QTLs related to plant height has been reported that are also associated to plant yield. Important QTLs associated with grain yield have also been located (McCartney et al. 2005; Heidari et al. 2011).

9.7 Conclusion

Agricultural production has been under threat to climate change and its various effects. Different parts of the world are facing extreme weather conditions; salinity and drought conditions have decreased the cultivable land globally. Crop yields have been significantly affected, and some studies showed their extent up to 50%. Plant breeders are playing their role in the development of such cultivars which can face abiotic conditions. For this purpose, they are using different technologies, such as QTL (quantitative trait loci), in particularly those regions that are controlling those traits which are associated with tolerance to abiotic stress. Using different suitable molecular markers, that is, RFLP, RAPD, AFLP, and SNPs, enabled plant breeders to screen genotypes based on the presence of desirable characters, so that they can be used in plant breeding programmes for introgression of genes and development of stress-tolerant cultivars.

REFERENCES

Abbasi GH, Ijaz M, Akhtar J, Anwar-Ul-Haq M, Jamil M, AliS AR, Khan, HN (2016) Profiling of anti-oxidative enzymes and lipid peroxidation in leaves of salt tolerant and salt sensitive maize hybrids under NaCl and cd stress. Sains Malay 45:177–184.

Adnan M, Fahad S, Khan IA, Saeed M, Ihsan MZ, Saud S, Riaz M, Wang D, Wu C (2019) Integration of poultry manure and phosphate solubilizing bacteria improved availability of Ca bound P in calcareous soils. 3 Biotech 9:368.

Adnan M, Fahad S, Muhammad Z, Shahen S, Ishaq AM, Subhan D, Zafar-ul-Hye M, Martin LB, Raja MMN, Beena S, Saud S, Imran A, Zhen Y, Martin B, Jiri H, Rahul D (2020) Coupling phosphate-solubilizing bacteria with phosphorus supplements improve maize phosphorus acquisition and growth under lime induced salinity stress. Plants 9:900. https://doi.org/10.3390/plants9070900.

Adnan M, Shah Z, Sharif M, Rahman H (2018b) Liming induces carbon dioxide (CO_2) emission in PSB inoculated alkaline soil supplemented with different phosphorus sources. Environ Sci Poll Res 25:9501–9509.

Adnan M, Zahir S, Fahad S, Arif M, Mukhtar A, Imtiaz AK, Ishaq AM, Abdul B, Hidayat U, Muhammad A, Inayat-Ur R, Saud S, Muhammad ZI, Yousaf J, Amanullah, Hafiz MH, Wajid N (2018a) Phosphate-solubilizing bacteria nullify the antagonistic effect of soil calcification on bioavailability of phosphorus in alkaline soils. Sci Rep 8:4339. https://doi.org/10.1038/s41598-018-22653-7.

Ahamed KU, Nahar K, Fujita M, Hasanuzzaman M (2010) Variation in plant growth, tiller dynamics and yield components of wheat (*Triticum aestivum* L.) due to high temperature stress. AAB Bioflux 2.

Ahmad S, Kamran M, Ding R, Meng X, Wang H, Ahmad I, Fahad S, Han Q (2019) Exogenous melatonin confers drought stress by promoting plant growth, photosynthetic capacity and antioxidant defense system of maize seedlings. PeerJ 7:e7793. http://doi.org/10.7717/peerj.7793.

Ahuja I, de Vos RC, Bones AM, Hall RD (2010) Plant molecular stress responses face climate change. Trends Plant Sci 15:664–674.

Akram R, Turan V, Hammad HM, Ahmad S, Hussain S, Hasnain A, Maqbool MM, Rehmani MIA, Rasool A, Masood N, Mahmood F, Mubeen M, Sultana SR, Fahad S, Amanet K, Saleem M, Abbas Y, Akhtar HM, Waseem F, Murtaza R, Amin A, Zahoor SA, ul Din MS, Nasim W (2018a) Fate of organic and inorganic pollutants in paddy soils. In: Hashmi MZ, Varma A (eds) Environmental pollution of paddy soils, soil biology. Springer International Publishing Ag, Switzerland, pp 197–214.

Akram R, Turan V, Wahid A, Ijaz M, Shahid MA, Kaleem S, Hafeez A, Maqbool MM, Chaudhary HJ, Munis, MFH, Mubeen M, Sadiq N, Murtaza R, Kazmi DH, Ali S, Khan N, Sultana SR, Fahad S, Amin A, Nasim W (2018b) Paddy land pollutants and their role in climate change. In: Hashmi MZ, Varma A (eds) Environmental pollution of paddy soils, soil biology. Springer International Publishing Ag, Switzerland, pp 113–124.

Aksoy E, Demirel U, Öztürk ZN, Çalişkan S, Çalişkan ME (2015) Recent advances in potato genomics, transcriptomics, and transgenics under drought and heat stresses: a review. Turk J Bot 39:920–940.

Ali S, Xu Y, Ahmad I, Jia Q, Ma X, Ullah H, Alam M, Adnan M, Daur I, Ren X, Cai T. (2018) Tillage and deficit irrigation strategies to improve winter wheat production through regulating root development under simulated rainfall conditions. Agric Water Manage 30:44–54.

Amjid MW, Malik TA, Shakeel A, Wahid A (2015) QTL Mapping for relative leaf water contents, cell membrane stability and excised leaf water loss under drought by using EST–SSR markers in *Gossypium hirsutum*. Int J Agric Biol 17.

Ammar MHM (2004) Molecular mapping for salt tolerance in rice. Doctoral dissertation, Indian Agricultural Research Institute, New Delhi.

Andaya VC, Mackill DJ (2003) Mapping of QTLs associated with cold tolerance during the vegetative stage in rice. J Exp Bot 54:2579–2585.

Andaya VC, Tai TH (2006) Fine mapping of the qCTS12 locus, a major QTL for seedling cold tolerance in rice. Theor Appl Genet 113:467–475.

Anjum SA, Xie XY, Wang LC, Saleem MF, Man C, Lei W (2011) Morphological, physiological and biochemical responses of plants to drought stress. Afr J Agr Res 6:2026–2032.

Araus JL, Bort J, Steduto P, Villegas D, Royo C (2003) Breeding cereals for Mediterranean conditions: ecophysiological clues for biotechnology application. Ann Appl Biol 142:129–141.

Arends D, Prins P, Jansen RC, Broman KW (2010) R/qtl: high-throughput multiple QTL mapping. Bioinformatics 26:2990–2992.

Arraouadi S, Badri M, Abdelly C, Huguet T, Aouani ME (2012) QTL mapping of physiological traits associated with salt tolerance in *Medicago truncatula* recombinant inbred lines. Genomics 99:118–125.

Arraouadi S, Chardon F, Huguet T, Aouani ME, Badri M (2011) QTLs mapping of morphological traits related to salt tolerance in *Medicago truncatula*. Acta Physiol Plant 33:917–926.

Ashraf M, Foolad MR (2007) Roles of glycine betaine and proline in improving plant abiotic stress resistance. Environ Exp Bot 59:206–216.

Ashraf M, Foolad MR (2013) Crop breeding for salt tolerance in the era of molecular markers and marker assisted selection. Plant Breed 132:10–20.

Aziz K, Daniel KYT, Fazal M, Muhammad ZA, Farooq S, Fan W, Fahad S, Ruiyang Z (2017a) Nitrogen nutrition in cotton and control strategies for greenhouse gas emissions: a review. Environ Sci Pollut Res 24:23471–23487. https://doi.org/10.1007/s11356-017-0131-y.

Aziz K, Daniel KYT, Muhammad ZA, Honghai L, Shahbaz AT, Mir A, Fahad S (2017b) Nitrogen fertility and abiotic stresses management in cotton crop: a review. Environ Sci Pollut Res 24:14551–14566. https://doi.org/10.1007/s11356-017-8920-x.

Baseer M, Adnan M, Fazal M, Fahad S, Muhammad S, Fazli W, Muhammad A, Jr. Amanullah, Depeng W, Saud S, Muhammad N, Muhammad Z, Fazli S, Beena S, Mian AR, Ishaq AM (2019) Substituting urea by organic wastes for improving maize yield in alkaline soil. J Plant Nutr https://doi.org/10.1080/01904167.2019.1659344.

Batool N, Ilyas N, Shahzad A, Hauser BA, Arshad M (2018) Quantitative trait loci (QTLs) mapping for salt stress tolerance in wheat at germination stage. Pak J Agr Sci 55.

Bennett D, Reynolds M, Mullan,D, Izanloo A, Kuchel H, Langridge P, Schnurbusch T (2012) Detection of two major grain yield QTL in bread wheat (*Triticum aestivum* L.) under heat, drought and high yield potential environments. Theor Appl Genet 125:1473–1485.

Bernardo R (2008) Molecular marker and selection for complex traits in plants: learning from the last 20 years. Crop Sci 48:1649–1664.

Bernier J, Serraj R, Kumar A, Venuprasad R, Impa S, Gowda V, Oane R, Spaner D, Atlin G (2009) The large–effect drought–resistance QTL qtl12.1 increases water uptake in upland rice. Field Crops Res 110: 139–146.

Bettey M, Finch-Savage WE, King GJ, Lynn JR (2000) Quantitative genetic analysis of seed vigour and pre-emergence seedling growth traits in *Brassica oleracea*. New Phytol 148:277–286.

Bizimana JB, Luzi-Kihupi A, Murori RW, Singh RK (2017) Identification of quantitative trait loci for salinity tolerance in rice (*Oryza sativa* L.) using IR29/Hasawi mapping population. J Genet 96:571–582.

Bo K, Ma Z, Chen J, Weng Y (2015) Molecular mapping reveals structural rearrangements and quantitative trait loci underlying traits with local adaptation in semi-wild Xishuangbanna cucumber (*Cucumis sativus* L. var. xishuangbannanesis Qi et Yuan). Theor Appl Gen 128:25–39.

Bonilla P, Dvorak J, Mackill D, Deal K, Gregorio G (2002) RFLP and SSLP mapping of salinity tolerance genes in chromosome 1 of rice (*Oryza sativa* L.) using recombinant inbred lines. Philipp Agric Sci 85:68–76.
Bradshaw HD (1996) Molecular genetics of populus. In: Biology of populous and its implications for management and conservation. Canadian Science Publishing (NRC Research Press), Ottawa, Canada, pp 183–199.
Bray EA, Bailey-Serres J, Weretilnyk (2000) Responses to abiotic stresses. Biochem Mol Biol P 1158–1249.
Broman KW, Wu H, Sen S, Churchill GA (2003) R/qtl: QTL mapping in experimental crosses. Bioinformatics 19:889–890.
Cao L, Zhao J, Zhan X, Li D, He L, Cheng S (2003) Mapping QTLs for heat tolerance and correlation between heat tolerance and photosynthetic rate in rice. Rice Sci 17:223–227.
Castonguay Y, Markhart AH (1992) Leaf gas exchange in water stressed common bean and tepary bean. Crop Sci 32:980–986.
Cattivelli L, Baldi P, Crosatti C, Di Fonzo N, Faccioli P, Grossi M, Mastrangelo AM, Pecchioni N, Stanca AM (2002) Chromosome regions and stress-related sequences involved in resistance to abiotic stress in Triticeae. Plant Mol Biol 48:649–665.
Chaves MM, Flexas J, Pinheiro C (2009) Photosynthesis under drought and salt stress: regulation mechanisms from whole plant to cell. Ann Bot 103:551–560.
Chen Z, Zhou M, Newman IA, Mendham NJ, Zhang G, Shabala S (2007) Potassium and sodium relations in 17alinized barley tissues as a basis of differential salt tolerance. Funct Plant Biol 34:150–162.
Christopher J, Christopher M, Jennings R, Jones S, Fletcher S, Borrell A, Manschadi AM, Jordan D, Mace E, Hammer G (2013) QTL for root angle and number in a population developed from bread wheats (*Triticum aestivum*) with contrasting adaptation to water–limited environments. Theor Appl Genet 126:1563–1574.
Cuartero J, Bolarín MC, Asíns MJ, Moreno V (2006) Increasing salt tolerance in the tomato. J Exp Bot 57:1045–1058.
Deng N, Ling X, Sun Y, Zhang C, Fahad S, Peng S, Cui K, Nie L, Huang J (2015) Influence of temperature and solar radiation on grain yield and quality in irrigated rice system. Eur J Agron 64:37–46.
Depeng W, Fahad S, Saud S, Muhammad K, Aziz K, Mohammad NK, Hafiz MH, Wajid N (2018) Morphological acclimation to agronomic manipulation in leaf dispersion and orientation to promote "Ideotype" breeding: evidence from 3D visual modeling of "super" rice (*Oryza sativa* L.). Plant Physiol Biochem 135. https://doi.org/10.1016/j.plaphy.2018.11.010.
Edwards MD, Stuber CW, Wendel JF (1987) Molecular-marker-facilitated investigations of quantitative-trait loci in maize. I. Numbers, genomic distribution and types of gene action. Genetics 116:113–125.
Eleuch L, Jilal A, Grando S, Ceccarelli S, von Korff Schmising M, Tsujimoto H, Hajer A, Daaloul A, Baum M (2008) Genetic diversity and association analysis for salinity tolerance, heading date and plant height of barley germplasm using simple sequence repeat markers. J Integr Plant Biol 50:1004–1014.
Elshafei AA, Saleh M, Al-Doss AA, Moustafa KA, Al-Qurainy FH, Barakat MN (2013) Identification of new SRAP markers linked to leaf chlorophyll content, flag leaf senescence and cell membrane stability traits in wheat under water-stressed condition. Aust J Crop Sci 7:887–893.
El-Soda M, Kruijer W, Malosetti M, Koornneef M, Aarts MG (2015) Quantitative trait loci and candidate genes underlying genotype by environment interaction in the response of A rabidopsis thaliana to drought. Plant Cell Environ 38:585–599.
Fahad S, Adnan M, Hassan S, Saud S, Hussain S, Wu C, Wang D, Hakeem KR, Alharby HF, Turan V, Khan MA, Huang J (2019b) Rice responses and tolerance to high temperature, In: Hasanuzzaman M, Fujita M, Nahar K, Biswas JK (eds) Advances in rice research for abiotic stress tolerance. Woodhead Publ Ltd, Cambridge, England, pp 201–224.
Fahad S, Bajwa AA, Nazir U, Anjum SA, Farooq A, Zohaib A, Sadia S, NasimW, Adkins S, Saud S, Ihsan MZ, Alharby H, Wu C, Wang D, Huang J (2017) Crop production under drought and heat stress: plant responses and management options. Front Plant Sci 8:1147. https://doi.org/10.3389/fpls.2017.01147.
Fahad S, Bano A (2012) Effect of salicylic acid on physiological and biochemical characterization of maize grown in saline area. Pak J Bot 44:1433–1438.
Fahad S, Chen Y, Saud S, Wang K, Xiong D, Chen C,Wu C, Shah F, Nie L, Huang J (2013) Ultraviolet radiation effect on photosynthetic pigments, biochemical attributes, antioxidant enzyme activity and hormonal contents of wheat. J Food Agri Environ 11:1635–1641.
Fahad S, Hussain S, Bano A, Saud S, Hassan S, Shan D, Khan FA, Khan F, Chen Y, Wu C, Tabassum MA,

Chun MX, Afzal M, Jan A, Jan MT, Huang J (2014a) Potential role of phytohormones and plant growth-promoting rhizobacteria in abiotic stresses: consequences for changing environment. Environ Sci Pollut Res 22:4907–4921. https://doi.org/10.1007/s11356-014-3754-2.

Fahad S, Hussain S, Matloob A, Khan FA, Khaliq A, Saud S, Hassan S, Shan D, Khan F, Ullah N, Faiq M, Khan MR, Tareen AK, Khan A, Ullah A, Ullah N, Huang J (2014b) Phytohormones and plant responses to salinity stress: A review. Plant Growth Regul 75:391–404. https://doi.org/10.1007/s10725-014-0013-y.

Fahad S, Hussain S, Saud S, Hassan S, Chauhan BS, Khan F et al. (2016a) Responses of rapid viscoanalyzer profile and other rice grain qualities to exogenously applied plant growth regulators under high day and high night temperatures. PLoS One 11:e0159590. https://doi.org/10.1371/journal.pone.0159590.

Fahad S, Hussain S, Saud S, Hassan S, Ihsan Z, Shah AN,Wu C, Yousaf M, Nasim W, Alharby H, Alghabari F, Huang J (2016c) Exogenously applied plant growth regulators enhance the morphophysiological growth and yield of rice under high temperature. Front Plant Sci 7:1250. https://doi.org/10.3389/fpls.2016.01250.

Fahad S, Hussain S, Saud S, Hassan S, Tanveer M, Ihsan MZ, Shah AN, Ullah A, Nasrullah KF, Ullah S, AlharbyH NW, Wu C, Huang J (2016d) A combined application of biochar and phosphorus alleviates heat-induced adversities on physiological, agronomical and quality attributes of rice. Plant Physiol Biochem 103:191–198.

Fahad S, Hussain S, Saud S, Khan F, Hassan S, Jr A, Nasim W, Arif M, Wang F, Huang J (2016b) Exogenously applied plant growth regulators affect heat-stressed rice pollens. J Agron Crop Sci 202:139–150.

Fahad S, Hussain S, Saud S, Tanveer M, Bajwa AA, Hassan S, Shah AN, Ullah A,Wu C, Khan FA, Shah F, Ullah S, Chen Y, Huang J (2015a) A biochar application protects rice pollen from high-temperature stress. Plant Physiol Biochem 96:281–287.

Fahad S, Muhammad ZI, Abdul K, Ihsanullah D, Saud S, Saleh A, Wajid N, Muhammad A, Imtiaz AK, Chao W, Depeng W, Jianliang H (2018) Consequences of high temperature under changing climate optima for rice pollen characteristics-concepts and perspectives. Arch Agron Soil Sci 64:1479–1488. https://doi.org/10.1080/03650340.2018.1443213.

Fahad S, Nie L, Chen Y, Wu C, Xiong D, Saud S, Hongyan L, Cui K, Huang J (2015b) Crop plant hormones and environmental stress. Sustain Agric Rev 15:371–400.

Fahad S, Rehman A, Shahzad B, Tanveer M, Saud S, Kamran M, Ihtisham M, Khan SU, Turan V, Rahman MHU (2019a) Rice responses and tolerance to metal/metalloid toxicity. In: Hasanuzzaman M, Fujita M, Nahar K, Biswas JK (eds) Advances in rice research for abiotic stress tolerance. Woodhead Publ Ltd, Cambridge, England, pp 299–312.

Fan Y, Shabala S, Ma Y, Xu R, Zhou M (2015) Using QTL mapping to investigate the relationships between abiotic stress tolerance (drought and salinity) and agronomic and physiological traits. BMC Genom 16:43.

Farhana G, Ishfaq A, Muhammad A, Dawood J, Fahad S, Xiuling L, Depeng W, Muhammad F, Muhammad F, Syed AS (2020) Use of crop growth model to simulate the impact of climate change on yield of various wheat cultivars under different agro-environmental conditions in Khyber Pakhtunkhwa, Pakistan. Arabian J Geosci 13:112. https://doi.org/10.1007/s12517-020-5118-1.

FAO (2013) Food and Agriculture Organization of the united nations, land resources. http://www.fao.org/economic/ess/ess-home/en/#.UcF1tY3E_Tp.

Farhat A, Hafiz MH, Wajid I, Aitazaz AF, Hafiz FB, Zahida Z, Fahad S, Wajid F, Artemi C (2020) A review of soil carbon dynamics resulting from agricultural practices. J Environ Manage 268:110319.

Farooq M, Hussain M, Wahid A, Siddique KHM (2012) Drought stress in plants: an overview. In: Aroca R (ed) Plant responses to drought stress. Springer, Berlin, Heidelberg, pp 1–33.

Fowler DB, N'Diaye A, Laudencia-Chingcuanco D, Pozniak CJ (2016) Quantitative trait loci associated with phenological development, low-temperature tolerance, grain quality, and agronomic characters in wheat (*Triticum aestivum* L.). PLoS One 11:152–185.

Fowler S, Thomashow MF (2002) Arabidopsis transcriptome profiling indicates that multiple regulatory pathways are activated during cold acclimation in addition to the CBF cold response pathway. Plant Cell 14:1675–1690.

Flowers TJ (2004) Improving crop salt tolerance. J Exp Bot 55:307–319.

Flowers TJ, Koyama ML, Flowers SA, Sudhakar C, Singh KP, Yeo AR (2000) QTL: their place in engineering tolerance of rice to salinity. J Exp Bot 51:99–106.

Fracheboud Y, Jompuk C, Ribaut JM, Stamp P, Leipner J (2004) Genetic analysis of cold-tolerance of photosynthesis in maize. Plant Mol Biol 56:241–253.

Frary A, Göl D, Keleş D, Ökmen B, Pınar H, Şığva HÖ, Yemenicioğlu A, Doğanlar S (2010) Salt tolerance in *Solanum pennellii*: antioxidant response and related QTL. BMC Plant Biol 10:58.

Fritsche-Neto R, Do Vale JC, Lanes Écmd, Resende Mdvd, Miranda GV (2012) Genome-wide selection for tropical maize root traits under conditions of nitrogen and phosphorus stress. Acta Sci Agron 34:389–395.

Fujino K, Sekiguchi H, Sato T, Kiuchi H, Nonoue Y, Takeuchi Y, Ando T, Lin SY, Yano M (2004) Mapping of quantitative trait loci controlling low-temperature germinability in rice (*Oryza sativa* L.). Theor Appl Genet 108:794–799.

Gelderman H (1975) Investigations on inheritance of quantitative characters in animals by gene markers. I. methods. Theor Appl Genet 46:319–330.

Genc Y, Oldach K, Verbyla AP, Lott G, Hassan M, Tester M, Wallwork H, McDonald GK (2010) Sodium exclusion QTL associated with improved seedling growth in bread wheat under salinity stress. Theor Appl Genet 121:877–894.

Gill MB, Zeng F, Shabala L, Zhang G, Yu M, Demidchik V, Shabala S, Zhou M (2019) Identification of QTL related to ROS formation under hypoxia and their association with waterlogging and salt tolerance in barley. Int J Mol Sci 20:699.

Gregorio GB (1997) Tagging salinity tolerance genes in rice using amplified fragment length polymorphism (AFLP). PhD Thesis, University of the Philippines.

Guo P, Baum M, Varshney RK, Graner A, Grando S, Ceccarelli S (2008) QTLs for chlorophyll and chlorophyll fluorescence parameters in barley under post–flowering drought. Euphytica 163:203–214.

Habib ur Rahman M, Ashfaq A, Aftab W, Manzoor H, Fahd R, Wajid I, Md. Aminul I, Vakhtang S, Muhammad A, Asmat U, Abdul W, Syeda RS, Shah S, Shahbaz K, Fahad S, Manzoor H, Saddam H, Wajid N (2017) Application of CSM-CROPGRO-Cotton model for cultivars and optimum planting dates: evaluation in changing semi-arid climate. Field Crops Res 238:139–152. http://dx.doi.org/10.1016/j.fcr.2017.07.007.

Hafiz MH, Abdul K, Farhat A, Wajid F, Fahad S, Muhammad A, Ghulam MS, Wajid N, Muhammad M, Hafiz FB (2020b) Comparative effects of organic and inorganic fertilizers on soil organic carbon and wheat productivity under arid region. Commun Soil Sci Plant Anal 51:1406–1422. https://doi.org/10.1080/00103624.2020.1763385.

Hafiz MH, Farhat A, Ashfaq A, Hafiz FB, Wajid F, Carol Jo W, Fahad S, Gerrit H (2020a) Predicting kernel growth of maize under controlled water and nitrogen applications. Int J Plant Prod 14:609–620. https://doi.org/10.1007/s42106-020-00110-8.

Hafiz MH, Farhat A, Shafqat S, Fahad S, Artemi C, Wajid F, Chaves CB, Wajid N, Muhammad M, Hafiz FB (2018) Offsetting land degradation through nitrogen and water management during maize cultivation under arid conditions. Land Degrad Dev 1–10. https://doi.org/10.1002/ldr.2933.

Hafiz MH, Muhammad A, Farhat A, Hafiz FB, Saeed AQ, Muhammad M, Fahad S, Muhammad A (2019) Environmental factors affecting the frequency of road traffic accidents: a case study of sub-urban area of Pakistan. Environ Sci Pollut Res 26:11674–11685. https://doi.org/10.1007/s11356-019-04752-8.

Hafiz MH, Wajid F, Farhat A, Fahad S, Shafqat S, Wajid N, Hafiz FB (2016) Maize plant nitrogen uptake dynamics at limited irrigation water and nitrogen. Environ Sci Pollut Res 24:2549–2557. https://doi.org/10.1007/s11356-016-8031-0.

Haq TU, Gorham J, Akhtar J, Akhtar N, Steele KA (2010) Dynamic quantitative trait loci for salt stress components on chromosome 1 of rice. Funct Plant Biol 37:634–645.

Hasanuzzaman M, Nahar K, Alam M, Roychowdhury R, Fujita M (2013) Physiological, biochemical, and molecular mechanisms of heat stress tolerance in plants. Int J Mol Sci 14:9643–9684.

Hayat S, Hayat Q, Alyemeni MN, Wani AS, Pichtel J, Ahmad A (2012) Role of proline under changing environments: a review. Plant Signal Behav 7:1456–1466.

Heidari B, Sayed-Tabatabaei BE, Saeidi G, Kearsey M, Suenaga K (2011) Mapping QTL for grain yield, yield components, and spike features in a doubled haploid population of bread wheat. Genome 54:517–527.

Hesham FA, Fahad S (2020) Melatonin application enhances biochar efficiency for drought tolerance in maize varieties: modifications in physio-biochemical machinery. Agron J 1–22.

Honsdorf N, March TJ, Berger B, Tester M, Pillen K (2014) High-throughput phenotyping to detect drought tolerance QTL in wild barley introgression lines. PloS One 9:97047.
Horie T, Schroeder J (2004) Sodium transporters in plants. Diverse genes and physiological functions. Plant Physiol 136:2457–2462.
Huang S, Spielmeyer W, Lagudah ES, James RA, Platten JD, Dennis ES, Munns R (2006) A sodium transporter (HKT7) is a candidate for Nax1, a gene for salt tolerance in durum wheat. Plant Physiol 142:1718–1727.
Hund A, Fracheboud Y, Soldati A, Frascaroli E, Salvi S, Stamp P (2004) QTL controlling root and shoot traits of maize seedlings under cold stress. Theor Appl Genet 109:618–629.
Hurkman WJ, Vensel WH, Tanaka CK, Whitehand L, Altenbach SB (2009) Effect of high temperature on albumin and globulin accumulation in the endosperm proteome of the developing wheat grain. J Cereal Sci 49:12–23.
Hussain SS (2006) Molecular breeding for abiotic stress tolerance. Drought Perspect Proc Pak Acad Sci 43:189–210.
Hussain MA, Fahad S, Rahat S, Muhammad FJ, Muhammad M, Qasid A, Ali A, Husain A, Nooral A, Babatope SA, Changbao S, Liya G, Ibrar A, Zhanmei J, Juncai H (2020) Multifunctional role of brassinosteroid and its analogues in plants. Plant Growth Regul 92:141–156. https://doi.org/10.1007/s10725-020-00647-8.
Hussain S, Khalid MF, Saqib M, Ahmad S, Zafar W, Rao MJ, Morillon R, Anjum MA (2018) Drought tolerance in citrus rootstocks is associated with better antioxidant defense mechanism. Acta Physiol Plant 40:135.
Hussain B, Lucas SJ, Ozturk L, Budak H (2017) Mapping QTLs conferring salt tolerance and micronutrient concentrations at seedling stage in wheat. Sci Rep 7:156–162.
Idrissi O, Udupa SM, De Keyser E, McGee RJ, Coyne CJ, Saha GC, Muehlbauer FJ, Van Damme P, De Riek J (2016) Identification of quantitative trait loci controlling root and shoot traits associated with drought tolerance in a lentil (*Lens culinaris* Medik.) recombinant inbred line population. Front Plant Sci 7:1174.
Ilyas M, Mohammad N, Nadeem K, Ali H, Aamir HK, Kashif H, Fahad S, Aziz K, Abid U (2020) Drought tolerance strategies in plants: a mechanistic approach. J Plant Growth Regul. https://doi.org/10.1007/s00344-020-10174-5.
Iqbal MM, Murtaza G, Saqib ZA, Rashid A (2015) Growth and physiological responses of two rice varieties to applied lead in normal and salt-affected soils. Int J Agric Biol 17.
Islam MR, Hassan L, Salam MA, Collard BCY, Singh RK, Gregorio GB (2011) QTL mapping for salinity tolerance at seedling stage in rice. Emir J Food Agr 137–146.
Islam M, Ontoy J, Subudhi PK (2019) Meta-analysis of quantitative trait loci associated with seedling-stage salt tolerance in rice (*Oryza sativa* L.). Plants 8:33.
Jamil A, Riaz S, Ashraf M, Foolad MR (2011) Gene expression profiling of plants under salt stress. Crit Rev Plant Sci 30:435–458.
Jan M, Anwar-ul-Haq M, Adnan NS, Muhammad Y, Javaid I, Xiuling L, Depeng W, Fahad S (2019) Modulation in growth, gas exchange, and antioxidant activities of salt-stressed rice (*Oryza sativa* L.) genotypes by zinc fertilization. Arabian J Geosci 12:775. https://doi.org/10.1007/s12517-019-4939-2.
Jansen RC, Stam P (1994) High resolution of quantitative traits into multiple loci via interval mapping. Genetics 136:1447–1455.
Jiang J, Zhuang JY, Fan YY, Bo, Shen (2009) Mapping of QTLs for leaf malondialdehyde content associated with stress tolerance in rice. Rice Sci 16:72–74.
Joehanes R, Nelson JC (2008) Qgene 4.0, an extensible Java QTL-analysis platform. Bioinformatics 24:2788–2789.
Johnson GR, Wheeler NC, Strauss SH (2000) Financial feasibility of marker-aided selection in Douglas-fir. Can J For Res 30:1942–1952.
Jompuk C, Fracheboud Y, Stamp P, Leipner J (2005) Mapping of quantitative trait loci associated with chilling tolerance in maize (*Zea mays* L.) seedlings grown under field conditions. J Exp Bot 56:1153–1163.
Jones H (2004) What is water use efficiency? In: Bacon MA (ed) Water use efficiency in plant biology. Blackwell, Oxford.
Kalladan R, Worch S, Rolletschek H, Harshavardhan VT, Kuntze L, Seiler C, Sreenivasulu N, Röder MS (2013) Identification of quantitative trait loci contributing to yield and seed quality parameters under terminal drought in barley advanced backcross lines. Mol Breed 32:71–90.

Kamaran M, Wenwen C, Irshad A, Xiangping M, Xudong Z, Wennan S, Junzhi C, Shakeel A, Fahad S, Qingfang H, Tiening L (2017) Effect of paclobutrazol, a potential growth regulator on stalk mechanical strength, lignin accumulation and its relation with lodging resistance of maize. Plant Growth Regul 84:317–332. https://doi.org/10.1007/s10725-017-0342-8.
Kao CH (2015) Mechanisms of salt tolerance in rice plants: Na+ transporters. Crop Environ Bioinform 12:113–119.
Kao CH, Zeng ZB, Teasdale RD (1999) Multiple interval mapping for quantitative trait loci. Genetics 152:1203–1216.
Kasim WA, Osman ME, Omar MN, El-Daim IAA, Bejai S, Meijer J (2013) Control of drought stress in wheat using plant-growth-promoting bacteria. J Plant Growth Regul 32:122–130.
Kasuga M, Liu Q, Miura S, Yamaguchi-Shinozaki K, Shinozaki K (1999) Improving plant drought, salt, and freezing tolerance by gene transfer of a single stress-inducible transcription factor. Nat Biotechnol 17:287–291.
Kebede H, Subudhi PK, Rosenow DT, Nguyen HT (2001) Quantitative trait loci influencing drought tolerance in grain sorghum (*Sorghum bicolor* L. Moench). Theor Appl 103:266–276.
Khan M (2015) QTL mapping: a tool for improvement in crop plants. Res J Recent Sci 4:2.
Kilasi NL, Singh J, Vallejos CE, Ye C, Jagadish SK, Kusolwa P, Rathinasabapathi B (2018) Heat stress tolerance in rice (*Oryza sativa* L.): identification of quantitative trait loci and candidate genes for seedling growth under heat stress. Front Plant Sci 9:1578.
Kim SI, Tai TH (2011) Evaluation of seedling cold tolerance in rice cultivars: a comparison of visual ratings and quantitative indicators of physiological changes. Euphytica 178:437–447.
Knoll J, Gunaratna N, Ejeta G (2008) QTL analysis of early-season cold tolerance in sorghum. Theor Appl Genet 116:577–587.
Koumoto T, Saito N, Aoki N, Iwasaki T, Kawai S, Yokoi S, Shimono, H (2016) Effects of salt and low light intensity during the vegetative stage on susceptibility of rice to male sterility induced by chilling stress during the reproductive stage. Plant Prod Sci 19:497–507.
Kuroki M, Saito K, Matsuba S, Yokogami N, Shimizu H, Ando I, Sato Y (2007) A quantitative trait locus for cold tolerance at the booting stage on rice chromosome 8. Theor Appl Genet 115:593–600.
Lander ES, Botstein D (1987) Homozygosity mapping: a way to map human recessive traits with the DNA of inbred children. Science 236:1567–1570.
Larcher W (2001) Ökophysiologie der Pflanzen. Eugen Ulmer, Stuttgart.
Lee SY, Ahn JH, Cha YS, Yun DW, Lee MC, Ko JC, Lee KS, Eun MY (2006) Mapping of quantitative trait loci for salt tolerance at the seedling stage in rice. Mol Cell 21.
Lehmann S, Funck D, Szabados L, Rentsch D (2010) Proline metabolism and transport in plant development. Amino Acids 39:949–962.
Li RH, Guo PP, Baumz M, Grando S, Ceccarelli S (2006) Evaluation of chlorophyll content and fluorescence parameters as indicators of drought tolerance in barley. Agr Sci China 5:751–757.
Li Z, Mu P, Li C, Zhang H, Li Z, Gao Y, Wang X (2005) QTL mapping of root traits in a doubled haploid population from a cross between upland and lowland japonica rice in three environments. Theor Appl Genet 110:1244–1252.
Li H, Ye G, Wang J (2007) A modified algorithm for the improvement of composite interval mapping. Genetics 175:361–374.
Lin HX, Zhu MZ, Yano M, Gao JP, Liang ZW, Su WA, Hu XH, Ren ZH, Chao DY (2004) QTLs for Na+ and K+ uptake of the shoots and roots controlling rice salt tolerance. Theor Appl Genet 108:253–260.
Liu Y, Subhash C, Yan J, Song C, Zhao J, Li J (2011) Maize leaf temperature responses to drought: thermal imaging and quantitative trait loci (QTL) mapping. Environ Exp Bot 71:158–165.
Lobell DB, Schlenker W, Costa-Roberts J (2011) Climate trends and global crop production since 1980. Science 333:616–620.
Lou Q, Chen L, Sun, Z, Xing Y, Li J, Xu X, Mei H, Luo L (2007) A major QTL associated with cold tolerance at seedling stage in rice (*Oryza sativa* L.). Euphytica 158:87–94.
Ma L, Zhou E, Huo N, Zhou R, Wang G, Jia J (2007) Genetic analysis of salt tolerance in a recombinant inbred population of wheat (*Triticum aestivum* L.). Euphytica 153:109–117.
Maccaferri M, Sanguineti MC, Corneti S, Ortega JLA, Salem MB, Bort J, DeAmbrogio E, del Moral LFG, Demontis A, El-Ahmed A, Maalouf F (2008) Quantitative trait loci for grain yield and adaptation of durum wheat (*Triticum durum* Desf.) across a wide range of water availability. Genetics 178:489–511.

Mahender A, Anandan A, Pradhan SK (2015) Early seedling vigour, an imperative trait for direct-seeded rice: an overview on physio-morphological parameters and molecular markers. Plant 241:1027–1050.
Malik S, Malik TA, Engineering G (2015) Genetic mapping of potential Qtls associated with drought tolerance in wheat. J Anim Plant Sci 25:1032–1040.
Manly KF, Cudmore Jr RH, Meer JM (2001) Map manager QTX, cross-platform software for genetic mapping. Mamm Genome 12:930–932.
Mano Y, Takeda K (1996) Genetical studies on salt tolerance at germination in recombinant inbred, isogenic, and doubled haploid lines of barley (*Hordeum vulgare* L.). Research Institute for Bioresources Okayama University 4:79–88.
Mano Y, Takeda K (1997) Mapping quantitative trait loci for salt tolerance at germination and the seedling stage in barley (Hordeum vulgare L.). Euphytica 94: 263–272.
Manschadi AM, Hammer GL, Christopher JT, de Well P (2008) Genotypic variation in seedling root architecture (traits and implications for drought adaptation in wheat (*T. Aestivum* L.). Plant Soil 303:115–129.
Mansfeld BN, Grumet R (2018) QTLseqr: An R package for bulk segregant analysis with next-generation sequencing. Plant Genome 11:180006.
Mardani Z, Rabiei B, Sabouri H, Sabouri A (2013) Mapping of QTLs for germination characteristics under non–stress and drought stress in rice. Rice Sci 20:391–399.
Mason RE, Mondal S, Beecher FW, Hays DB (2011) Genetic loci linking improved heat tolerance in wheat (*Triticum aestivum* L.) to lower leaf and spike temperatures under controlled conditions. Euphytica 180:181–194.
Mason RE, Mondal S, Beecher FW, Pacheco A, Jampala B, Ibrahim AM, Hays DB (2010) QTL associated with heat susceptibility index in wheat (*Triticum aestivum* L.) under short-term reproductive stage heat stress. Euphytica 174:423–436.
McCartney CA, Somers DJ, Humphreys DG, Lukow O, Ames N, Noll J, Cloutier S, McCallum BD (2005) Mapping quantitative trait loci controlling agronomic traits in the spring wheat cross RL4452 × ‘AC Domain’. Genome 48:870–883.
Michelmore R (1995) Molecular approaches to manipulation of disease resistance genes. Annu Rev Phytopathol 33:393–427.
Ming-zhe Y, Jian-fei W, Hong-you C, Hu-qu, Z, Hong-sheng Z (2005) Inheritance and QTL mapping of salt tolerance in rice. Rice Sci 12:25–32.
Mittler R (2006) Abiotic stress, the field environment and stress combination. Trend Plant Sci 11:15–19.
Mittler R, Blumwald E (2010) Genetic engineering for modern agriculture: challenges and perspectives. Annu Rev Plant Biol 61:443–462.
Moffat AS (2002) Finding new ways to protect drought-stricken plants. Science 296:1226–1229.
Mohan M, Nair S, Bhagwat A, Krishna TG, Yano M, Bhatia CR, Sasaki T (1997) Genome mapping, molecular markers and marker-assisted selection in crop plants. Mol Breed 3:87–103.
Mora JRH, Micheletti D, Bink M, Van de Weg E, Cantín C, Nazzicari N, Caprera A, Dettori MT, Micali S, Banchi E, Campoy JA (2017) Integrated QTL detection for key breeding traits in multiple peach progenies. BMC Genom 18:404.
Morton MJ, Awlia M, Al-Tamimi N, Saade S, Pailles Y, Negrão S, Tester M (2019) Salt stress under the scalpel–dissecting the genetics of salt tolerance. Plant J 97:148–163.
Mubeen M, Ashfaq A, Hafiz MH, Muhammad A, Hafiz UF, Mazhar S, Muhammad Sami ul Din, Asad A, Amjed A, Fahad S, Wajid N (2020) Evaluating the climate change impact on water use efficiency of cotton-wheat in semi-arid conditions using DSSAT model. J Water Climate Chang 11:1661–1675. https://doi.org/10.2166/wcc.2019.179/622035/jwc2019179.pdf.
Muhammad Z, Abdul MK, Abdul MS, Kenneth BM, Muhammad S, Shahen S, Ibadullah J, Fahad S (2019) Performance of *Aeluropus lagopoides* (mangrove grass) ecotypes, a potential turfgrass, under high saline conditions. Environ Sci Pollut Res 26:13410–13421. https://doi.org/10.1007/s11356-019-04838-3.
Munns R (2005) Genes and salt tolerance: bringing them together. New Phytol 167:645–663.
Munns R, Tester M (2008) Mechanisms of salinity tolerance. Annu Rev Plant Biol 59:651–681.
Murtaza G, Murtaza B, Kahlon UZ, Yaseen M (2017) A comparative study of different amendments on amelioration of saline-sodic soils irrigated with water having different EC: SAR ratios. Commun Soil Sci Plant Anal 48:2630–2641.
Nadeem MA, Nawaz MA, Shahid MQ, Doğan Y, Comertpay G, Yıldız M, Hatipoğlu R, Ahmad F, Alsaleh A,

Labhane N, Özkan H (2018) DNA molecular markers in plant breeding: current status and recent advancements in genomic selection and genome editing. Biotechnol Biotechnol Equip 32:261–285.
Neale DB, Savolainen O (2004) Association genetics of complex traits in conifers. Trend Plant Sci 9:325–330.
Nguyen DT, Nguyen TKL, Pham QC, Nguyen TH, Tran QT, Dao XH, Nguyen HT (2004) Mapping QTLs associated with root traits related to drought resistance in Vietnamese upland rice. Resil Crop Water Limit Environ 234.
Ozden M, Demirel U, Kahraman A (2009) Effects of proline on antioxidant system in leaves of grapevine (*Vitis vinifera* L.) exposed to oxidative stress by H_2O_2. Sci Hort 119:163–168.
Paliwal R, Röder MS, Kumar U, Srivastava JP, Joshi AK (2012) QTL mapping of terminal heat tolerance in hexaploid wheat (*T. aestivum* L.). Theor Appl Genet 125:561–575.
Passioura JB (2012) Phenotyping for drought tolerance in grain crops: when is it useful to breeders? Funct Plant Biol 39:851–859.
Pekosz A, He B, Lamb RA (1999) Reverse genetics of negative-strand RNA viruses: closing the circle. PNAS 96:8804–8806.
Peltzer D, Dreyer E, Polle A (2002) Temperature dependencies of antioxidative enzymes in two contrasting species. Plant Physiol Biochem 40:141–150.
Piramila BHM, Prabha AL, Nandagopalan V, Stanley AL (2012) Effect of heat treatment on germination, seedling growth and some biochemical parameters of dry seeds of black gram. Int J Pharm Phytopharmacol Res 1:194–202.
Prasad SR, Bagali PG, Hittalmani S, Shashidhar HE (2000) Molecular mapping of quantitative trait loci associated with seedling tolerance to salt stress in rice (*Oryza sativa* L.). Curr Sci 162–164.
Price AH, Townend J, Jones MP, Audebert A, Courtois B (2002) Mapping QTLs associated with drought avoidance in upland rice grown in the Philippines and West Africa. Plant Mol Biol 48:683–695.
Qian Q, Zeng D, He P, Zheng X, Chen Y, Zhu L (2000) QTL analysis of the rice seedling cold tolerance in a double haploid population derived from anther culture of a hybrid between indica and japonica rice. Chinese Sci Bull 45:448–453.
Quarrie SA, Pekic Quarrie S, Radosevic R, Rancic D, Kaminska A, Barnes JD, Leverington M, Ceoloni C, Dodig D (2006) Dissecting a wheat QTL for yield present in a range of environments: from the QTL to candidate genes. J Exp Bot 57:2627–2637.
Rehman M, Fahad S, Saleem MH, Hafeez M, Habib ur Rahman M, Liu F, Deng G (2020) Red light optimized physiological traits and enhanced the growth of ramie (*Boehmeria nivea* L.). Photosynthetica 58:922–931.
Reinheimer JL, Barr AR, Eglinton JK (2004) QTL mapping of chromosomal regions conferring reproductive frost tolerance in barley (*Hordeum vulgare* L.). Theor Appl Genet 109:1267–1274.
Ren ZH, Gao JP, Li LG, Cai XL, Huang W, Chao DY, Zhu MZ, Wang ZY, Luan S, Lin HX (2005) A rice quantitative trait locus for salt tolerance encodes a sodium transporter. Nat Genet 37:1141.
Ren Z, Zheng Z, Chinnusamy V, Zhu J, Cui X, Iida K, Zhu JK (2010) RAS1, a quantitative trait locus for salt tolerance and ABA sensitivity in Arabidopsis. Proc Natl Acad Sci 107:5669–5674.
Reynolds MP, Dreccer F, Trethowan R (2007) Drought adaptive traits derived from wheat wild relatives and landraces. J Exp Biol 58:177–186.
Richards RA (1996) Defining selection criteria to improve yield under drought. Plant Growth Regul 20:157–166.
Rizhsky L, Liang H, Shuman J, Shulaev V, Davletova S, Mittler R (2004) When defense pathways collide. The response of Arabidopsis to a combination of drought and heat stress. Plant Physiol 134:1683–1696.
Rohde R, Muller RA, Jacobsen R, Muller E, Perlmutter S, Rosenfeld A, Wurtele J, Groom D, Wickham C (2013) A new estimate of the average earth surface land temperature spanning 1753 to 2011. Geoinfor Geostat Overview 7:2.
Roostaei M, Mohammadi SA, Amrid A, Majidi E, Nachit M, Haghparast R (2011) Chlorophyll fluorescence parameters and drought tolerance in a mapping population of winter bread wheat in the highlands of Iran. Russ J Plant Physiol 58:351–358.
Sabouri H, Rezai AM, Moumeni A, Kavousi A, Katouzi M, Sabouri A (2009) QTLs mapping of physiological traits related to salt tolerance in young rice seedlings. Biol Plant 53:657–662.
Saeed M, Guo WZ, Ullah I, Tabbasam N, Zafar Y, Mehboob R, Zhang TZ (2011) QTL mapping for

physiology, yield and plant architecture traits in cotton (Gossypium hirsutum L.) grown under well–watered versus water–stress conditions. Electron J Biotechnol 14:1–13.
Saito K, Miura K, Nagano K, Hayano-Saito Y, Araki H, Kato A (2001) Identification of two closely linked quantitative trait loci for cold tolerance on chromosome 4 of rice and their association with anther length. Theor Appl Genet 103:862–868.
Saito K, Miura K, Nagano K, Hayano-Saito Y, Saito A, Araki H, Kato A (1995) Chromosomal location of quantitative trait loci for cool tolerance at the booting stage in rice variety Norin-PL8. Breed Sci 45:337–340.
Sajjad H, Muhammad M, Ashfaq A, Waseem A, Hafiz MH, Mazhar A, Nasir M, Asad A, Hafiz UF, Syeda RS, Fahad S, Depeng W, Wajid N (2019) Using GIS tools to detect the land use/land cover changes during forty years in Lodhran district of Pakistan. Environ Sci Pollut Res 27:39676–39692. https://doi.org/10.1007/s11356-019-06072-3.
Sakai A, Larcher W (1987) Frost survival of plants: response and adaptation to freezing stress. In Billings WD, Golley F, Lange OL, Olson JS, Remmert H (eds) Ecological studies. Springer, Berlin, New York.
Saleem MH, Fahad S, Adnan M, Mohsin A, Muhammad SR, Muhammad K, Qurban A, Inas AH, Parashuram B, Mubassir A, Reem MH (2020a) Foliar application of gibberellic acid endorsed phytoextraction of copper and alleviates oxidative stress in jute (*Corchorus capsularis* L.) plant grown in highly copper-contaminated soil of China. Environ Sci Pollution Res 27:37121–37133. https://doi.org/10.1007/s11356-020-09764-3.
Saleem MH, Fahad S, Shahid UK, Mairaj D, Abid U, Ayman ELS, Akbar H, Analía L, Lijun L (2020c) Copper-induced oxidative stress, initiation of antioxidants and phytoremediation potential of flax (Linum usitatissimum L.) seedlings grown under the mixing of two different soils of China. Environ Sci Poll Res 27:5211–5221. https://doi.org/10.1007/s11356-019-07264-7.
Saleem MH, Rehman M, Fahad S, Tung SA, Iqbal N, Hassan A, Ayub A, Wahid MA, Shaukat S, Liu L, Deng G (2020b) Leaf gas exchange, oxidative stress, and physiological attributes of rapeseed (*Brassica napus* L.) grown under different light-emitting diodes. Photosynthetica 58:836–845.
Sandhu N, Singh A, Dixit S, Cruz MTS, Maturan PC, Jain RK, Kumar A (2014) Identification and mapping of stable QTL with main and epistasis effect on rice grain yield under upland drought stress. BMC Genet 15:63.
Sanghera GS, Wani SH, Hussain W, Singh NB (2011) Engineering cold stress tolerance in crop plants. Curr Genom 12:30–43.
Sato S, Kamiyama M, Iwata T, Makita N, Furukawa H, Ikeda H (2006) Moderate increase of mean daily temperature adversely affects fruit set of *Lycopersicon esculentum* by disrupting specific physiological processes in male reproductive development. Ann Bot 97:731–738.
Saud S, Chen Y, Fahad S, Hussain S, Na L, Xin L, Alhussien SA (2016) Silicate application increases the photosynthesis and its associated metabolic activities in Kentucky bluegrass under drought stress and post-drought recovery. Environ Sci Pollut Res 23:17647–17655. https://doi.org/10.1007/s11356-016-6957-x.
Saud S, Chen Y, Long B, Fahad S, Sadiq A (2013) The different impact on the growth of cool season turf grass under the various conditions on salinity and drought stress. Int J Agric Sci Res 3:77–84.
Saud S, Fahad S, Cui G, Chen Y, Anwar S (2020) Determining nitrogen isotopes discrimination under drought stress on enzymatic activities, nitrogen isotope abundance and water contents of Kentucky bluegrass. Sci Rep 10:6415. https://doi.org/10.1038/s41598-020-63548-w.
Saud S, Fahad S, Yajun C, Ihsan MZ, Hammad HM, Nasim W, Amanullah Jr, Arif M, Alharby H (2017) Effects of nitrogen supply on water stress and recovery mechanisms in kentucky bluegrass plants. Front. Plant Sci 8:983. https://doi.org/10.3389/fpls.2017.00983.
Saud S, Li X, Chen Y, Zhang L, Fahad S, Hussain S, Sadiq A, Chen Y (2014) Silicon application increases drought tolerance of Kentucky bluegrass by improving plant water relations and morph physiological functions. SciWorld J 2014:1–10. https://doi.org/10.1155/2014/368694.
Savin R, Stone PJ, Nicolas ME, Wardlaw IF (1997) Effects of heat stress and moderately high temperature on grain growth and malting quality of barley. Aust J Agric Res 48:615–624.
Sayed MA, Schumann H, Pillen K, Naz AA, Léon J (2012) AB–QTL analysis reveals new alleles associated to proline accumulation and leaf wilting under drought stress conditions in barley (*Hordeum vulgare* L.). BMC Genet 13:61.
Shafi MI, Adnan M, Fahad S, Fazli W, Ahsan K, Zhen Y, Subhan D, Zafar-ul-Hye M, Martin B, Rahul D

(2020) Application of single superphosphate with humic acid improves the growth, yield and phosphorus uptake of wheat (*Triticum aestivum* L.) in calcareous soil. Agron 10:1224. https://doi.org/10.3390/agronomy10091224.

Shah F, Lixiao N, Kehui C, Tariq S, Wei W, Chang C, Liyang Z, Farhan A, Fahad S, Huang J (2013) Rice grain yield and component responses to near 2°C of warming. Field Crop Res 157:98–110.

Shahzad S, Chaudhry UK, Anwar B, Saboor A, Yousaf MF, Saeed F, Yaqoob S (2016) Drought stress effect on morphological and physiological characteristics of different varieties of annual verbena (Verbena hybrid). J Biod Environ Sci 9:32–46.

Shahzad S, Khan MY, Zahir ZA, Asghar HN, Chaudhry UK (2017) Comparative effectiveness of different carriers to improve the efficacy of bacterial consortium for enhancing wheat production under salt affected field conditions. Pak J Bot 49: 1523–1530.

Shalata A, Mittova V, Volokita M, Guy M, Tal M (2001) Response of the cultivated tomato and its wild salt tolerant relative Lycopersiconpennellii to salt dependent oxidative stress: the root antioxidative system. Physiol Plant 112:487–494.

Sharma DK, Torp AM, Rosenqvist E, Ottosen CO, Andersen SB (2017) QTLs and potential candidate genes for heat stress tolerance identified from the mapping populations specifically segregating for Fv/Fm in wheat. Front Plant Sci 8:1668.

Shinada H, Iwata N, Sato T, Fujino K (2014) QTL pyramiding for improving of cold tolerance at fertilization stage in rice. Breed Sci 63:483–488.

Siahsar BA, Narouei M (2010) Mapping QTLs of physiological traits associated with salt tolerance in "Steptoe"×"Morex" doubled haploid lines of barley at seedling stage. J Food Agric Environ 7:751–759.

Soliman WS, Fujimori M, Tase K, Sugiyama SI (2011) Oxidative stress and physiological damage under prolonged heat stress in C3 grass *Lolium perenne*. Grassl Sci 57:101–106.

Solis J, Gutierrez A, Mangu V, Sanchez E, Bedre R, Linscombe S, Baisakh N (2018) Genetic mapping of quantitative trait loci for grain yield under drought in rice under controlled greenhouse conditions. Front Chem 5:129.

Soller M, Brody T, Genizi A (1976) On the power of experimental designs for the detection of linkage between marker loci and quantitative loci in crosses between inbred lines. Theor Appl Genetic 47:35–39.

Steponkus PL, Uemura M, Joseph RA, Gilmour SJ, Thomashow MF (1998) Mode of action of the COR15a gene on the freezing tolerance of Arabidopsis thaliana. Proc Natl Acad Sci USA 95:14570–14575.

Stuber CW, Lincoln SE, Wolff DW, Helentjaris T, Lander ES (1992) Identification of genetic factors contributing to heterosis in a hybrid from two elite maize inbred lines using molecular markers. Genetic 132:823–839.

Subhan D, Zafar-ul-Hye M, Fahad S, Saud S, Martin B, Tereza H, Rahul D (2020) Drought stress alleviation by ACC deaminase producing *Achromobacter xylosoxidans* and *Enterobacter cloacae*, with and without timber waste biochar in maize. Sustain 12:6286. https://doi.org/10.3390/su12156286.

Suh JP, Jeung, JU, Lee JI, Choi YH, Yea JD, Virk PS, Mackill DJ, Jena KK (2010) Identification and analysis of QTLs controlling cold tolerance at the reproductive stage and validation of effective QTLs in cold-tolerant genotypes of rice (*Oryza sativa* L.). Theor Appl Genet 120:985–995.

Suzuki M, Latshaw S, Sato Y, Settles AM, Koch KE, Hannah LC, Kojima M, Sakakibara H, McCarty DR (2008) The maize Viviparous8 locus, encoding a putative ALTERED MERISTEM PROGRAM1-like peptidase, regulates abscisic acid accumulation and coordinates embryo and endosperm development. Plant Physiol 146:1193–1206.

Suzuki N, Rivero RM, Shulaev V, Blumwald E, Mittler R (2014) Abiotic and biotic stress combinations. New Phytol 203:32–43.

Tabata M, Hirabayashi H, Takeuchi Y, Ando I, Iida Y, Ohsawa R (2007) Mapping of quantitative trait loci for the occurrence of white-back kernels associated with high temperatures during the ripening period of rice (*Oryza sativa* L.). Breed Sci 57:47–52.

Tahmasebi S, Heidari B, Pakniyat H, McIntyre CL (2016) Mapping QTLs associated with agronomic and physiological traits under terminal drought and heat stress conditions in wheat (*Triticum aestivum* L.). Genome 60:26–45.

Takehisa H, Ueda T, Fukuta Y, Obara M, Abe T, Yano M, Yamaya T, Kameya T, Higashitani A, Sato T

(2006) Epistatic interaction of QTLs controlling leaf bronzing in rice (*Oryza sativa* L.) grown in a saline paddy field. Breed Sci 56:287–293.

Takeuchi Y, Hayasaka H, Chiba B, Tanaka I, Shimano T, Yamagishi M, Nagano K, Sasaki T, Yano M (2001) Mapping quantitative trait loci controlling cool-temperature tolerance at booting stage in temperate japonica rice. Breed Sci 51:191–197.

Talukder SK, Babar MA, Vijayalakshmi K, Poland J, Prasad PVV, Bowden R, Fritz A (2014) Mapping QTL for the traits associated with heat tolerance in wheat (*Triticum aestivum* L.). BMC Genet 15:97.

Tardieu F (2005) Plant tolerance to water deficit: physical limits and possibilities for progress. Comptes Rendus Geosci 337:57–67.

Tariq M, Ahmad S, Fahad S, Abbas G, Hussain S, Fatima Z, Nasim W, Mubeen M, ur Rehman MH, Khan MA, Adnan M (2018) The impact of climate warming and crop management on phenology of sunflower-based cropping systems in Punjab, Pakistan. Agric For Met 256:270–282.

Thomson MJ, de Ocampo M, Egdane J, Rahman MA, Sajise AG, Adorada DL, Tumimbang-Raiz E, Blumwald E, Seraj ZI, Singh RK, Gregorio GB, Ismail AM (2010) Characterizing the Saltol quantitative trait locus for salinity tolerance in rice. Rice 3:148–160.

Tripathy JN, Zhang J, Robin S, Nguyen TT, Nguyen HT (2000) QTLs for cell-membrane stability mapped in rice (*Oryza sativa* L.) under drought stress. Theor Appl Genet 100:1197–1202.

Tuberosa R, Salvi S (2006) Genomics-based approaches to improve drought tolerance of crops. Trend Plant Sci 11:405–412.

Tuna AL, Kaya C, Ashraf M, Altunlu H, Yokas I, Yagmur B (2007) The effects of calcium sulphate on growth, membrane stability and nutrient uptake of tomato plants grown under salt stress. Environ Exp Bot 59:173–178.

Uga Y, Kitomi Y, Yamamoto E, Kanno N, Kawai S, Mizubayashi T, Fukuoka S (2015) A QTL for root growth angle on rice chromosome 7 is involved in the genetic pathway of deeper rooting 1. Rice 8:8.

Uz Zaman Q, Zubair A, Muhammad Y, Muhammad ZI, Abdul K, Fahad S, Safder B, Ramzani PMA, Muhammad N (2017) Zinc biofortification in rice: leveraging agriculture to moderate hidden hunger in developing countries. Arch Agron Soil Sci 64:147–161. https://doi.org/10.1080/03650340.2017.1338343.

Vijayalakshmi K, Fritz AK, Paulsen GM, Bai G, Pandravada S, Gill BS (2010) Modeling and mapping QTL for senescence-related traits in winter wheat under high temperature. Mol Breed 26:163–175.

Wahid F, Fahad S, Subhan D, Adnan M, Zhen Y, Saud S, Manzer HS, Martin B, Tereza H, Rahul D (2020) Sustainable management with mycorrhizae and phosphate solubilizing bacteria for enhanced phosphorus uptake in calcareous soils. Agri 10:334. https://doi.org/10.3390/agriculture10080334.

Wahid A, Gelani S, Ashraf M, Foolad MR (2007) Heat tolerance in plants: an overview. Environ Exp Bot 61:199–223.

Wainaina CM, Makihara D, Nakamura M, Ikeda A, Suzuki T, Mizukami Y, Nonoyama T, Doi K, Kikuta M, Samejima H, Menge DM (2018) Identification and validation of QTLs for cold tolerance at the booting stage and other agronomic traits in a rice cross of a Japanese tolerant variety, Hananomai, and a NERICA parent, WAB56-104. Plant Prod Sci 21:132–143.

Wajid N, Ashfaq A, Asad A, Muhammad T, Muhammad A, Muhammad S, Khawar J, Ghulam MS, Syeda RS, Hafiz MH, Muhammad IAR, Muhammad ZH, Habib ur Rahman M, Veysel T, Fahad S, Suad S, Aziz K, Shahzad A (2017) Radiation efficiency and nitrogen fertilizer impacts on sunflower crop in contrasting environments of Punjab. Pak Environ Sci Pollut Res 25:1822–1836. https://doi.org/10.1007/s11356-017-0592-z.

Wang SCJB (2007) Windows QTL cartographer 2.5. http://statgen.ncsu.edu/qtlcart/WQTLCart.html.

Wang X, Cai J, Liu F, Jin M, Yu H, Jiang D, Wollenweber B, Dai T, Cao W (2012) Pre-anthesis high temperature acclimation alleviates the negative effects of post-anthesis heat stress on stem stored carbohydrates remobilization and grain starch accumulation in wheat. J Cereal Sci 55:331–336.

Wang H, Chen G, Zhang H, Liu B, Yang Y, Qin L, Chen E, Guan Y (2014) Identification of QTLs for salt tolerance at germination and seedling stage of *Sorghum bicolor* L. Moench. Euphytica 196:117–127.

Warland JS, McDonald MR, McKeown AM (2006) Annual yields of five crops in the family Brassicacae in southern Ontario in relation to weather and climate. Can J Plant Sci 86:1209–1215.

Wasson AP, Richards RA, Chatrath R, Misra SC, Sai Prasad SV, Rebetzke GJ, Kirkegaard JA, Christopher J, Watt M (2012) Traits and selection strategies to improve root systems and water uptake in water–limited wheat crop. J Exp Bot 63:3485–3498.

Welcker C, Boussuge B, Bencivenni C, Ribaut JM, Tardieu F (2006) Are source and sink strengths genetically linked in maize plants subjected to water deficit? A QTL study of the responses of leaf growth and of anthesis-silking interval to water deficit. J Exp Bot 58:339–349.

Wilcox PL, Carson SD, Richardson TE, Ball RD, Horgan GP, Carter P (2001) Cost-benefit analysis of marker based selection in seed orchard production populations of *Pinus radiata*. Can J Plant Sci 31:2213–2224.

Wu C, Kehui C, She T, Ganghua L, Shaohua W, Fahad S, Lixiao N, Jianliang H, Shaobing P, Yanfeng D (2020) Intensified pollination and fertilization ameliorate heat injury in rice (*Oryza sativa* L.) during the flowering stage. Field Crops Res 252:107795.

Wu C, Tang S, Li G, Wang S, Fahad S, Ding Y (2019) Roles of phytohormone changes in the grain yield of rice plants exposed to heat: a review. PeerJ 7:e7792. https://doi.org/10.7717/peerj.7792.

Xie F, Guo L, Ren G, Hu P, Wang F, Xu J, Li X, Qiu F, dela Paz MA (2012) Genetic diversity and structure of indica rice varieties from two heterotic pools of southern China and IRRI. Plant Genet Resour 10:186–193.

Xie G, Kato H, Sasaki K, Imai R (2009) A cold-induced thioredoxin h of rice, OsTrx23, negatively regulates kinase activities of OsMPK3 and OsMPK6 in vitro. FEBS Lett 583:2734–2738.

Xiong L, Zhu JK (2002) Salt tolerance. Arabidopsis Book 1:e00048.

Xu YF, An DG, Liu DC, Zhang AM, Xu HX, Li B (2012) Mapping QTLs with epistatic effects and QTL× treatment interactions for salt tolerance at seedling stage of wheat. Euphytica 186:233–245.

Xu JL, Lafitte HR, Gao YM, Fu BY, Torres R, Li ZK (2005) QTLs for drought escape and tolerance identified in a set of random introgression lines of rice. Theor Appl Genet 111:1642–1650.

Xu Y, Zhang H, Hu J, Wang X, Huang M, Wang H (2018) Further QTL mapping for yield component traits using introgression lines in rice (*Oryza sativa* L.) under drought field environments. Euphytica 214:33.

Xu Z, Zhou G, Shimizu H (2010) Plant responses to drought and rewatering. Plant Signal Behav 5:649–654.

Xue D, Huang Y, Zhang X, Wei K, Westcott S, Li C, Chen M, Zhang G, Lance R (2009) Identification of QTLs associated with salinity tolerance at late growth stage in barley. Euphytica 169:187–196.

Yang DL, Jing RL, Chang XP, Li W (2007) Quantitative trait loci mapping for chlorophyll fluorescence and associated traits in wheat (*Triticum aestivum*). J Integ Plant Biol 49:646–654.

Yang Z, Zhang Z, Zhang T, Fahad S, Cui K, Nie L, Peng S, Huang J (2017) The effect of season-long temperature increases on rice cultivars grown in the central and southern regions of China. Front Plant Sci 8:1908. https://doi.org/10.3389/fpls.2017.01908.

Ye C, Argayoso MA, Redoña ED, Sierra SN, Laza MA, Dilla CJ, Mo Y, Thomson MJ, Chin J, Delaviña CB, Diaz GQ (2012) Mapping QTL for heat tolerance at flowering stage in rice using SNP markers. Plant Breed 131:33–41.

Zafar-ul-Hye M, Muhammad N, Subhan D, Fahad S, Rahul D, Mazhar A, Ashfaq AR, Martin B, Ji˘rí H, Zahid HT, Muhammad N (2020a) Alleviation of cadmium adverse effects by improving nutrients uptake in bitter gourd through cadmium tolerant rhizobacteria. Environ 7:54. https://doi.org/10.3390/environments7080054.

Zafar-ul-Hye M, Tahzeeb-ul-Hassan M, Muhammad A, Fahad S,, Martin B, Tereza D, Rahul D, Subhan D (2020b) Potential role of compost mixed biochar with rhizobacteria in mitigating lead toxicity in spinach. Scientific Rep 10:12159. https://doi.org/10.1038/s41598-020-69183-9.

Zahida Z, Hafiz FB, Zulfiqar AS, Ghulam MS, Fahad S, Muhammad RA, Hafiz MH, Wajid N, Muhammad S (2017) Effect of water management and silicon on germination, growth, phosphorus and arsenic uptake in rice. Ecotoxicol Environ Saf 144:11–18.

Zhang M, Cao Y, Wang Z, Wang ZQ, Shi J, Liang X, Song W, Chen Q, Lai J, Jiang C (2018) A retrotransposon in an HKT1 family sodium transporter causes variation of leaf Na+ exclusion and salt tolerance in maize. New Phytol 217:1161–1176.

Zhang T, Yang L, Jang KF, Huang M, Sun Q, Chen WF, Zheng JK (2008) QTL mapping for teat tolerance of the tassel period of rice. Mol Plant Breed 6:867–873.

Zhi-Hong Z, Li S, Wei L, Wei C, Ying-Guo Z (2005) A major QTL conferring cold tolerance at the early seedling stage using recombinant inbred lines of rice (*Oryza sativa* L.). Plant Sci 168:527–534.

Zhou GF, Johnson P, Ryan PR, Delhaize E, Zhou MX (2012) Quantitative trait loci for salinity tolerance in barley (Hordeum vulgare L.). Mol Breed 29: 427–439.

Zhou J, Wang J, Yu JQ, Chen Z (2014) Role and regulation of autophagy in heat stress responses of tomato plants. Front Plant Sci 5:174.

10

Adapting Climate Change and Variability for Current and Future Production Systems of Maize

Dr. Ishfaq Ahmad[1], Engr. Dr. Muhammad Mohsin Waqas[2], Miss Farhana Gul[3], and Dr. Muhammad Mumtaz[4]

[1]Centre for Climate Research and Development (CCRD), COMSATS University Islamabad, Pakistan

[2]Department of Agricultural Engineering, Khwaja Fareed University of Engineering and Information Technology, Rahim Yar Khan, Pakistan

[3]Department of Agriculture, University of Swabi, Pakistan

[4]Department of Management Sciences, University of Haripur, Pakistan

10.1 Introduction

Maize is important cereal crop with highest productivity after wheat and rice (Milind and Isha 2013). It is a versatile crop that is grown in 166 countries across the globe including tropical, subtropical, and temperate regions (Ranum et al. 2014). The production area of maize continues to increase due to improved technologies and increased demand (Ahmad et al. 2018a). However, the impact of climate change has shown to negatively affect the maize (Lobell et al. 2013; Westengen and Brysting 2014; Ahmed et al. 2018). Substantial changes in temperatures and precipitation cause the reduction in crop yield (Bannayan et al. 2004; Abdrabbo et al. 2013). Thus, the adaptations are the key strategies to offset negative impacts of climate change.

Future projections showed that temperature is expected to rise by 2.5 °C up to 2050 (IPCC 2013a). Intense and uncertain precipitation will cause the destruction of ecosystem due to floods and water scarcity due to complete absence of rainfall (Richardson et al. 2018). Extreme weather events such as floods, droughts, thunderstorms, and cyclones are projected to rise in future, which will negatively affect the yield (Lesk et al. 2016). Rise in temperature during crop growing season accelerates the growth rate that shortens the length of growth period (Asseng et al. 2011). High temperature reduces the yield due to shortening of growth period, less interception of light, and less grain number per unit area (Devasirvatham et al. 2012).

Climate change is affecting the biophysical, socioeconomic, and agriculture sector (Fahad and Bano 2012; Lobell et al. 2012; Fahad et al. 2013; Asseng et al. 2014; Fahad et al. 2014a,b; Fahad et al. 2015a,b; Fahad et al. 2016a,b,c,d; Fahad et al. 2017; Ahmad et al. 2018b; Fahad et al. 2018; Rahman et al. 2018; Fahad et al. 2019a,b; Hesham and Fahad 2020). There is need to take action for mitigating the negative effects of climate change (Smit and Wandel 2006; Hoffmann and Sgro 2011). The changes in agricultural practices are the effective adaptation strategies (Burke and Emerick 2016). However, the current production system will change in future due to shortage of water, soil degradation, urbanization, and the economic development (Kriegler et al. 2012). Thus, the Representative Agricultural Pathways (RAPs) are developed, which are the socioeconomic scenarios and can be used to represent the future agriculture system (Antle et al. 2017; Riahi et al. 2017). The RAPs are the future biophysical and socioeconomics pathways which are useful to develop the adaptation strategies under changing climate.

Various approaches are used widely for adaptation of the current production system such as modification in management practices, the efficient use of fertilizers, and irrigations (Mózner et al. 2012). However, the adaptation strategies such as RAPs with development of climate-resilient cultivars are development of future production system (Sutcliffe et al. 2016). The adaptability approaches such as altering the planting dates and optimum application of irrigation water are among others (Ahmad et al. 2019). The fertilizers also provide a key role in adapting to the adverse effects of changing climate. The Information Technological (IT) techniques such as crop modelling, remote sensing, and robotics in agricultures are considered the novel approaches to combat the impact of climate change (Ospina and Heeks 2010).

10.2 Climate-Change Projections

Climate projections are sets of simulations by climate models using different CO_2 concentrations and related forcing functions (Collins et al. 2013; Fahad et al. 2016a,b,c,d; Fahad et al. 2018). The current level of CO_2 concentration is near to 400 ppm in 2018 and expected to rise by 794–1142 ppm by 2100 without any abatement scenarios (Collins et al. 2013). The global temperature in high emission scenarios is expected to increase by 2.6 to 4.8 °C throughout the 21th century with increased concentration of CO_2 (IPCC 2013b). This extent of rise in temperature is expected to be more at land surface than at the oceans. The projected rise in temperature will cause more hot events and fewer cold events at daily and seasonal scale (Trenberth 2011).

The precipitation will increase by 1–3% by 1 °C rise in temperature in the 21th century (IPCC 2013b). The spatial pattern of precipitation will change and cause the heavy floods and water scarcity issues (Marengo and Espinoza 2016). The annual evaporation is related to evapotranspiration and evaporation increases due to rise in temperature. Future projections further indicated that water-holding capacity in the air will be increased by 7% due to rise in 1 °C (IPCC 2013b). The dry air takes more water and causes high and intense rainfall.

The projected changes in climatic parameters affected the maize growth and productivity. However, the response of increased concentration of CO_2 has minimal effect due to C4 plant (Hatfield et al. 2011). The leaf photosynthesis increased by 3%, grain yield and biomass by 4% due to doubling of the CO_2 concentration, whereas the stomatal conductance decreased by 34%. The response to physiological activity was less in C4 plants as compared to C3, due to rise in CO_2 concentration (Leakey et al. 2006). The rise in temperature in future will decrease the yield by 27% till the middle of century (Ahmed et al. 2018).

10.3 Representative Agricultural Pathways (RAPs)

The RAPs are the plausible pathways and future projections for socioeconomic and biophysical conditions that are used for climate change impact assessment. The RAPs are useful scenarios that can be used by the scientist to estimate the range of uncertainty in climate impacts and adaptation strategies. The RAPs are needed to develop the protocol-based approach to assess the vulnerabilities in climate impacts and in production system such as land degradation and shortage of water. The use of RAPs with Representative Concentrations pathways (RCPs) provides the detailed insight and magnitude of climate impacts.

The RAPs are developed for regional and local through historical data, global economic projections, and series of consultative sessions with the scientist of various disciplines such as soil, water and crop management, pathologist, and plant breeder (Parry et al. 2004). The RAPs are then finalized and quantified for regional integrated assessment. The knowledge experts are agreed that current production system will be changed in future. For instance, there will be shortage of water due to climate change, and intense and heavy rainfall will cause the soil degradation and nutrient loss. Thus, the use of RAPs in climate analyses is useful for accurate estimation of climate impacts and development of adaptation strategies (Kriegler et al. 2012).

10.4 Impacts of Climate Change and Variability on Maize Productivity

10.4.1 Impacts of Climate Change and Variability on Phenology

The development of leaves or collar leaf at vegetative stage of crop and grains at reproductive stages defines the phenology of maize (Darby and Lauer 2004). The critical growth stages of maize crop are described by Abendroth et al. (2011). Temperature is a key variable for development of phenology, and each plant has a specific range of lower, upper, and optimum temperature for successful growth (Gupta and Gyanpur 2012). The range of temperature for vegetative stage of maize is 8 to 38 °C with an optimum of 34 °C, and for reproductive stage, the range is 8–30 °C (Badu-Apraku et al. 1983; Kiniry and Bonhomme 1991). The lower temperature limits the growth of crop, and in crop model, the lower temperature is assumed as 10 °C (Muchow et al. 1990). The pollens are very sensitive to variation in temperature, as the temperature higher than 35 °C affects the pollen viability (Sánchez et al. 2014). Higher temperature increased the water pressure deficit, resulting in disruption of pollination process in maize in the form of movement of tassel to silk (Cakir 2004).

The increase in temperature accelerates the growth rate and decreases the length of growing period due to rapid accumulation of degree days (Lobell et al. 2013; Asseng et al. 2014). Ahmed et al. (2018) found that as temperature is increased by 2 °C, the days to anthesis and maturity decreased by 10 days; however, when temperature is increased from 0 to 8 °C, the days to anthesis decreased by 24 days and maturity days by 29 days (Figure 10.1). They further found that lower temperature extends the growing season length due to more accumulation of degree days and interception of light (Figure 10.1).

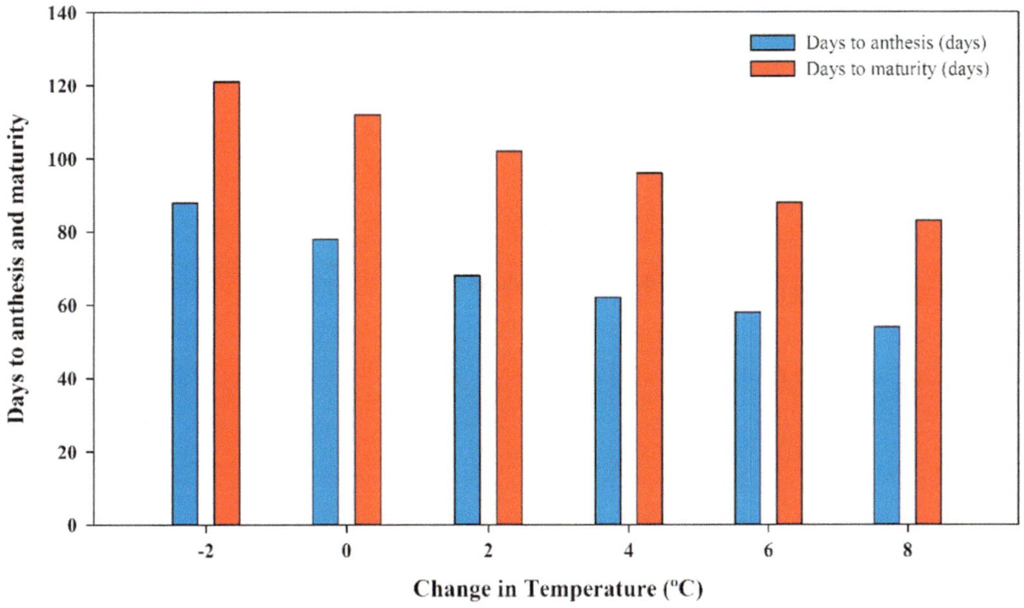

FIGURE 10.1 Effect of Change in Temperature on Phenology of Maize.

10.4.2 Impacts of Climate Change and Variability on Yield

Extreme weather conditions such as drought, floods, and intense rainfall have adverse effect on crop yield (Lobell and Gourdji 2012). Maize is the third important food crop in the world (Isaacson 2005; Cairns et al. 2013; Ahmad et al. 2018a). However, warming weather conditions have shown to affect the maize yield negatively (Lobell and Burke 2008; Ahmad et al. 2018b). It is projected that maize yield will decline by 14% at mid-century (2040–2069) and 33% at the end of century (2070–2099) in the central Africa under A1B and B1 scenarios (Msowoya et al. 2016). The reduction in maize yield could be due to shortening of growing period with higher temperature (Ringler et al. 2010). The ARP4 assessment report of IPCC showed that 1 °C rise in temperature in European countries has decreased the crop yield by 5%.

The impacts of climate change on maize yield showed that projected rise in temperature in different countries reduced the yield. However, positive impacts of climate change were also noted in Chine (Table 10.2). Higher projected rise in temperature was recorded in Pakistan and Romania, which is causing reduction in yield of 27% in Pakistan and 12% in Romania under mid-century scenarios. The rise in temperature of 3.5 to 4 °C was recorded in France with a huge reduction in yield of 20% from 2016 to 2035 (Table 10.1). A decline in yield of 7.4% was noted in the USA due to rise in temperature of 1.2 to 1.5 °C from 2008 to 2030. However, with the same scenarios, the yield is increased by 10.7% in China (Table 10.1).

10.4.2.1 Effect of Rise in Temperature, CO_2 Concentration, and Precipitation on Maize Yield: A Case Study of Pakistan

A study by Ahmed et al. (2018) indicated that an increase in 2 °C temperature is responsible for decline in the maize yield by 16%. Further, rise in temperature from 2 to 4 °C will likely reduce the yield by 25%. The huge reduction in yield of 86% was recorded when temperature is increased by 8 °C (reference). However, the decrease in temperature of 2 °C showed higher maize yield (Figure 10.2). The higher yield could be due to extended growth period, which could cause more light interception, resulting an increase in yield. The projected decline in yield due to rise in temperature could shorten growth period as shown in Figure 10.1.

The increase in CO_2 concentration also increased the yield. A rise in CO_2 level from 360 to 450 ppm increases the yield by 17%. However, further rise in CO_2 from 450 to 540 ppm increases the yield by 11%. The yield is increased by 4% when CO_2 increased from 630 to 720 ppm (Figure 10.3). The lower increase in yield at high concentration of CO_2 could be fertilization effect (Manderscheid et al. 2014). The increase in precipitation slightly increased the yield. A rise of 25% precipitation increased the yield by 1.5%, while further rise in precipitation from 50 to 70% increased the yield by 0.85% (Figure 10.4). The less increase in yield was due to cultivation of maize in irrigated condition. It is further reported that an increase of precipitation from 125 to 150% caused the reduction in yield of 0.33% (Figure 10.4).

TABLE 10.1
Impacts of Climate Change on Yield of Maize in Different Countries

Sr. No	Country	ΔY (%)	Duration/Century	ΔT (°C)	Reference
	USA	−7.4	2008–2030	1.2–1.5	Li et al. (2011)
	China	+10.7	2008–2030	1.46	Li et al. (2011)
	Brazil	−20	2041–2070	3–3.7	Martins et al. (2019)
	Mexico	−16	2020–2050	2.3–3	
	Argentina	−16.5	2020–2050	3–3.8	Magrin et al. (2007)
	France	−20	2016–2035	3.5–4	Hawkins et al. (2013)
	India	−13	2010–2030	1.8–2	Singh et al. (2010)
	Pakistan	−27	2040–2070	3.4–3.8	Ahmed et al. (2018)
	Romania	−12	2041–2071	3.9–4	Cuculeanu et al. (1999)
	Central Africa	−14	2040–2070	1.3–2.2	Msowoya et al. (2016)

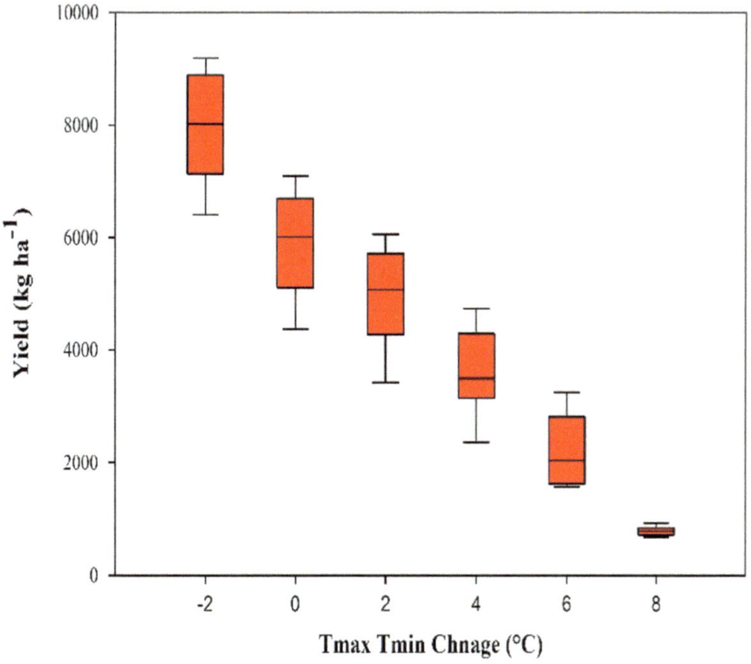

FIGURE 10.2 Effect of Rise in Temperature on Maize Yield.

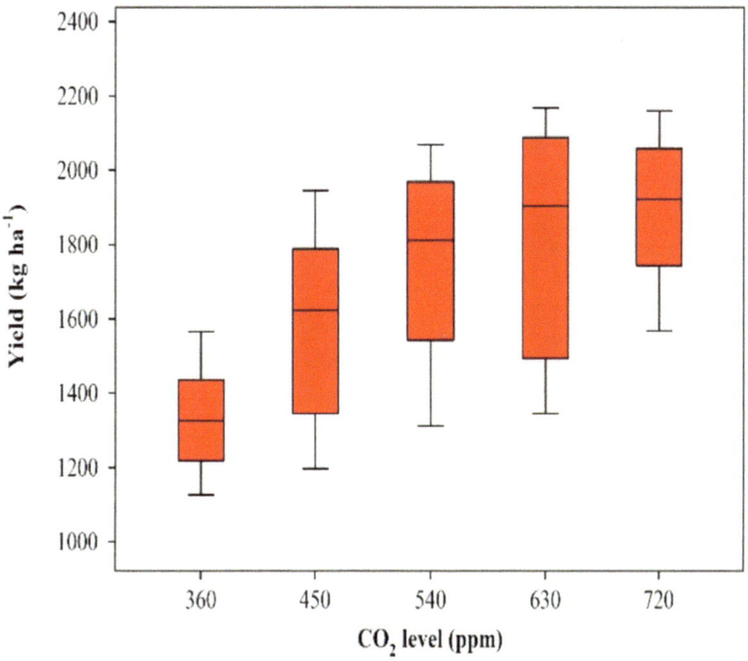

FIGURE 10.3 Effect of Increase in Concentration of CO_2 on Maize Yield.

The reduction in yield was due to leaching of nutrients (NO_3) from soil, leading to decrease in nitrogen uptake from soil and decrease in the yield (Figure 10.5). As precipitation increased the leaching of NO_3, it reduced the nitrogen uptake (Figure 10.5).

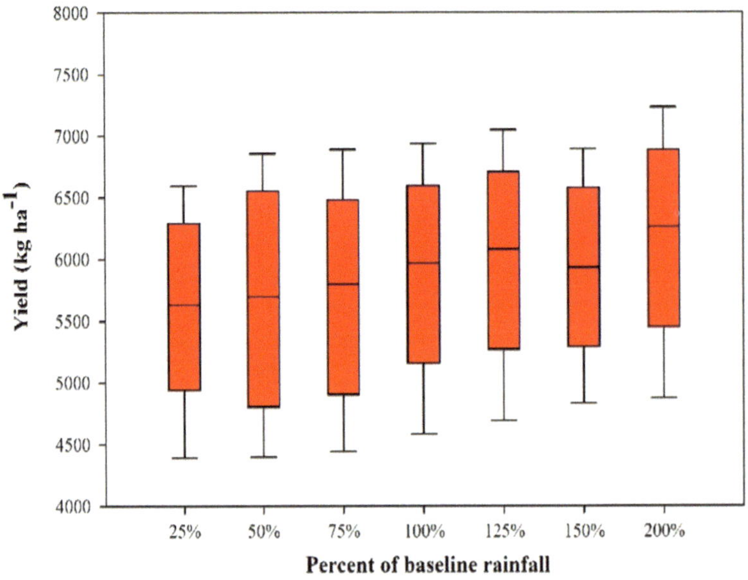

FIGURE 10.4 Effect of Per Cent Increase in Rainfall on Maize Yield.

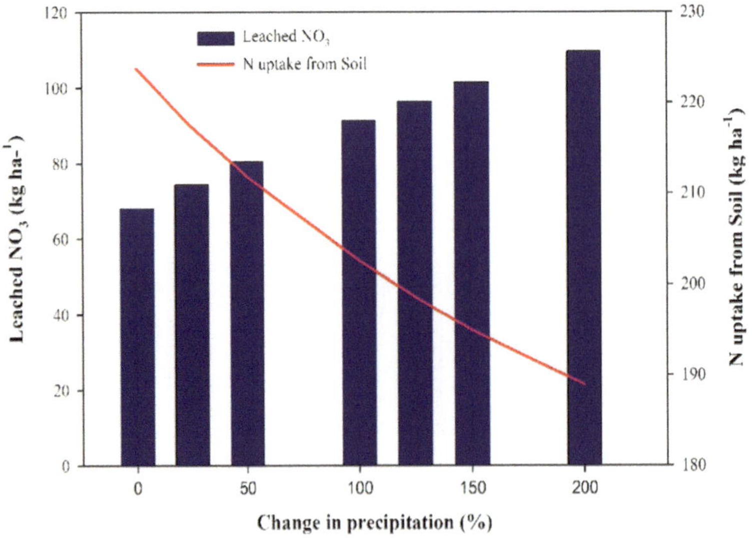

FIGURE 10.5 Effect of Per Cent Increase in Rainfall of Nitrogen Uptake and Leaching.

10.5 Climate-Change Adaptations

10.5.1 Cultural Management

Various studies have been conducted to investigate the suitable adaptation strategies to offset the negative impacts of climate change. Modification in cultural managements as shown in Table 10.2 lessens the danger of climate change and variability and helps in better adaptability of crop. The selection of suitable planting time according to environmental condition plays a key role in maximizing the crop productivity (Olesen et al. 2012). The precise management of fertilizer (Table 10.2) reduces the

TABLE 10.2

Adapting the Impact of Climate Change Through Cultural Management

Sr. No.	Cultural Management	References
1	Modification in planting time	Ahmad et al. (2018b)
2	Precise management of fertilizer	Henderson et al. (2018)
3	Efficient management of irrigation	Deligios et al. (2019)
4	Optimum planting density	Tokatlidis (2013)
5	Minimum tillage	Van Kessel et al. (2013)
6	Plantation of cover crops	Kaye and Quemada (2017)
7	Integrated soil fertility management (composting, manuring)	Battisti et al. (2018)
8	Crop rotation (Soybean–maize)	Duku et al. (2018)

fertilizer use and emission intensity which results in increase in crop productivity. The efficient management of irrigation reduces the water use, run off, and wastage of water and energy which is required for irrigating the crop (Ahmad et al. 2019). The number of plant per unit area is very important for maize. Therefore, maintenance of recommended plant population (8–10 plants m^{-2}) is important for enhancing the crop production.

The use of minimum tillage (Table 10.2) increased the water retention and improved the organic matter/carbon storage, which can reduce the loss of crop due to water stress and energy required for irrigation. The planting of cover crop with maize improves the soil fertility, water infiltration, and organic carbon and decreases the nitrate emission and input cost (Table 10.2). The management of soil fertility through application of animal manure, compost, and green manuring in maize field improved the water-holding capacity and reduces the soil-borne diseases. Manuring also increased the carbon storage and reduced the use of synthetic fertilizer, which could cause the reduction of emission. The rotation of maize with soybean (Table 10.2) prevents the soil erosion that contributes in fixation of CO_2 that reduced the use of nitrogen fertilizer.

10.5.2 Climate-Resilient Cultivars Development

The resilient cultivars should be developed to tolerate the climatic stresses such as heat and drought. The development of climate-resilient cultivars plays a crucial role for copping the climate change and variability (Shiferaw et al. 2014). The heat and drought cultivars could be developed.

10.5.2.1 Heat- and Drought-Tolerant Varieties

The development of heat- and drought-tolerant cultivars increases the production under dry and heat stress period. The resilient cultivars consume less energy for irrigation and synthetic fertilizers and chemical application and are tolerant to moisture stress and climate shock. The stress-tolerant genes could be introduced through breeding. The crop modelling developed virtual climate-resilient cultivars showed an increase in maize yield of 5–25% in various regions of Africa (Setimela et al. 2018). Ahmed et al. (2018) developed heat-tolerant cultivars through modelling approach. The future rise in temperature was used in the model and then genetic coefficients of crop were adjusted to re-gain the phenology of crop, which showed an increase in yield of 6.11%.

10.5.2.2 Early Maturing Varieties

The development of new or cultivation of existing early maturity varieties of maize increased the yield, income, and profit due to less input costs. By utilizing early maturating cultivars, the resources such as fertilizers and soil moisture can be utilized efficiently (Bavec and Bavec 2002). Due to uncertainty in precipitation pattern, it is difficult to cultivate the crop at recommended sowing time. In such conditions, medium or late maturating cultivars delay the sowing to next crop. In this situation, the cultivation of early maturity vacated is a suitable option to avoid the terminal heat stress and drought (Cairns et al. 2013).

10.5.3 ICTs Techniques for Combating Climate Change

Information and communication technologies (ICTs) are the computer-based technologies and software application for management and communicating the information (Magawata 2014). Three actions are taken while using ICT: to record, transform, and communicate the data and knowledge (Meera et al. 2004). The remote sensing is an effective technology for monitoring and assessment of natural disasters such as drought and heat waves. The use of remote sensing is also helpful for reducing the risk through different planning. The Geographic Information Systems (GIS), Global Positioning System (GPS), and Global Navigation Satellite Systems (GNSS) are innovative adaptation approaches for disaster risk management through geospatial mappings and analyses of carbon stock (Yong et al. 2002).

The variable rate technologies (VRTs) for irrigation, seeding, and fertilizer application (Table 10.3) are modern techniques for efficient utilization of resources. The fertilizer and irrigation are applied according to requirement of crop. The senor senses the status of crop and sends a signal to receiver to apply appropriate amount of inputs. The Internet of Things (IoTs) is the new approach in which devices are connected to off/on switch of internet. The sensors are installed at field that collects the real-time data of soil moister, temperature, and satellite imagery for monitoring of crop. The devices are integrated and supply the input according to deficiency (Gondchawar and Kawitkar 2016).

The early warning system should be developed for weather prediction. General circulation model (GCMs) and regional climate models (RCMs) are appropriate techniques to climate change projection in near term, which are helpful for provision of scientific basis for climate change adaptations. Crop models such as Decision Support System for Agrotechnology Transfer (DSSAT) and Model of Agricultural Adaptation to Climatic Variation (MAACV) can be used for climate change adaptation at farm or regional scale. The comprehensive assessment of climate impacts and development of appropriate agronomic adaptations for sustainable crop production are possible through these models (Table 10.3).

10.6 Conclusions

- Climate change is negatively affecting the maize production. The projected rise in temperature reduced the grain yield through accelerating the phenological development.

TABLE 10.3

ICT Techniques as Adaptations to Climate Change

Sr. No.	ICTs Techniques	References
1	Use of remote sensing	Liaghat and Balasundram (2010)
2	Use of GIS, GPs, and GNSS	Djaja et al. (2017)
3	UAVs for precise application of inputs	Zhang and Kovacs (2012)
4	Variable-rate irrigation (VRI)	Hedley et al. (2010)
5	Internet of Things (IoTs)	Stewart et al. (2017)
6	Sensor for optimization of inputs	Srbinovska et al. (2015)
7	Variable rate seeding	Sun and Wang (2009)
8	Weather modelling for weather updates	
9	Crop modelling for input optimization	Braga and Jones (2004)
10	Nitrogen modelling/variable rate technology (VRT)	Bullock et al. (2009)

- The maize-producing countries such as France and Brazil showed a reduction in yield of 20% due to rise in temperature of 3–4 °C.
- The model simulations showed that a rise of 2 °C temperature causes the 10-day early maturity and reduces the yield by 16%. The increased in CO_2 concentration (450–630 ppm) slightly increased the yield, but further increase caused the reduction in yield due to fertilization effect.
- Climate change adaptations: modification in production technology, development of climate resilient, early maturing cultivars, and new innovation ICTs techniques could avoid the extreme impacts of climate change.
- It is imperative to utilize the newly established models and scientific backing while establishing new climate change adaptation policies for agriculture sector (maize), especially at the local scale. However, it is challenging for the developing and most valuable countries battling climate change to incorporate such technological advancement and improved scientific knowledge due to their low adaptive capabilities and financial constraints. On the heels of Paris Agreement and international commitments to curb climate change, international cooperation, transference of innovative techniques, and flow of climate, finances to the most vulnerable regions battling climate change are highly desirable and can be helpful in addressing the global challenge of climate change.

REFERENCES

Abdrabbo MAA, Hashem FA, Elsayed ML, Abul-Soud MA, Farag AA, Hamada MM, Refaie KM (2013) Evaluation of CSM-CERES-maize model for simulating maize production in northern delta of Egypt. Life Sci J 10:3179–3192.

Abendroth LJ, Elmore RW, Boyer MJ, Marley SK (2011) Corn growth and development. PMR 1009. Iowa State University Extension and Outreach, Ames, Iowa.

Ahmad I, Saeed U, Fahad M, Ullah A, Habib ur Rahman M, Ahmad A, Judge J (2018a) Yield forecasting of spring maize using remote sensing and crop modeling in Faisalabad-Punjab Pakistan. Ind Soc Remote Sens 46:1701–1711. https://doi.org/10.1007/s12524-018-0825-8.

Ahmad I, Wajid SA, Ahmad A, Cheema MJM, Judge J (2018b) Assessing the impact of thermo-temporal changes on the productivity of spring maize under semi-arid environment. Int J Agric Biol 20:2203–2210.

Ahmad I, Wajid SA, Ahmad A, Cheema MJM, Judge J (2019) Optimizing irrigation and nitrogen requirements for maize through empirical modeling in semi-arid environment. Environ Sci Pollut Res 26:1227–1237. https://doi.org/10.1007/s11356-018-2772-x.

Ahmed I, Habib ur Rahman M, Ahmed S, Hussain J, Ullah A, Judge J (2018) Assessing the impact of climate variability on maize using simulation modeling under semi-arid environment of Punjab, Pakistan. Environ Sci Pollut Res 25:28413–28430.

Antle JM, Mu JE, Zhang H, Capalbo SM, Diebel PL, Eigenbrode SD, Kruger CE, Stöckle CO, Wulfhorst JD, Abatzoglou JT (2017) Design and use of representative agricultural pathways for integrated assessment of climate change in US Pacific Northwest cereal-based systems. Front Ecol Evol 5:99.

Asseng S, Ewert F, Martre P, et al. (2014) Rising temperatures reduce global wheat production. Nat Clim Change 5:143–147. https://doi.org/10.1038/nclimate2470.

Asseng S, Foster IAN, Turner NC (2011) The impact of temperature variability on wheat yields. Global Change Biol 17:997–1012.

Badu-Apraku B, Hunter RB, Tollenaar M (1983) Effect of temperature during grain filling on whole plant and grain yield in maize (*Zea mays* L.). Can J Plant Sci 63:357–363.

Bannayan M, Hoogenboom G, Crout NMJ (2004) Photothermal impact on maize performance: a simulation approach. Ecol Model 180:277–290. https://doi.org/10.1016/j.ecolmodel.2004.04.022.

Battisti R, Sentelhas PC, Parker PS, et al. (2018) Assessment of crop-management strategies to improve soybean resilience to climate change in Southern Brazil. Crop Pasture Sci 69:154–162.

Bavec F, Bavec M (2002) Effects of plant population on leaf area index, cob characteristics and grain yield of early maturing maize cultivars (FAO 100–400). Eur J Agron 16:151–159.

Braga RP, Jones JW (2004) Using optimization to estimate soil inputs of crop models for use in site-specific management. Trans ASAE 47:1821.

Bullock DS, Ruffo ML, Bullock DG, Bollero GA (2009) The value of variable rate technology: an information-theoretic approach. Am J Agric Econ 91:209–223.

Burke M, Emerick K (2016) Adaptation to climate change: evidence from US agriculture. Am Econ J Econ Policy 8:106–140.

Cairns JE, Hellin J, Sonder K, et al. (2013) Adapting maize production to climate change in sub-Saharan Africa. Food Secur 5:345–360.

Cakir R (2004) Effect of water stress at different development stages on vegetative and reproductive growth of corn. Field Crop Res 89:1–16.

Collins M, Knutti R, Arblaster J, et al. (2013) Long-term climate change: projections, commitments and irreversibility. In: Climate change 2013-The physical science basis: Contribution of Working Group I to the Fifth Assessment Report of the Intergovernmental Panel on Climate Change. Cambridge University Press, Cambridge, pp 1029–1136.

Cuculeanu V, Marica A, Simota C (1999) Climate change impact on agricultural crops and adaptation options in Romania. Clim Res 12:153–160.

Darby H, Lauer J (2004) Plant physiology—critical stages in the life of a corn plant. F Corn (online). http://www.mn.nrcs.usda.gov/technical/ecs/pest/planningaids. Accessed 5 January 2007.

Deligios PA, Chergia AP, Sanna G, et al. (2019) Climate change adaptation and water saving by innovative irrigation management applied on open field globe artichoke. Sci Total Environ 649:461–472.

Devasirvatham V, Tan DKY, Gaur PM, et al. (2012) High temperature tolerance in chickpea and its implications for plant improvement. Crop and Pasture Sci 63:419–428.

Djaja K, Putera R, Rohman AF, et al. (2017) The integration of geography information system (GIS) and global navigation satellite system-real time kinematic (GNSS-RTK) for land use monitoring. Int J Geomate 13:31–34.

Duku C, Zwart SJ, Hein L (2018) Impacts of climate change on cropping patterns in a tropical, sub-humid watershed. PLoS One 13:e0192642.

Fahad S, Adnan M, Hassan S, Saud S, Hussain S, Wu C, Wang D, Hakeem KR, Alharby HF, Turan V, Khan MA, Huang J (2019b) Rice responses and tolerance to high temperature, In: Hasanuzzaman M, Fujita M, Nahar K, Biswas JK (eds) Advances in rice research for abiotic stress tolerance. Woodhead Publ Ltd, Cambridge, England, pp 201–224.

Fahad S, Bajwa AA, Nazir U, Anjum SA, Farooq A, Zohaib A, Sadia S, NasimW, Adkins S, Saud S, Ihsan MZ, Alharby H, Wu C, Wang D, Huang J (2017) Crop production under drought and heat stress: plant responses and management options. Front Plant Sci 8:1147. https://doi.org/10.3389/fpls.2017.01147.

Fahad S, Bano A (2012) Effect of salicylic acid on physiological and biochemical characterization of maize grown in saline area. Pak J Bot 44:1433–1438.

Fahad S, Chen Y, Saud S, Wang K, Xiong D, Chen C,Wu C, Shah F, Nie L, Huang J (2013) Ultraviolet radiation effect on photosynthetic pigments, biochemical attributes, antioxidant enzyme activity and hormonal contents of wheat. J Food Agri Environ 11:1635–1641.

Fahad S, Hussain S, Bano A, Saud S, Hassan S, Shan D, Khan FA, Khan F, Chen Y, Wu C, Tabassum MA, Chun MX, Afzal M, Jan A, Jan MT, Huang J (2014a) Potential role of phytohormones and plant growth-promoting rhizobacteria in abiotic stresses: consequences for changing environment. Environ Sci Pollut Res 22:4907–4921. https://doi.org/10.1007/s11356-014-3754-2.

Fahad S, Hussain S, Matloob A, Khan FA, Khaliq A, Saud S, Hassan S, Shan D, Khan F, Ullah N, Faiq M, Khan MR, Tareen AK, Khan A, Ullah A, Ullah N, Huang J (2014b) Phytohormones and plant responses to salinity stress: A review. Plant Growth Regul 75:391–404. https://doi.org/10.1007/s10725-014-0013-y.

Fahad S, Hussain S, Saud S, Hassan S, Chauhan BS, Khan F et al. (2016a) Responses of rapid viscoanalyzer profile and other rice grain qualities to exogenously applied plant growth regulators under high day and high night temperatures. PLoS One 11:e0159590. https://doi.org/10.1371/journal.pone.0159590.

Fahad S, Hussain S, Saud S, Hassan S, Ihsan Z, Shah AN,Wu C, Yousaf M, Nasim W, Alharby H, Alghabari F, Huang J (2016c) Exogenously applied plant growth regulators enhance the morphophysiological growth and yield of rice under high temperature. Front Plant Sci 7:1250. https://doi.org/10.3389/fpls.2016.01250.

Fahad S, Hussain S, Saud S, Hassan S, Tanveer M, Ihsan MZ, Shah AN, Ullah A, Nasrullah KF, Ullah S, AlharbyH NW, Wu C, Huang J (2016d) A combined application of biochar and phosphorus alleviates

heat-induced adversities on physiological, agronomical and quality attributes of rice. Plant Physiol Biochem 103:191–198.
Fahad S, Hussain S, Saud S, Khan F, Hassan S, Jr A, Nasim W, Arif M, Wang F, Huang J (2016b) Exogenously applied plant growth regulators affect heat-stressed rice pollens. J Agron Crop Sci 202:139–150.
Fahad S, Hussain S, Saud S, Tanveer M, Bajwa AA, Hassan S, Shah AN, Ullah A,Wu C, Khan FA, Shah F, Ullah S, Chen Y, Huang J (2015a) A biochar application protects rice pollen from high-temperature stress. Plant Physiol Biochem 96:281–287.
Fahad S, Muhammad ZI, Abdul K, Ihsanullah D, Saud S, Saleh A, Wajid N, Muhammad A, Imtiaz AK, Chao W, Depeng W, Jianliang H (2018) Consequences of high temperature under changing climate optima for rice pollen characteristics-concepts and perspectives. Arch Agron Soil Sci 64:1479–1488. https://doi.org/10.1080/03650340.2018.1443213.
Fahad S, Nie L, Chen Y, Wu C, Xiong D, Saud S, Hongyan L, Cui K, Huang J (2015b) Crop plant hormones and environmental stress. Sustain Agric Rev 15:371–400.
Fahad S, Rehman A, Shahzad B, Tanveer M, Saud S, Kamran M, Ihtisham M, Khan SU, Turan V, Rahman MHU (2019a) Rice responses and tolerance to metal/metalloid toxicity. In: Hasanuzzaman M, Fujita M, Nahar K, Biswas JK (eds) Advances in rice research for abiotic stress tolerance. Woodhead Publ Ltd, Cambridge, England, pp 299–312.
Gondchawar N, Kawitkar RS (2016) IoT based smart agriculture. Int J Adv Res Comput Commun Eng 5:838–842.
Gupta KS, Gyanpur SRN (2012) Sustainability of scientific maize cultivation practices in Uttar Pradesh, India. J Agric Sci Technol 8:1089–1098.
Hatfield JL, Boote KJ, Kimball BA, et al. (2011) Climate impacts on agriculture: implications for crop production. Agron J 103:351–370.
Hawkins E, Fricker TE, Challinor AJ, et al. (2013) Increasing influence of heat stress on French maize yields from the 1960s to the 2030s. Glob Chang Biol 19:937–947.
Hedley CB, Bradbury S, Ekanayake J, et al. (2010) Spatial irrigation scheduling for variable rate irrigation. In: Proceedings of the New Zealand Grassland Association. New Zealand Grassland Association, pp 97–102.
Henderson B, Cacho O, Thornton P, et al. (2018) The economic potential of residue management and fertilizer use to address climate change impacts on mixed smallholder farmers in Burkina Faso. Agric Syst 167:195–205.
Hesham FA, Fahad S (2020) Melatonin application enhances biochar efficiency for drought tolerance in maize varieties: modifications in physio-biochemical machinery. Agron J 1–22.
Hoffmann AA, Sgro CM (2011) Climate change and evolutionary adaptation. Nature 470:479.
IPCC (2013a) Climate Change 2013: The physical science basis. Contribution of Working Group I to the Fifth Assessment Report of the intergovernmental panel on climate change.
IPCC (2013b) Climate change 2013. The Physical Science Basis. Contribution of Working Group I to the Fifth Assessment Report of the Intergovernmental Panel on Climate Change.
Isaacson C (2005) The change of the staple diet of black South Africans from sorghum to maize (corn) is the cause of the epidemic of squamous carcinoma of the oesophagus. Med Hypotheses 64:658–660.
Kaye JP, Quemada M (2017) Using cover crops to mitigate and adapt to climate change. A review. Agron Sustain Dev 37:4.
Kiniry JR, Bonhomme R (1991) Predicting maize phenology. In: Hodges T (ed) Predicting crop phenology. CRC Press, Boca Raton, FL, USA, pp 115–132.
Kriegler E, O'Neill BC, Hallegatte S, et al. (2012) The need for and use of socio-economic scenarios for climate change analysis: a new approach based on shared socio-economic pathways. Glob Environ Chang 22:807–822.
Leakey ADB, Uribelarrea M, Ainsworth EA, et al. (2006) Photosynthesis, productivity, and yield of maize are not affected by open-air elevation of CO_2 concentration in the absence of drought. Plant Physiol 140:779–790.
Lesk C, Rowhani P, Ramankutty N (2016) Influence of extreme weather disasters on global crop production. Nature 529:84.
Li X, Takahashi T, Suzuki N, Kaiser HM (2011) The impact of climate change on maize yields in the United States and China. Agric Syst 104:348–353.

Liaghat S, Balasundram SK (2010) A review: the role of remote sensing in precision agriculture. Am J Agric Biol Sci 5:50–55.
Lobell DB, Burke MB (2008) Why are agricultural impacts of climate change so uncertain? The importance of temperature relative to precipitation. Environ Res Lett 3:34007.
Lobell DB, Gourdji SM (2012) The influence of climate change on global crop productivity. Plant Physiol 160:1686–1697.
Lobell DB, Hammer GL, McLean G, et al. (2013) The critical role of extreme heat for maize production in the United States. Nat Clim Chang 3:497.
Lobell DB, Sibley A, Ivan Ortiz-Monasterio J (2012) Extreme heat effects on wheat senescence in India. Nat Clim Chang 2:186–189. https://doi.org/10.1038/nclimate1356.
Magawata D (2014) Information and communication technology: A variable tool for mitigating climate change and improving crop production. Int Lett Nat Sci 2:50–65.
Magrin GO, Travasso MI, Díaz RA, Rodríguez RO (2007) Vulnerability of the agricultural systems of Argentina to climate change. Clim Res 9:31–36.
Manderscheid R, Erbs M, Weigel H-J (2014) Interactive effects of free-air CO_2 enrichment and drought stress on maize growth. Eur J Agron 52:11–21.
Marengo JA, Espinoza JC (2016) Extreme seasonal droughts and floods in Amazonia: causes, trends and impacts. Int J Climatol 36:1033–1050.
Martins MA, Tomasella J, Dias CG (2019) Maize yield under a changing climate in the Brazilian Northeast: impacts and adaptation. Agric Water Manag 216:339–350.
Meera SN, Jhamtani A, Rao DUM (2004) Information and communication technology in agricultural development: a comparative analysis of three projects from India. Network Paper No 135.
Milind P, Isha D (2013) Zea maize: a modern craze. Int Res J Pharm 4:39–43.
Mózner Z, Tabi A, Csutora M (2012) Modifying the yield factor based on more efficient use of fertilizer—The environmental impacts of intensive and extensive agricultural practices. Ecol Indic 16:58–66.
Msowoya K, Madani K, Davtalab R, et al. (2016) Climate change impacts on maize production in the warm heart of Africa. Water Resour Manag 30:5299–5312.
Muchow RC, Sinclair TR, Bennett JM (1990) Temperature and solar radiation effects on potential maize yield across locations. Agron J 82:338–343.
Olesen JE, Børgesen CD, Elsgaard L, et al. (2012) Changes in time of sowing, flowering and maturity of cereals in Europe under climate change. Food Addit Contam A 29:1527–1542.
Ospina A, Heeks R (2010) Linking ICTs and climate change adaptation. Institute for Development Policy and Management, University of Manchester, Manchester, UK.
Parry ML, Rosenzweig C, Iglesias A, et al. (2004) Effects of climate change on global food production under SRES emissions and socio-economic scenarios. Glob Environ Chang 14:53–67.
Rahman MH, Ahmad A, Wang X, et al. (2018) Multi-model projections of future climate and climate change impacts uncertainty assessment for cotton production in Pakistan. Agric For Meteorol 253–254:94–113. https://doi.org/10.1016/j.agrformet.2018.02.008.
Ranum P, Peña-Rosas JP, Garcia-Casal MN (2014) Global maize production, utilization, and consumption. Ann N Y Acad Sci 1312:105–112.
Riahi K, Van Vuuren DP, Kriegler E, et al. (2017) The shared socioeconomic pathways and their energy, land use, and greenhouse gas emissions implications: an overview. Glob Environ Chang 42:153–168.
Richardson KJ, Lewis KH, Krishnamurthy PK, et al. (2018) Food security outcomes under a changing climate: impacts of mitigation and adaptation on vulnerability to food insecurity. Clim Change 147:327–341.
Ringler C, Zhu T, Cai X, et al. (2010) Climate change impacts on food security in sub-Saharan Africa: insights from comprehensive climate change scenarios. International Food Policy Research Institute, USA.
Sánchez B, Rasmussen A, Porter JR (2014) Temperatures and the growth and development of maize and rice: a review. Glob Chang Biol 20:408–417.
Setimela PS, Zaman-Allah M, Ndoro O (2018) Performance of elite drought tolerant maize varieties: across eastern and southern Africa. ASA, CSSA & SSSA Annual Meetings, Tampa, Florida.
Shiferaw B, Tesfaye K, Kassie M, et al. (2014) Managing vulnerability to drought and enhancing livelihood

resilience in sub-Saharan Africa: technological, institutional and policy options. Weather Clim Extreme 3:67–79.

Singh S, Kattarkandi B, Deka S, Choudhary R (2010) Impact of climatic variability and climate change on maize productivity in north India. Int J Adv Agric Sci 2:5–9.

Smit B, Wandel J (2006) Adaptation, adaptive capacity and vulnerability. Glob Environ Chang 16:282–292.

Srbinovska M, Gavrovski C, Dimcev V, et al. (2015) Environmental parameters monitoring in precision agriculture using wireless sensor networks. J Clean Prod 88:297–307.

Stewart J, Stewart R, Kennedy S (2017) Internet of things—Propagation modelling for precision agriculture applications. In: 2017 Wireless Telecommunications Symposium (WTS). IEEE, pp 1–8.

Sun L, Wang F (2009) Research on variable rate seeding and fertilization technology. J Northeast Agric Univ 3:12–25.

Sutcliffe C, Dougill AJ, Quinn CH (2016) Evidence and perceptions of rainfall change in Malawi: do maize cultivar choices enhance climate change adaptation in sub-Saharan Africa? Region Environ Chang 16:1215–1224.

Tokatlidis IS (2013) Adapting maize crop to climate change. Agron Sustain Dev 33:63–79.

Trenberth KE (2011) Changes in precipitation with climate change. Clim Res 47:123–138.

Van Kessel C, Venterea R, Six J, et al. (2013) Climate, duration, and N placement determine N2O emissions in reduced tillage systems: a meta-analysis. Glob Chang Biol 19:33–44.

Westengen OT, Brysting AK (2014) Crop adaptation to climate change in the semi-arid zone in Tanzania: the role of genetic resources and seed systems. Agric Food Secur 3:1–12. https://doi.org/10.1186/2048-7010-3-3.

Yong H, Hui F, Lei F (2002) Information processing system for precision agriculture based on GPS and GIS. Life S J 4:3179–3192.

Zhang C, Kovacs JM (2012) The application of small unmanned aerial systems for precision agriculture: a review. Precis Agric 13:693–712.

11

Climate Change and Plants: Biodiversity, Growth, and Interaction

Dong Qin, Fuchun Xie, and Junwei Huo
College of Horticulture & Landscape Architecture, Northeast Agricultural University, Harbin, Heilongjiang Province, China, 150030

11.1 Introduction

In the 21st century, one of the most vehemently debated environmental issues is climate change (Saud et al. 2013; Shah et al. 2013; Saud et al. 2014; Burhan and Zafar 2015; Hafiz et al. 2016; Saud et al. 2016; Aziz et al. 2017a,b; Habib ur Rahman et al. 2017; Kamaran et al. 2017; Saud et al. 2017; Uz Zaman et al. 2017; Wajid et al. 2017; Yang et al. 2017; Zahida et al. 2017; Adnan et al. 2018a,b; Akram et al. 2018a,b; Depeng et al. 2018; Hafiz et al. 2018; Tariq et al. 2018; Adnan et al. 2019; Ahmad et al. 2019; Baseer et al. 2019; Hafiz et al. 2019; Jan et al. 2019; Muhammad et al. 2019; Sajjad et al. 2019; Wu et al. 2019; Adnan et al. 2020; Farhana et al. 2020; Farhat et al. 2020; Hafiz et al. 2020a,b; Ilyas et al. 2020; Mubeen et al. 2020; Rehman 2020; Saleem et al. 2020a,b,c; Saud et al. 2020; Shafi et al. 2020; Subhan et al. 2020; Wahid et al. 2020; Wu et al. 2020; Zafar-ul-Hye et al. 2020a,b). The climate change and global warming and more recently global cooling are now part of our life. Climate change has come upon us in a relatively short space of time and is accelerating with alarming speed. It is perhaps the most serious problem that the civilized world has had to face (Fahad and Bano 2012; Fahad et al. 2013; Fahad et al. 2014a,b; Fahad et al. 2015a,b; Ur Rehman et al. 2015; Fahad et al. 2016a,b,c,d; Fahad et al. 2017; Fahad et al. 2018; Fahad et al. 2019a,b; Hesham and Fahad 2020). The temperature and atmospheric carbon dioxide are increasing, and warming is decreasing frost, snow, and ice cover. Rain may increase in some areas, particularly high latitudes, but decrease in others. Also cloud and therefore solar radiation will change (David 2005).

Strong scientific consensus has highlighted that climate changes are occurring mainly due to anthropogenic activities (IPCC 2007a,b) with clear evidence of an observed increase in average global temperatures and changes in patterns of precipitation (Fauchereau et al. 2003). Although still largely unresolved, the effect new climate is likely to have upon complete establishment, an expected tectonic shift of climatic belts has provoked extensive amount of research over the past two decades. Where, when, and how these changes would happen remain largely unsettled (IPCC 2007a), but these are expected to have far-reaching impacts on marine and terrestrial ecosystems in the decades to come.

Plausible climate change scenarios include higher atmospheric CO_2 concentrations, higher temperatures, changes in precipitation patterns, droughts, and other weather extremes that will severely affect all forms of plant life both natural as well as cultivated (Jablonski et al. 2002; Morecroft and Paterson 2006; Ziska and Bunce 2006; Figure 11.1).

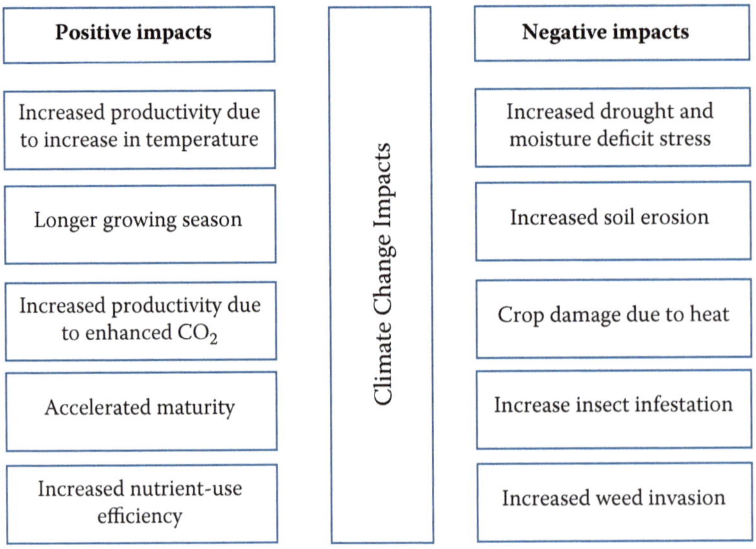

FIGURE 11.1 Impacts of Climate Change (from Burhan and Zafar 2015).

11.2 The Main Climate Changes

11.2.1 Atmospheric Carbon Dioxide (CO_2) Concentration

Global emissions of radiatively active gases, including CO_2, have increased rapidly since 1750, a trend that is likely to accelerate if increase in global emissions cannot be curbed effectively. The atmospheric CO_2 concentration has substantially increased from preindustrial levels of 285 ppm to current level of 384 ppm mainly due to anthropogenic activities and is predicted to approach to 448 ppm by the year 2050 (IPCC 2007b), and will reach to 600 ppm at current annual rate of increase by the end of this century. Such an increase in the CO_2 levels will certainly affect the globally important process of photosynthesis, which sustains the life on this planet.

The primary effects of rising CO_2 on plants have been well documented and include reduction in stomatal conductance and transpiration, improved water-use efficiency, higher rates of photosynthesis, and increased light-use efficiency (Drake and González-Meler 1997). So, increased CO_2 levels may increase plant growth and productivity, especially to some plants grown in water-limited areas or when there are less than normal rainfall amounts (Hatfield et al. 2011).

Elevated CO_2 concentration affects the photosynthetic process directly and evokes a wide range of morphological, anatomical, and biochemical responses in plants. These responses depend mainly on photosynthetic pathways operative in plant species and thus vary accordingly. The most common responses in plants include growth rate changes (Poorter et al. 1997), change in the pattern of allocation of resources (Bazzaz 1990), water-use efficiency (Eamus 1991), and the rate of nutrient uptake (Jackson and Reynolds 1996).

11.2.2 Globe Warming

The increase in levels of atmospheric CO_2 will result in an increase in average global surface temperatures by 0.6 ± 0.2 °C. Climate simulation models predict that surface air temperature will increase by 1.1–2.9 °C ('low scenario') or 2.4–6.4 °C ('usual business scenario') relative to 1990 by the end of the 21st century (IPCC 2007a,b).

Temperature being one of the main climatic factors determining how efficiently plants will grow, its increase can be both beneficial and detrimental to plant productivity, but general trends indicate a reduction in yield of many plants especially cereals (Fahad et al. 2014a,b; Fahad et al. 2016a,b,c,d; Fahad et al. 2017; Fahad et al. 2018). A faster growth will be obvious due to warming in plants that are below their optimum temperature; however, at the same time, in some crops such as cereals, warming reduces time for proper grain development, thereby affecting yield (Hatfield et al. 2008).

An increase in temperature will also cause northward movement of optimum latitude for many crops, whereas a decrease in temperature would shift the same towards the equator, causing a shift in the growing areas of the world which will increase plant productivity in some areas, particularly in temperate regions, and at the same time will drastically decrease it in equatorial regions (Hardy 2003).

An increase in local temperature over a range of 1–3 °C will increase crop productivity, but above this range, a decrease in same will occur due to its adverse effects on photosynthesis (Nobel 2005; IPCC 2007a,b), transpiration, and phenological stages (Mitchell et al. 1993).

The mean temperature for most regions across the world is increasing according to the climate change models. The response of crops to temperature mainly depends on the specific optimum temperature for photosynthesis, growth, and yield (Conroy et al. 1994).

11.3 Climate Change and Plant Biodiversity

The researchers have focused their research on the predictions of responses of biodiversity to climate change (Salamin et al. 2010; Hakeem 2015), and the response of biodiversity to climate change plays significant role in planning and alteration of the thinking of scientists to stop the biological changes due to climate change and decrease the climate change impacts on biodiversity (Parmesan et al. 2011).

Climate change can decrease genetic diversity of populations due to quick migration and directional selection which may in turn impact changes in ecosystem resilience and functioning (Meyers and Bull 2002; Botkin et al. 2007). The climate change has induced phonological shifts in pollinating insects and flowering plants which have caused a mismatch between pollinating insect population and flowering plants which has caused extinction of both pollinators and plant populations (Rafferty and Ives 2010).

Climate change may induce variation in plant communities at the higher degrees of biodiversity which are expected to affect biome integrity. It has been forecasted by the Millennium Ecosystem Assessment (MES) that 5–20% shifts for Earth's terrestrial ecosystems will take place, in particular in the cool conifer forests, savannahs and boreal forests, tundra, and shrub land (Sala et al. 2005).

Reduction in fruit-tree diversity in both agricultural and forest (native or endemic) vegetation is likely to occur as a long-term consequence of global warming. Temperate fruit trees are likely to be more affected by climate change than trees in the subtropics and tropics; however, among temperate fruit trees, cultivars might be less or more adapted to changing climatic conditions (Ramírez and Kallarackal 2015).

Genetic resources of fruits are facing a serious threat of extinction due to climate change, large-scale urbanization, and developmental projects (Malik et al. 2010). Tropical fruit trees may respond to climate change through phenotypic plasticity, adaptive evolution, migration to suitable sites, or extinction (Sthapit et al. 2012). Moreover, selection of appropriate rootstocks in various fruit crops, for example, mango and guava, to suit the changed climatic conditions could be one of the solutions to temperature change (Sthapit et al. 2012).

There are two categories for plants to adapt to climate changes, mechanisms and responses. There are two mechanisms, that is, genetic and plastic. In genetic mechanisms, plants may adapt genetically to changing environment by mutation or selection, and in plastic mechanisms, plants are provided with short-term responses in behaviour, whether or not the plants will be able to adopt the mechanisms to cope with changing climate (Lavergne et al. 2010). There are three types of responses, that is, spatial, temporal, and self. The first two are easily observed and well-reputed responses to climate change and self-correspond to changes which are less obvious physiological and behavioural responses which enable plants to adapt to environmental changes in the same space and time paradigms.

11.4 Plant Response to the Climate Changes and Interaction

Increased temperature, aberrant precipitation, and a host of other related factors are expected to cause a global climate change that would adversely affect life on this planet. Otherwise, plants have also evolved over time to respond to the changing climates.

11.4.1 Plant Response to CO_2

The increase in CO_2 concentration generally enhances plant growth and development by stimulating photosynthesis and improving water-use efficiency (Ramírez and Kallarackal 2015). Plants have been classified mainly into three groups, namely C3, C4, and CAM plants, on the basis of the photosynthetic pathway operational in them, and the variation in the physiological response of plants to CO_2 enrichment mainly stems from CO_2-fixing enzymes involved, namely ribulose-1,5- bisphosphate carboxylase/oxygenase (Rubisco), phosphoenolpyruvate carboxylase PEP carboxylase, and to some extent carbonic anhydrase.

The present ambient atmospheric CO_2 concentration is a limiting factor for C3 plants when other factors affecting plant growth (light, water, temperature, nutrients) are optimal (Downton et al. 1980), and any increase in its concentration will be beneficial to these plants. Rubisco, which is the primary enzyme involved in leaf photosynthesis in C3 plants, is not CO_2 saturated at present-day atmospheric conditions, and can bind to either CO_2 or O_2. An increase in the concentration of CO_2 enables Rubisco to compete better with O_2 for binding sites of the Rubisco protein and thus can fix more CO_2 due to the increased CO_2:O_2 ratio (Ziska and Bunce 2006). Thus, CO_2 enrichment will momentarily enhance the relative growth rate by increasing net carbon uptake (Lambers et al. 1998), thereby increasing biomass and yield (Kimball 1983).

The plants with C3 photosynthetic pathway may benefit in dry matter production of high CO_2 in three ways: (1) enhancement of leaf expansion by increasing leaf thickness, leaf area, and leaf area index (LAI) due to accumulation of nonstructural carbohydrates (Lambers et al. 1998), (2) increase in the photosynthetic rate per unit leaf area, and (3) increase in water-use efficiency. Elevated levels of CO_2 accelerate growth and also bring in changes in inherent established timings of stages of plant development (Bowes 1993).

Experiments under idealized conditions revealed that a doubling of atmospheric CO_2 concentration increased photosynthesis by 30–50% in C3 plant species (Ainsworth and Long 2005) and crop yield by 10–20% (Gifford 2004; Ainsworth and Long 2005).

11.4.2 Plant Response to Elevated and High Temperature

Elevated temperature accelerates various aspects of plant metabolism (Larcher 2003; Fahad et al. 2016a,b,c,d; Fahad et al. 2017) and can affect crop yield primarily by causing rapid crop development (Badeck et al. 2004; Fahad et al. 2016a,b,c,d; Fahad et al. 2017), thereby shortening duration of various aspects of plant metabolism (Larcher 2003), through changing the rate and timing of physiological processes, including organ development rate, respiration, and senescence (Farrar and Williams 1991). A slight temperature increase will lead to increased plant growth if the ambient air temperature is below optimum for photosynthesis, but can have a negative effect on growth if the same is close to the maximum (Baker and Allen 1993; Fahad et al. 2016a,b,c,d; Fahad et al. 2017).

The optimum temperature for photosynthesis of C4 crop is higher than that of C3 crop, while C4 crops show declines in photosynthesis at high temperature (Crafts and Salvucci 2002). An exponential increase in the air saturation vapour pressure will occur due to the increase in air temperature resulting in an increased vapour pressure deficit (VPD) between air and the leaf that will reduce water-use efficiency (WUE), as a result of which plants will lose more water per unit carbon gain. High temperatures for extended periods not only reduce plant biomass, but also adversely affect reproductive efforts of crops during critical reproductive periods, most predominantly during fertilization and flowering which can lead to severe sterility, reduced pollen germination and viability, and fruit abortion adversely affecting seeds, thereby increasing risk of complete crop failure.

Plants respond to an increase in temperature by adopting various strategies. Plants mitigate overheating by leaf drooping, leaf rolling, and vertical orientation (Nobel 2005) or by transient wilting (Chiariello et al. 1987); all these mechanisms, however, may lead to decreased photosynthesis. Plants under high temperatures will tend to produce a robust root system to increase water uptake and smaller leaves to offset the leaf water loss (Gliessman 1998).

An increase in mean temperature will result in decreased crop yield in subtropical and tropical regions (Parry et al. 2007) because of lower levels of soil moisture and increase in evapotranspiration rendering agricultural lands in tropics unsuitable for cropping (Bals et al. 2008).

11.5 Horticultural Crops' Response to the Climate Changes

The value of horticultural crops is derived from not only the quantity but also the quality of the harvested product. Perennial fruit crops such as apple, banana, grape, and citrus are important components of human diet providing fibre, vitamins, antioxidants, and nutrients such as manganese and calcium that can improve human health but are often limited in cereals and other major food crops. They are also high value agricultural commodities that are important for the local as well as the global economy.

Climate change brings both challenges and opportunities to agriculture for sustaining food and energy supplies while protecting and maintaining the environment and natural resources essential for global food production (Glenn et al. 2013). Like all other crops, perennial horticultural crops are sensitive to temperature, water availability, solar radiation, air pollution, and CO_2.

In some locations, the increasing CO_2 in the atmosphere has given rise to higher productivity, while at the same time controversy remains as to whether the increasing temperature due to CO_2 will sustain this productivity. Increased atmospheric CO_2 generally increases growth rate and yield, resulting in a higher accumulation of biomass, fruit production, and quality in both temperate and tropical fruit trees (Idso and Kimball 1997; Centritto et al. 1999a,b; Kimball et al. 2007).

The value of a horticultural crop is determined and limited at many points before and during the growing season because the value is based on not only production quantities (i.e., biomass, yield) but also size, colour, chemical composition, firmness, and other measurable criteria. Several examples exist that illustrate the sensitivities and the complexity of responses of perennial crop systems to climate change and climate variability (Lobell et al. 2006). For instance, extreme cold can kill fruit flower bud in the dormant winter, and warming periods can deacclimate buds, making them susceptible to spring frost. High temperature can reduce cell division resulting in small fruits in the spring and can cause sunburn damage in summer, which reduces production efficiency, that is, pack-out at harvest (Caprio and Quamme 1999). High temperatures not only alter maturity, fruit firmness, colour development, but also decrease the suitability of fruit for short- or long-term storage (Woolf and Ferguson 2000).

Stomatal conductance to water vapour (gs) in general was reduced at elevated CO_2 by 23% in cherry, citrus, and peach, while leaf water-use efficiency (WUE) will be increased by 58%. This is likely because of an increase in tree leaf area, offsetting the increase in WUE per leaf area under elevated CO_2.

On average, aboveground biomass increased by 60% in elevated CO_2 (NCA 2012); however, the ratio of root/shoot remained similar in apple (Chen et al. 2002) and citrus (Kimball et al. 2007), while it slightly increased in cherry (Druta 2001). In summary, perennial horticultural crops exhibit physiological and growth responses that are similar to trees in forest and other unmanaged ecosystems. The CO_2 fertilization effect may be amplified and sustained longer for perennial horticultural crops when the nutrients and water are amply supplied and management such as pruning and thinning is practiced to facilitate the prolonged CO_2 effects.

Increased CO_2 concentrations are expected to increase crop biomass, but this largely depends on the photosynthetic mechanism of the plant as well as on the prevailing climate conditions, particularly temperature and vapour pressure deficit (VPD) (Leakey et al., 2009). In addition, CO_2 fertilization effects can be negated or can even enhance the negative response of crops to high temperatures (Vara Prasad et al. 2006). In addition, intermittent drought is a significant limiting factor in most rained perennial systems (Van Asten et al. 2011), but little is known about the interactions between different types of drought, changes in VPD, and high temperatures particularly in tropical species.

The ozone concentration was significantly increased with a mean trend of 0.26 ppb per year, corresponding to an average increase of 5 ppb between 1987 and 2004 (Jaffe and Ray 2007). Current levels at background sites range between 20 and 45 ppb, depending on location, elevation, and distance to emission sources (Vingarzan 2004). According to Fuhrer (2009), there is likely an increase in worldwide ozone in the next century.

When elevated ozone is combined with elevated CO_2, yield loss is typically considerably less than with ozone alone. The protective effect of CO_2 is primarily due to reduced stomatal conductance reducing ozone flux into the leaf, and this mechanism is associated with elevated CO_2. Consequently, elevated ozone can also diminish the stimulating effect on yield of elevated CO_2, and the CO_2 protection from ozone effects also becomes less effective with increasing temperature (Fuhrer 2009).

Higher growing season temperatures will likely reduce crop quality. Under high temperatures, banana leaves fold downwards across the petiole. This reduces leaf temperatures by 7–8°C and the amount of water transpired (Turner 1998; Turner et al. 2007). Turner et al. (2007) showed that fruit damage due to high temperatures can be avoided through the use of paper layers in between the bunch and the bunch cover.

Japanese citrus production will likely move northwards due to higher temperatures (Sugiura and Yokozawa 2004). By 2060, the favourable areas for 'Satsuma' mandarin production will possibly move from the southern coastal sites to inland areas of western and southern Japan. Potential for northwards expansion of US citrus production was small because results indicated that in 2030 and 2090, northern sites of current marginal production would continue to have lower fruit yield, higher risk of crop loss due to freezing, and lower water availability than the southern sites.

Global warming has caused temperate fruit tree phenology to change in various parts of the world (Ramírez and Kallarackal 2015). The chilling hours, which is a major determinant in tree phenology in temperate regions, have come down, causing considerable reduction in yield in several species. In the tropics, precipitation is a major factor regulating the phenology and yield in fruit trees. There is a need to develop phenological models in order to estimate the impact of climate change on plant development in different regions of the world. Past records on phenological data from around the world have shown that the flowering of fruit trees has advanced by a few days or weeks compared to their reproductive behaviour a century ago.

Weather variability in the spring will affect flowering and the potential for frost damage. Areas with reduced winter chilling will require either adapted cultivars with less chilling requirement or chemical methods to mimic chilling hours. Research on technology to protect flowers from freezing could stabilize production. Chemical agents to break dormancy have not been developed beyond the use of hydrogen cyanamide, and further improvements will be needed for all the deciduous crops. Mitigating freeze damage is approached in two nonexclusive areas of research: freeze avoidance and freeze tolerance. Many physical mechanisms of freeze avoidance have been identified and include ice nucleators, anti-nucleators, preferential ice accumulation sites, supercooling, and ice barriers (Gusta and Wisniewski 2012). However, no commercial product has been developed that consistently avoids freeze damage.

Mechanisms associated with freeze tolerance are associated with biochemical adaptations under genetic control. Effective biochemical changes associated with freeze tolerance include compositional changes in cell membranes, increased osmotic adjustment, cryoprotective compounds, antioxidant defence systems, and cold-induced proteins (Gusta and Wisniewski 2012). Modern molecular biology, genomics, proteomics, and metabolomics will provide markers and genes for adapted cultivars with freeze tolerance and perhaps avoidance.

11.6 Conclusion

Climate change effects are complex phenomena involving atmospheric gases, temperature, and other extreme weather. As carbon dioxide increases, the climate warms, with potentially disastrous outcomes. To build and maintain a sustainable society, it is necessary to understanding the mechanisms of plants' response to the climate changes. So plantation of horticultural (perennials) plants that are at greater risk can also help in mitigating the climate change effects by absorbing more radiation than annual or seasonal crops. With the development of modern molecular biology, genomics, proteomics, and metabolomics, we will be able to accurately reveal the mechanisms that plants respond to climate changes and, in practice, to better deal with these changes.

REFERENCES

Adnan M, Fahad S, Khan IA, Saeed M, Ihsan MZ, Saud S, Riaz M, Wang D, Wu C (2019) Integration of poultry manure and phosphate solubilizing bacteria improved availability of Ca bound P in calcareous soils. 3 Biotech 9:368.

Adnan M, Fahad S, Muhammad Z, Shahen S, Ishaq AM, Subhan D, Zafar-ul-Hye M, Martin LB, Raja MMN, Beena S, Saud S, Imran A, Zhen Y, Martin B, Jiri H, Rahul D (2020) Coupling phosphate-solubilizing bacteria with phosphorus supplements improve maize phosphorus acquisition and growth under lime induced salinity stress. Plants 9:900. https://doi.org/10.3390/plants9070900.

Adnan M, Shah Z, Sharif M, Rahman H (2018b) Liming induces carbon dioxide (CO_2) emission in PSB inoculated alkaline soil supplemented with different phosphorus sources. Environ Sci Poll Res 25:9501–9509.

Adnan M, Zahir S, Fahad S, Arif M, Mukhtar A, Imtiaz AK, Ishaq AM, Abdul B, Hidayat U, Muhammad A, Inayat-Ur R, Saud S, Muhammad ZI, Yousaf J, Amanullah, Hafiz MH, Wajid N (2018a) Phosphate-solubilizing bacteria nullify the antagonistic effect of soil calcification on bioavailability of phosphorus in alkaline soils. Sci Rep 8:4339. https://doi.org/10.1038/s41598-018-22653-7.

Ahmad S, Kamran M, Ding R, Meng X, Wang H, Ahmad I, Fahad S, Han Q (2019) Exogenous melatonin confers drought stress by promoting plant growth, photosynthetic capacity and antioxidant defense system of maize seedlings. PeerJ 7:e7793. http://doi.org/10.7717/peerj.7793.

Ainsworth EA, Long SP (2005) What have we learned from 15 years of free-air CO_2 enrichment (FACE)? A meta-analytic review of the responses of photosynthesis, canopy properties and plant production to rising CO_2. New Phytol 165:351–372.

Akram R, Turan V, Hammad HM, Ahmad S, Hussain S, Hasnain A, Maqbool MM, Rehmani MIA, Rasool A, Masood N, Mahmood F, Mubeen M, Sultana SR, Fahad S, Amanet K, Saleem M, Abbas Y, Akhtar HM, Waseem F, Murtaza R, Amin A, Zahoor SA, ul Din MS, Nasim W (2018a) Fate of organic and inorganic pollutants in paddy soils. In: Hashmi MZ, Varma A (eds) Environmental pollution of paddy soils, soil biology. Springer International Publishing Ag, Switzerland, pp 197–214.

Akram R, Turan V, Wahid A, Ijaz M, Shahid MA, Kaleem S, Hafeez A, Maqbool MM, Chaudhary HJ, Munis, MFH, Mubeen M, Sadiq N, Murtaza R, Kazmi DH, Ali S, Khan N, Sultana SR, Fahad S, Amin A, Nasim W (2018b) Paddy land pollutants and their role in climate change. In: Hashmi MZ, Varma A (eds) Environmental pollution of paddy soils, soil biology. Springer International Publishing Ag Switzerland, pp 113–124.

Aziz K, Daniel KYT, Fazal M, Muhammad ZA, Farooq S, Fan W, Fahad S, Ruiyang Z (2017a) Nitrogen nutrition in cotton and control strategies for greenhouse gas emissions: a review. Environ Sci Pollut Res 24:23471–23487. https://doi.org/10.1007/s11356-017-0131-y.

Aziz K, Daniel KYT, Muhammad ZA, Honghai L, Shahbaz AT, Mir A, Fahad S (2017b) Nitrogen fertility and abiotic stresses management in cotton crop: a review. Environ Sci Pollut Res 24:14551–14566. https://doi.org/10.1007/s11356-017-8920-x.

Badeck FW, Bondeau A, Bottcher K, Doktor D, Lucht W, Schaber J, Sitch S (2004) Responses of spring phenology to climate change. New Phytol 162:295–309.

Baker JT, Allen LH (1993) Contrasting crop species responses to CO_2 and temperature: rice, soybean and citrus. Vegetation 105:239–260.

Bals C, Harmeling S, Windfuhr M (2008) Climate change, food security and the right to adequate food. GermanWatch

Baseer M, Adnan M, Fazal M, Fahad S, Muhammad S, Fazli W, Muhammad A, Jr. Amanullah, Depeng W, Saud S, Muhammad N, Muhammad Z, Fazli S, Beena S, Mian AR, Ishaq AM (2019) Substituting urea by organic wastes for improving maize yield in alkaline soil. J Plant Nutr https://doi.org/10.1080/01904167.2019.1659344.

Bazzaz FA (1990) The response of natural ecosystems to the rising global CO_2 levels. Annu Rev Ecol Syst 21:167–196.

Botkin DB, Saxe H, Araujo MB, Betts R, Bradshaw RHW, Cedhagen T, Chesson P, Dawson TP, Etterson JR, Faith DR, Ferrier S, Guisan A, Hansen AS, Hilbert DW, Loehle C, Margules C, New M, Sobel MJ, Stockwell DRB (2007) Forecasting the effects of global warming on biodiversity. Bioscience 57:227–236.

Bowes G (1993) Facing the inevitable: plants and increasing atmospheric CO_2. Annu Rev Plant Physiol Plant Mol Biol 44:309–332.

Burhan A, Zafar AR (2015) Climate change and plants. In Hakeem KR (ed) Crop production and global environmental issues. https://doi.org/10.1007/978-3-319-23162-4_20.

Caprio JM, Quamme HA (1999) Weather conditions associated with apple production in the Okanagan Valley of British Columbia. Can J Plant Sci 79:129–137.

Centritto M, Lee HSJ, Jarvis PG (1999a) Increased growth in elevated CO_2: An early, short-term response? Global Change Biol 5:623–633.

Centritto M, Lee HSJ, Jarvis PG (1999b) Interactive effects of elevated CO_2 and drought on cherry (*Prunus avium*) seedlings—I. Growth, whole-plant water use efficiency and water loss. New Phytol 141:129–140.

Chen K, Hu GQ, Lenz F (2002) Effects of doubled atmospheric CO_2 concentration on apple trees. II. Dry mass production. Gartenbauwissenschaft 67:28–33.

Chiariello NR, Field CB, Mooney HA (1987) Midday wilting in a tropical pioneer tree. Funct Ecol 1:3–11.

Conroy JP, Seneweera S, Basra AS, Rogers G, Nissen-Wooller B (1994) Influence of rising atmospheric CO_2 concentrations and temperature on growth, yield and grain quality of cereal crops. Aust J Plant Physiol 21:741–758.

Crafts BSJ, Salvucci ME (2002) Sensitivity of photosynthesis in a C_4 plant, maize, to heat stress. Plant Physiol 129:1773–1780.

David WL (2005) Plant responses to climate change: impacts and adaptation. In: Omasa K, Nouchi I, De LJ (eds) Plant responses to air pollution and global change. Springer-Verlag, Tokyo

Depeng W, Fahad S, Saud S, Muhammad K, Aziz K, Mohammad NK, Hafiz MH, Wajid N (2018) Morphological acclimation to agronomic manipulation in leaf dispersion and orientation to promote "Ideotype" breeding: evidence from 3D visual modeling of "super" rice (*Oryza sativa* L.). Plant Physiol Biochem 135. https://doi.org/10.1016/j.plaphy.2018.11.010.

Downton WJS, Bjorkman O, Pike C (1980) Consequences of increased atmospheric concentrations of carbon dioxide for growth and photosynthesis of higher plants. In: Pearman GI (ed) Carbon dioxide and climate: Australian research. Australian Academy of Sciences, Canberra, pp 143–151.

Drake BG, González-Meler MA (1997) More efficient plants: a consequence of rising atmospheric CO_2? Annu Rev Plant Physiol Plant Mol Biol 48:609–639.

Druta A (2001) Effect of long term exposure to high CO_2 concentrations on photosynthetic characteristics of *Prunus avium* L. plants. Photosynthetica 39:289–297.

Eamus D (1991) The interaction of rising CO_2 and temperatures with water use efficiency. Plant Cell Environ 14:843–852.

Fahad S, Adnan M, Hassan S, Saud S, Hussain S, Wu C, Wang D, Hakeem KR, Alharby HF, Turan V, Khan MA, Huang J (2019b). Rice responses and tolerance to high temperature, In: Hasanuzzaman M, Fujita M, Nahar K, Biswas JK (eds) Advances in rice research for abiotic stress tolerance. Woodhead Publ Ltd, Cambridge, England, pp 201–224.

Fahad S, Bajwa AA, Nazir U, Anjum SA, Farooq A, Zohaib A, Sadia S, Nasim W, Adkins S, Saud S, Ihsan MZ, Alharby H, Wu C, Wang D, Huang J (2017) Crop production under drought and heat stress: plant responses and management options. Front Plant Sci 8:1147. https://doi.org/10.3389/fpls.2017.01147.

Fahad S, Bano A (2012) Effect of salicylic acid on physiological and biochemical characterization of maize grown in saline area. Pak J Bot 44:1433–1438.

Fahad S, Chen Y, Saud S, Wang K, Xiong D, Chen C,Wu C, Shah F, Nie L, Huang J (2013) Ultraviolet radiation effect on photosynthetic pigments, biochemical attributes, antioxidant enzyme activity and hormonal contents of wheat. J Food Agri Environ 11:1635–1641.

Fahad S, Hussain S, Bano A, Saud S, Hassan S, Shan D, Khan FA, Khan F, Chen Y, Wu C, Tabassum MA, Chun MX, Afzal M, Jan A, Jan MT, Huang J (2014a) Potential role of phytohormones and plant growth-promoting rhizobacteria in abiotic stresses: consequences for changing environment. Environ Sci Pollut Res 22:4907–4921. https://doi.org/10.1007/s11356-014-3754-2.

Fahad S, Hussain S, Matloob A, Khan FA, Khaliq A, Saud S, Hassan S, Shan D, Khan F, Ullah N, Faiq M, Khan MR, Tareen AK, Khan A, Ullah A, Ullah N, Huang J (2014b) Phytohormones and plant responses to salinity stress: A review. Plant Growth Regul 75:391–404. https://doi.org/10.1007/s10725-014-0013-y.

Fahad S, Hussain S, Saud S, Hassan S, Chauhan BS, Khan F, Ihsan MZ, Ullah A, Wu Chao, Bajwa AA, Alharby H, Amanullah, NasimW, Shahzad B, Tanveer M, Huang J (2016a) Responses of rapid viscoanalyzer profile and other rice grain qualities to exogenously applied plant growth regulators under high day and high night temperatures. PLoS One 11:e0159590. https://doi.org/10.1371/journal.pone.0159590.

Fahad S, Hussain S, Saud S, Hassan S, Ihsan Z, Shah AN,Wu C, Yousaf M, Nasim W, Alharby H, Alghabari F, Huang J (2016c) Exogenously applied plant growth regulators enhance the morphophysiological growth and yield of rice under high temperature. Front Plant Sci 7:1250. https://doi.org/10.3389/fpls.2016.01250.

Fahad S, Hussain S, Saud S, Hassan S, Tanveer M, Ihsan MZ, Shah AN, Ullah A, Nasrullah KF, Ullah S, AlharbyH NW, Wu C, Huang J (2016d) A combined application of biochar and phosphorus alleviates heat-induced adversities on physiological, agronomical and quality attributes of rice. Plant Physiol Biochem 103:191–198.

Fahad S, Hussain S, Saud S, Khan F, Hassan S, Jr A, Nasim W, Arif M, Wang F, Huang J (2016b) Exogenously applied plant growth regulators affect heat-stressed rice pollens. J Agron Crop Sci 202:139–150.

Fahad S, Hussain S, Saud S, Tanveer M, Bajwa AA, Hassan S, Shah AN, Ullah A,Wu C, Khan FA, Shah F, Ullah S, Chen Y, Huang J (2015a) A biochar application protects rice pollen from high-temperature stress. Plant Physiol Biochem 96:281–287.

Fahad S, Muhammad ZI, Abdul K, Ihsanullah D, Saud S, Saleh A, Wajid N, Muhammad A, Imtiaz AK, Chao W, Depeng W, Jianliang H (2018) Consequences of high temperature under changing climate optima for rice pollen characteristics-concepts and perspectives. Arch Agron Soil Sci 64:1479–1488. https://doi.org/10.1080/03650340.2018.1443213.

Fahad S, Nie L, Chen Y, Wu C, Xiong D, Saud S, Hongyan L, Cui K, Huang J (2015b) Crop plant hormones and environmental stress. Sustain Agric Rev 15:371–400.

Fahad S, Rehman A, Shahzad B, Tanveer M, Saud S, Kamran M, Ihtisham M, Khan SU, Turan V, Rahman MHU (2019a) Rice responses and tolerance to metal/metalloid toxicity. In: Hasanuzzaman M, Fujita M, Nahar K, Biswas JK (eds) Advances in rice research for abiotic stress tolerance. Woodhead Publ Ltd, Cambridge, England, pp 299–312.

Farhana G, Ishfaq A, Muhammad A, Dawood J, Fahad S, Xiuling L, Depeng W, Muhammad F, Muhammad F, Syed AS (2020) Use of crop growth model to simulate the impact of climate change on yield of various wheat cultivars under different agro-environmental conditions in Khyber Pakhtunkhwa, Pakistan. Arabian J Geosci 13:112. https://doi.org/10.1007/s12517-020-5118-1.

Farhat A, Hafiz MH, Wajid I, Aitazaz AF, Hafiz FB, Zahida Z, Fahad S, Wajid F, Artemi C (2020) A review of soil carbon dynamics resulting from agricultural practices. J Environ Manage 268:110319.

Farrar JF, Williams ML (1991) The effects of increased atmospheric carbon-dioxide and temperature on carbon partitioning, source-sink relations and respiration. Plant Cell Environ 14:819–830.

Fauchereau N, Trzaska M, Rouault M, Richard Y (2003) Rainfall variability and changes in Southern Africa during the 20th century in the global warming context. Nat Hazards 29:139–154.

Fuhrer J (2009) Ozone risk for crops and pastures in present and future climates. Naturwissen- schaften. 96:173–194.

Gifford RM (2004) The CO_2 fertilising effect—does it occur in the real world? New Phytol 163:221–225.

Gliessman SR (1998) Agroecology: ecological processes in sustainable agriculture. Sleeping Bear Press, Chelsea, MI

Gusta LV, Wisniewski W (2012) Understanding plant cold hardiness: an opinion. Physiol Plant 147:4–14. https://doi.org/10.1111/j.1399-3054.2012.01611.x.

Habib ur Rahman M, Ashfaq A, Aftab W, Manzoor H, Fahd R, Wajid I, Md. Aminul I, Vakhtang S, Muhammad A, Asmat U, Abdul W, Syeda RS, Shah S, Shahbaz K, Fahad S, Manzoor H, Saddam H, Wajid N (2017) Application of CSM-CROPGRO-Cotton model for cultivars and optimum planting dates: evaluation in changing semi-arid climate. Field Crops Res 238:139–152. http://dx.doi.org/10.1016/j.fcr.2017.07.007.

Hafiz MH, Abdul K, Farhat A, Wajid F, Fahad S, Muhammad A, Ghulam MS, Wajid N, Muhammad M, Hafiz FB (2020b) Comparative effects of organic and inorganic fertilizers on soil organic carbon and wheat productivity under arid region. Commun Soil Sci Plant Anal 51:1406–1422. https://doi.org/10.1080/00103624.2020.1763385.

Hafiz MH, Farhat A, Ashfaq A, Hafiz FB, Wajid F, Carol Jo W, Fahad S, Gerrit H (2020a) Predicting kernel growth of maize under controlled water and nitrogen applications. Int J Plant Prod 14:609–620. https://doi.org/10.1007/s42106-020-00110-8.

Hafiz MH, Farhat A, Shafqat S, Fahad S, Artemi C, Wajid F, Chaves CB, Wajid N, Muhammad M, Hafiz FB (2018) Offsetting land degradation through nitrogen and water management during maize cultivation under arid conditions. Land Degrad Dev 1–10. https://doi.org/10.1002/ldr.2933.

Hafiz MH, Muhammad A, Farhat A, Hafiz FB, Saeed AQ, Muhammad M, Fahad S, Muhammad A (2019) Environmental factors affecting the frequency of road traffic accidents: a case study of sub-urban area of Pakistan. Environ Sci Pollut Res 26:11674–11685. https://doi.org/10.1007/s11356-019-04752-8.

Hafiz MH, Wajid F, Farhat A, Fahad S, Shafqat S, Wajid N, Hafiz FB (2016) Maize plant nitrogen uptake dynamics at limited irrigation water and nitrogen. Environ Sci Pollut Res 24:2549–2557. https://doi.org/10.1007/s11356-016-8031-0.

Hakeem KR (2015) Crop production and global environmental issues. Springer International Publishing AG, Cham, pp 598.

Hardy JT (2003) Climate change: causes, effects, and solutions. Wiley, Chichester

Hatfield J, Boote K, Fay P, Hahn L, Izaurralde C, Kimball BA, Mader T, Morgan J, Ort D, Polley W, Thomson A, Wolfe D (2008) Agriculture. In: Backlund P, Janetos A, Schimel D, Hatfield J, Boote K, Fay P, Hahn L, Izaurralde C, Kimball BA, Mader T, et al. (eds) The effects of climate change on agriculture, land resources, water resources, and biodiversity in the United States. U.S. Department of Agriculture, Washington, DC, pp 21–74.

Hatfield JL, Boote KJ, Kimball BA, Ziska LH, Izaurralde RC, Ort D, Thomson AM, Wolfe D (2011) Climate impacts on agriculture: implications for crop production. Agron J 103:351–370.

Hesham FA, Fahad S (2020) Melatonin application enhances biochar efficiency for drought tolerance in maize varieties: modifications in physio-biochemical machinery. Agron J 1–22.

Houghton JT, Ding Y, Griggs DJ, Noguer M, van der Linden PJ, Dai X, Maskell K, Johnson CA. Hussain MA, Fahad S, Rahat S, Muhammad FJ, Muhammad M, Qasid A, Ali A, Husain A, Nooral A, Babatope SA, Changbao S, Liya G, Ibrar A, Zhanmei J, Juncai H (2020) Multifunctional role of brassinosteroid and its analogues in plants. Plant Growth Regul 92:141–156. https://doi.org/10.1007/s10725-020-0064 7-8.

Idso SB, Kimball BA (1997) Effects of long-term atmospheric CO_2 enrichment on the growth and fruit production of sour orange trees. Global Change Biol 3:89–96.

Ilyas M, Mohammad N, Nadeem K, Ali H, Aamir HK, Kashif H, Fahad S, Aziz K, Abid U (2020) Drought tolerance strategies in plants: a mechanistic approach. J Plant Growth Regul. https://doi.org/10.1007/s00344-020-10174-5.

impacts, adaptation, and vulnerability. Contribution of Working Group II to the third assessment report of the intergovernmental panel on climate change. Cambridge University Press, Cambridge, 1000 pp

Jablonski LM, Wang X, Curtis PS (2002) Plant reproduction under elevated CO_2 conditions: a meta-analysis of reports on 79 crop and wild species. New Phytol 156:9–26.

IPCC (2007a) Summary for policymakers. In: Solomon S et al (eds) Climate change 2007: the physical science basis. Contribution of Working Group I to the fourth assessment report of the intergovernmental panel on climate change. Cambridge University Press, Cambridge

IPCC (2007b) Climate change 2007: impacts, adaptation and vulnerability: contribution of Working Group II to the fourth assessment report of the Intergovernmental Panel on Climate Change. Cambridge University Press, Cambridge

Jackson RB, Reynolds HL (1996) Nitrate and ammonium uptake for single and mixed species communities grown at elevated CO_2. Oecologia 105:74–80.

Jaffe D, J Ray (2007) Increase in surface ozone at rural sites in the western US. Atmos Environ 41:5452–5463.

Jan M, Muhammad Anwar-ul-Haq, Adnan NS, Muhammad Y, Javaid I, Xiuling L, Depeng W, Fahad S, (2019) Modulation in growth, gas exchange, and antioxidant activities of salt-stressed rice (*Oryza sativa* L.) genotypes by zinc fertilization. Arabian J Geosci 12:775. https://doi.org/10.1007/s12517-019-4939-2.

Kamaran M, Wenwen C, Irshad A, Xiangping M, Xudong Z, Wennan S, Junzhi C, Shakeel A, Fahad S, Qingfang H, Tiening L (2017) Effect of paclobutrazol, a potential growth regulator on stalk mechanical strength, lignin accumulation and its relation with lodging resistance of maize. Plant Growth Regul 84:317–332. https://doi.org/10.1007/ s10725-017-0342-8.

Kimball BA (1983) Carbon dioxide and agricultural yields: an assemblage and analyses of 430 Prior Observations. Agron J 75:779.

Kimball BA, SB Idso, S Johnson, MC Rillig (2007) Seventeen years of carbon dioxide enrichment of sour orange trees: final results. Global Change Biol 13:2171–2183.

Lambers H, Chapin FS, Pons TL (1998) Plant physiological ecology. Springer, New York, NY

Larcher W (2003) Physiological plant ecology, 4th edn. Springer, Berlin

Lavergne S, Mouquet N, Thuiller W, Ronce O (2010) Biodiversity and climate change: integrating evolutionary and ecological responses of species and communities. Ann Rev Ecol Evol Syst 41:321–350.

Leakey ADB, Ainsworth EA, Bernacchi CJ, Rogers A, Long SP, Ort DR (2009) Elevated CO_2 effects on plant carbon, nitrogen, and water relations: six important lessons from FACE. J Exp Bot 60:2859–2876.

Lobell DB, Field CB, Cahill KN, Bonfils C (2006) Impacts of future climate change on California perennial crop yields: model projections with climate and crop uncertainties. Agr For Meteorol 141:208–218.

Malik SK, Chaudhury R, Dhariwal OP, Bhandari DC (2010) Genetic resources of tropical underutilized fruits in India. NBPGR, New Delhi

Meyers LA, Bull JJ (2002) Fighting change with change: adaptive variation in an uncertain world. Trend Ecol Evol 17:551–557.

Mitchell RAC, Mitchell VJ, Driscoll SP, Franklin J, Lawlor DW (1993) Effects of increased CO_2 concentration and temperature on growth and yield of winter wheat at two levels of nitrogen application. Plant Cell Environ 16:521–529.

Morecroft MD, Paterson JS (2006) Effects of temperature and precipitation changes on plant communities. In: Morison JIL, Morecroft MD (eds) Plant growth and climate change. Blackwell Publishing Ltd, Oxford, pp 146–164.

Mubeen M, Ashfaq A, Hafiz MH, Muhammad A, Hafiz UF, Mazhar S, Muhammad Sami ul Din, Asad A, Amjed A, Fahad S, Wajid N (2020) Evaluating the climate change impact on water use efficiency of cotton-wheat in semi-arid conditions using DSSAT model. J Water Climate Chang 11:1661–1675. https://doi.org/10.2166/wcc.2019.179/622035/jwc2019179.pdf.

Muhammad Z, Abdul MK, Abdul MS, Kenneth BM, Muhammad S, Shahen S, Ibadullah J, Fahad S (2019) Performance of *Aeluropus lagopoides* (mangrove grass) ecotypes, a potential turfgrass, under high saline conditions. Environ Sci Pollut Res 26:13410–13421. https://doi.org/10.1007/s11356-019-04838-3.

Nobel PS (2005) Physicochemical and environmental plant physiology, 3rd edn. Academic Press, Inc., San Diego, CA

NCA (2012) National climate assessment. Climate change and agriculture: effects and adaptation. USDA, U.S. Government Printing Office, Washington, DC

Parmesan C, Duarte CM, Poloczanska E, Richardson AJ, Singer MC (2011) Overstretching attribution. Nat Clim Change 1:2–4.

Parry ML, Canziani OF, Palutikof JP, Van der Linden PJ, Hanson CE (2007) Climate change: prior observations. Agronomy 75:779–788.

Poorter H, Van Berkel Y, Baxter R, Den Hertog J, Dijkstra P, Gifford RM, Griffin KL, Roumet C, Roy J, Wong SC (1997) The effects of elevated CO_2 on the chemical composition and construction costs of leaves of 27 C3 species. Plant Cell Environ 20:472–482.

Rafferty NE, Ives AR (2010) Effects of experimental shifts in flowering phenology on plant–pollinator interactions. Ecol Lett 14:69–74.

Ramírez F, Kallarackal J (2015) Responses of fruit trees to global climate change. Springer briefs in plant science. Springer International Publishing. https://doi.org/10.1007/978-3-319-14200-5_2.

Rehman M, Fahad S, Saleem MH, Hafeez M, Habib ur Rahman M, Liu F, Deng G (2020) Red light optimized physiological traits and enhanced the growth of ramie (*Boehmeria nivea* L.). Photosynthetica 58:922–931.

Sajjad H, Muhammad M, Ashfaq A, Waseem A, Hafiz MH, Mazhar A, Nasir M, Asad A, Hafiz UF, Syeda RS, Fahad S, Depeng W, Wajid N (2019) Using GIS tools to detect the land use/land cover changes during forty years in Lodhran district of Pakistan. Environ Sci Pollut Res 27:39676–39692. https://doi.org/10.1007/s11356-019-06072-3.

Sala OE, Detlef van Vuuren D, Pereira HM, Lodge D, Alder J, Cumming G, Dobson A, Wolters V, Xenopoulos MA (2005) Biodiversity across scenarios. In: Ecosystems and human well-being: scenarios, Vol. 2. Island Press, Washington, DC, pp 375–408.

Salamin N, Wüest RO, Lavergne S, Thuiller W, Pearman PB (2010) Assessing rapid evolution in a changing environment. Tren Ecol Evol 25:692–698.

Saleem MH, Fahad S, Adnan M, Mohsin A, Muhammad SR, Muhammad K, Qurban A, Inas AH, Parashuram B, Mubassir A, Reem MH (2020a) Foliar application of gibberellic acid endorsed phytoextraction of copper and alleviates oxidative stress in jute (*Corchorus capsularis* L.) plant grown in highly copper-contaminated soil of China. Environ Sci Pollution Res 27:37121–37133. https://doi.org/10.1007/s11356-020-09764-3.

Saleem MH, Fahad S, Shahid UK, Mairaj D, Abid U, Ayman ELS, Akbar H, Analía L, Lijun L (2020c) Copper-induced oxidative stress, initiation of antioxidants and phytoremediation potential of flax (*Linum usitatissimum* L.) seedlings grown under the mixing of two different soils of China. Environ Sci Poll Res 27:5211–5221. https://doi.org/10.1007/s11356-019-07264-7.

Saleem MH, Rehman M, Fahad S, Tung SA, Iqbal N, Hassan A, Ayub A, Wahid MA, Shaukat S, Liu L, Deng G (2020b) Leaf gas exchange, oxidative stress, and physiological attributes of rapeseed (*Brassica napus* L.) grown under different light-emitting diodes. Photosynthetica 58:836–845.

Saud S, Chen Y, Fahad S, Hussain S, Na L, Xin L, Alhussien SA (2016) Silicate application increases the photosynthesis and its associated metabolic activities in Kentucky bluegrass under drought stress and post-drought recovery. Environ Sci Pollut Res 23:17647–17655. https://doi.org/10.1007/s11356-016-6957-x.

Saud S, Chen Y, Long B, Fahad S, Sadiq A (2013) The different impact on the growth of cool season turf grass under the various conditions on salinity and drought stress. Int J Agric Sci Res 3:77–84.

Saud S, Fahad S, Cui G, Chen Y, Anwar S (2020) Determining nitrogen isotopes discrimination under drought stress on enzymatic activities, nitrogen isotope abundance and water contents of Kentucky bluegrass. Sci Rep 10:6415. https://doi.org/10.1038/s41598-020-63548-w.

Saud S, Fahad S, Yajun C, Ihsan MZ, Hammad HM, Nasim W, Amanullah Jr, Arif M, Alharby H (2017) Effects of nitrogen supply on water stress and recovery mechanisms in kentucky bluegrass plants. Front. Plant Sci 8:983. https://doi.org/10.3389/fpls.2017.00983.

Saud S, Li X, Chen Y, Zhang L, Fahad S, Hussain S, Sadiq A, Chen Y (2014) Silicon application increases drought tolerance of Kentucky bluegrass by improving plant water relations and morph physiological functions. SciWorld J 2014:1–10. https://doi.org/10.1155/2014/368694.

Shafi MI, Adnan M, Fahad S, Fazli W, Ahsan K, Zhen Y, Subhan D, Zafar-ul-Hye M, Martin B, Rahul D (2020) Application of single superphosphate with humic acid improves the growth, yield and phosphorus uptake of wheat (*Triticum aestivum* L.) in calcareous soil. Agron 10:1224. https://doi.org/10.3390/agronomy10091224.

Shah F, Lixiao N, Kehui C, Tariq S, Wei W, Chang C, Liyang Z, Farhan A, Fahad S, Huang J (2013) Rice grain yield and component responses to near 2 °C of warming. Field Crop Res 157:98–110.

Sthapit BR, Ramanatha Rao V, Sthapit SR (2012) Tropical fruit tree species and climate change. Biodiversity International, New Delhi

Subhan D, Zafar-ul-Hye M, Fahad S, Saud S, Martin B, Tereza H, Rahul D (2020) Drought stress alleviation by ACC deaminase producing *Achromobacter xylosoxidans* and *Enterobacter cloacae*, with and without timber waste biochar in maize. Sustain 12:6286. https://doi.org/10.3390/su12156286.

Sugiura T, Yokozawa M (2004) Impact of global warming on environments for apple and satsuma mandarin production estimated from changes of the annual mean temperature. J Japan Soc Hortic Sci 73:72–78.

Tariq M, Ahmad S, Fahad S, Abbas G, Hussain S, Fatima Z, Nasim W, Mubeen M, Habib ur Rehman M, Khan MA, Adnan M (2018). The impact of climate warming and crop management on phenology of sunflower-based cropping systems in Punjab, Pakistan. Agri and Forest Met 256:270–282.

Turner DW (1998) Ecophysiology of bananas: the generation and functioning of the leaf canopy. Acta Hortic 490:211–222.

Turner DW, Fortescue JA, Thomas CD (2007) Environmental physiology of the bananas (*Musa spp.*). Braz J Plant Physiol 19:463–484.

Ur Rehman M, Hassan Rather G, Gull Y, Mir MR, Mir MM, Waida UI, Hakeem KR (2015) Effect of climate change on horticultural crops. In Hakeem KR (ed) Crop production and global environmental issues, Springer, Cham. https://doi.org/10.1007/978-3-319-23162-4_9.

Uz Zaman Q, Zubair A, Muhammad Y, Muhammad ZI, Abdul K, Fahad S, Safder B, Ramzani PMA, Muhammad N (2017) Zinc biofortification in rice: leveraging agriculture to moderate hidden hunger in developing countries. Arch Agron Soil Sci 64:147–161. https://doi.org/10.1080/03650340.2017.1338343.

Van Asten PJA, Fermont AM, Taulya G (2011) Drought is a major yield loss factor for rained East African highland banana. Agric Water Manage 98:541–552.

Vara Prasad PV, Boote KJ, Allen LH (2006) Adverse high temperature effects on pollen viability, seed-set, seed yield and harvest index of grain-sorghum [*Sorghum bicolor* (L.) Moench] are more severe at elevated carbon dioxide due to higher tissue temperatures. Agric For Meteorol 139:237–251.

Vingarzan R (2004) A review of surface ozone background levels and trends. Atmos Environ 38:3431–3442.

Wajid N, Ashfaq A, Asad A, Muhammad T, Muhammad A, Muhammad S, Khawar J, Ghulam MS, Syeda RS, Hafiz MH, Muhammad IAR, Muhammad ZH, Habib ur Rahman M, Veysel T, Fahad S, Suad S, Aziz K, Shahzad A (2017) Radiation efficiency and nitrogen fertilizer impacts on sunflower crop in contrasting environments of Punjab. Pakistan Environ Sci Pollut Res 25:1822–1836. https://doi.org/10.1007/s11356-017-0592-z.

Wahid F, Fahad S, Subhan D, Adnan M,, Zhen Y, Saud S, Manzer HS, Martin B, Tereza H, Rahul D (2020) Sustainable management with mycorrhizae and phosphate solubilizing bacteria for enhanced phosphorus uptake in calcareous soils. Agri 10:334. https://doi.org/10.3390/agriculture10080334.

Woolf AB, Ferguson IB (2000) Postharvest responses to high fruit temperatures in the field. Postharvest Biol Technol 21:7–20.

Wu C, Kehui C, She T, Ganghua L, Shaohua W, Fahad S, Lixiao N, Jianliang H, Shaobing P, Yanfeng D (2020) Intensified pollination and fertilization ameliorate heat injury in rice (*Oryza sativa* L.) during the flowering stage. Field Crops Res 252:107795.

Wu C, Tang S, Li G, Wang S, Fahad S, Ding Y (2019) Roles of phytohormone changes in the grain yield of rice plants exposed to heat: a review. PeerJ 7:e7792. https://doi.org/10.7717/peerj.7792.

Yang Z, Zhang Z, Zhang T, Fahad S, Cui K, Nie L, Peng S, Huang J (2017) The effect of season-long temperature increases on rice cultivars grown in the central and southern regions of China. Front Plant Sci 8:1908. https://doi.org/10.3389/fpls.2017.01908.

Zafar-ul-Hye M, Muhammad N, Subhan D, Fahad S, Rahul D, Mazhar A, Ashfaq AR, Martin B, Jiˇrí H, Zahid HT, Muhammad N (2020a) Alleviation of cadmium adverse effects by improving nutrients uptake in bitter gourd through cadmium tolerant rhizobacteria. Environ 7:54. https://doi.org/10.3390/environments7080054.

Zafar-ul-Hye M, Tahzeeb-ul-Hassan M, Muhammad A, Fahad S, Martin B, Tereza D, Rahul D, Subhan D (2020b) Potential role of compost mixed biochar with rhizobacteria in mitigating lead toxicity in spinach. Scientific Rep 10:12159. https://doi.org/10.1038/s41598-020-69183-9.

Zahida Z, Hafiz FB, Zulfiqar AS, Ghulam MS, Fahad S, Muhammad RA, Hafiz MH, Wajid N, Muhammad S (2017) Effect of water management and silicon on germination, growth, phosphorus and arsenic uptake in rice. Ecotoxicol Environ Saf 144:11–18.

Ziska LH, Bunce JA (2006) Plant responses to rising atmospheric carbon dioxide. In: Morison JIL, Morecroft MD (eds) Plant growth and climate change. Blackwell Publishing Ltd, Oxford, pp 17–47.

Index

Q

R

S

T

V

W

Z